Molecular Modelling
Computational Chemistry Demystified

Molecular Modelling
Computational Chemistry Demystified

Molecular Modelling
Computational Chemistry Demystified

Peter Bladon
Interprobe Chemical Services, Lenzie, Kirkintilloch, Glasgow, UK

John E. Gorton
Gorton Systems, Glasgow, UK

Robert B. Hammond
Institute of Particle Science and Engineering, The University of Leeds, UK

RSCPublishing

ISBN: 978-1-84973-352-6

A catalogue record for this book is available from the British Library

© Bladon, Gorton and Hammond 2012

All rights reserved

Apart from fair dealing for the purposes of research for non-commercial purposes or for private study, criticism or review, as permitted under the Copyright, Designs and Patents Act 1988 and the Copyright and Related Rights Regulations 2003, this publication may not be reproduced, stored or transmitted, in any form or by any means, without the prior permission in writing of The Royal Society of Chemistry or the copyright owner, or in the case of reproduction in accordance with the terms of licences issued by the Copyright Licensing Agency in the UK, or in accordance with the terms of the licences issued by the appropriate Reproduction Rights Organization outside the UK. Enquiries concerning reproduction outside the terms stated here should be sent to The Royal Society of Chemistry at the address printed on this page.

The RSC is not responsible for individual opinions expressed in this work.

Published by The Royal Society of Chemistry,
Thomas Graham House, Science Park, Milton Road,
Cambridge CB4 0WF, UK

Registered Charity Number 207890

For further information see our web site at www.rsc.org

Preface
'Molecular Modelling - Computational Chemistry Demystified'

While there are many books that deal with *Molecular Modelling* or *Computational Chemistry*, few of them meet the needs of someone who is just starting out, and wishes to set up their own system. Problems arise right at the beginning with the bewildering array of acronyms to do with both chemistry and computing.

A student studying chemistry at a well endowed university, and needing to fulfil requirements in computational chemistry, would find the necessary equipment and programs already set up. All (s)he has to do is to follow the prescribed parts of the course and pass the tests. The same student, having graduated, and now proceeding to a higher degree would be able to use the same or similar facilities when needed. But there are those who, for various reasons, choose to work alone, or are obliged to do so. Catering for these is what this book is all about. Molecular modelling can provide an entry into chemistry, that in former years was available by experimentation, now unfortunately, but understandably, frowned on by authority. For someone in this category who is impatient and wants to get started right away, we offer a first chapter that allows just this.

Demystification is one of our goals. But it does not require that Molecular Modelling be trivialized. So we have tried to provide a book that would be valuable to people at many stages of learning about Chemistry. Is this possible? We believe it is, since even the business of writing the book and the software has provided us with insights into aspects of chemistry that were new to us.

"First catch your hare..." This quotation is wrongly attributed to Mrs Beeton in her recipe for jugged hare. In the present context we can paraphrase it "First buy/acquire/beg/borrow... your computer". We show in the second chapter how to use such a computer, that you maybe already have, and that is running a version of the Windows® Operating system, to run Molecular Modelling and Computational Chemistry programs. Not only that - the Interprobe programs (INTERCHEM, PRESTO, *etc*.) are included with this book, and sections are devoted to setting them up and using them!

The rest of the book is devoted to explaining the mystique and jargon that surrounds computational chemistry. There is a chapter devoted to modelling the crystalline state, a feature not catered for elsewhere. Other chapters deal with modelling biopolymers such as proteins that impinge on medicinal chemistry. There is an underlying heuristic aim throughout the book; problems for the reader to solve are provided.

The authors are conscious of the difficulties of explaining in detail how to perform tasks on a personal computer. The expectation of users today is that all the tasks will be started as the results of user - computer interaction. This means that *every* request from the computer and *every* response from the user shall be specified in detail. Such requirements can result in text that is difficult to read; we have done our best in this regard, and hope that the parts of the book that are not so circumscribed will make easier reading.

We have framed the instructional examples and the questions in the context of organic chemistry of recent decades, with an emphasis on structures. Modelling chemical *reactions* presents significant challenges and, it may be argued, can only be done in terms of starting, intermediate, and end *structures* and the stereochemistry and thermodynamics that relates them. To aid this we provide a chapter dealing with stereochemistry, and conformational analysis.

The book aims to be a practical guide to the use of computers in aiding structural chemistry, and when we have digressed into the chemistry itself, we have done so to explain the relevance of the modelling techniques or computations. The book is neither to be regarded as a textbook on crystallography nor a textbook on medicinal chemistry. However with these two topics, and the topic of protein structures, we hope to provide an introduction to fields of chemical science that are both challenging and worthy of serious study.

The reasons for issuing software with the book
In coupling the publication of this book with an offering of Open-Source software, we are making a bold statement. This comes from a belief that openness is the key to scientific progress. It is possible that a scientist in 2061, referring to one of the many papers concerning the docking of small molecule structures into protein cavities (a subject we touch upon in Chapter 9), would not be able to make any sense of the paper itself, and would have difficulty in reproducing the experiments. This would be in part due the complexity of the science, but a contributing factor would be the fact that the software being used is proprietary and secret. The case for software being open-source now largely rests on contemporary issues (who should own the software?), but we believe that this other aspect of future traceability is a more important reason for the algorithms and source code based on them to be completely disclosed. The present situation in computational chemistry is in stark contrast to that existing in conventional experimental work, where results obtained today can be (and regularly are) compared with results obtained in the last two centuries, if necessary by repeating the original experiments.

The ethical aspects of molecular modelling have been raised before. In a paper entitled *Guidelines for Publications in Molecular Modeling Related to Medicinal Chemistry*,[1] the members of Working Party on Computer Assisted Molecular Modeling, commissioned by IUPAC made several significant recommendations. In relation to proprietary algorithms they suggested that ".... the commercializing companies have the obligation to submit for publication articles giving at least general descriptions of algorithms and databases used, with sample calculation results to aid calibration and evaluation."

The reasons why the Interprobe software came about
The straightforward answer is economic necessity. Twenty five years ago commercially available modelling packages were expensive and of limited scope. Subsequently mergers, takeovers, company failures, and 'rationalisations' have reduced the number of choices for purchasing software without reducing costs. What has resulted from twenty years of development is a set of programs that fills most of the needs of entry- level molecular modelling. We would be foolish to claim that it is the best, but we hope that what is provided may stimulate purchasers/users to

experiment. We would also welcome constructive comments aimed at its improvement.

The software owes its origin to earlier versions that were designed to work on Digital Equipment Vax computers and on SGI Irix systems. These versions were developed in the Department of Pure and Applied Chemistry at The University of Strathclyde where one of the authors was employed. The version of the software that is supplied with this book, and that works on systems running Microsoft Windows operating systems, derives from these programs. It is a reworking of these earlier systems and has been developed independently and without any outside financial aid.

We understand that in some circumstances complete openness of software has disadvantages; the chief being that 'modified' versions become available! In this connection, the Interprobe software is 'protected' by the fact that the compilers used are not open software; if you wish to modify the code for and distribute it, this is possible only if you acquire the Salford software currently licensed by the firm Silverfrost. This firm does provide, at no cost, a demonstration version of the compilers *etc.* that does allow you to make modifications to our programs for your own use, *but not redistribute them*. Furthermore, this enabling software only allows the production of versions for Microsoft Windows operating systems. We do not hide the fact that we would have preferred, in an ideal world, to have a system working under Linux. Maybe that will come.

Chemistry needs computing needs computers need chemistry!
That chemistry depends on computation and therefore on computers is not much in doubt. But let it not be forgotten that the reverse is also true. Two facts stand out:
(1) All modern computers depend on devices that require highly purified elemental silicon. The key to this is the chemistry of silanes, an area of chemistry opened up by the work of Alfred Stock in the first third of the 20th Century, and at the time considered of only academic interest.
(2) Any laptop computer that you own will rely on a *liquid crystal* screen for a display, and most desktop machines will also have screens of this type. In 1924 Daniel Vorländer wrote on page 89 of his monograph on liquid crystals:[2]

"Man hat mir wohl die Frage gestellt, ob sich die kristallin-flüssigen Substanzen technisch verwerten lassen? Ich sehe keine Möglichkeit dazu"

["I have indeed asked myself whether liquid crystal substances are capable of technical application? I see no possibility of that."]

He expressed his honest opinion, but how wrong he was!

These are just two examples, but what better testimony and reasons could we want, for the continued support of blue-sky research?

[1] P. Grund, D. C. Berry, J. M. Blaney, and N. C. Cohen, *J. Med. Chem.*, 1988. **31**, 2230.

[2] D. Vorländer, *Chemische Kristallographie der Flüssigkeiten.*, Akademische Verlagsgesellschaft, Leipzig, 1924.

Acknowledgements

J.G. acknowledges the help from many students and colleagues who have given great help and simplified his convoluted explanations.

P.B. acknowledges the help of many former colleagues at the University of Strathclyde; in particular, Dr. Mark Dufton for stimulating interest in protein chemistry, Dr. Robin Breckenridge for collaboration in the early days of Interprobe and for subsequently keeping him on track on drug design politics and economics, Dr. David Pugh for putting him right when dealing with crystal structures, and Professor Douglas McGregor for guidance on the humane treatment of computers. The molecular mechanics that is used in the Interprobe software is based on code that was generously supplied by Dr. Armin Widmer of Novartis, Basel and to whom a great debt is due.

R.B.H. acknowledges the help and support of colleagues at the University of Leeds in particular, Dr. Christopher Hammond for permission to reproduce a Table from his book. He also wishes to thank his parents Dorothy and Bryan for their steadfast support at all times.

We are also grateful to the Royal Society for permission to quote from Dirac's paper that is referred to in Chapter 1.

Formal permission has not been obtained for our quotation (in the Preface) from Vorländer's monograph; investigation has failed to reveal who is the current owner of the copyright.

We are grateful to the Cambridge Crystallographic Data Centre for permission to access, reproduce, and comment on selected files from that repository.

Contents

Chapter 1		Introduction	1
	1.1	The Beginnings - Some History	1
	1.2	About the Book	1
		1.2.2 Some Caveats	3
	1.3	Getting Started	3
	1.4	Basics	6
	1.5	Atoms and Bonds	6
	1.6	Isomerism, Stereoisomerism, Configuration, and Conformation	6
	1.7	Low Dimensional Structural Information	8
	1.8	SMILES in INTERCHEM	8
	1.9	Where now?	9
		References and Endnotes: Chapter 1	10
Chapter 2		Computers for Molecular Modelling	13
	2.1	Caveats	13
		2.1.1 Is Special Computer Equipment Needed?	13
		2.1.3 Are there choices to be made?	13
	2.2	Choice of Operating System	13
	2.3	Choice of Hardware - Desktop or Laptop	14
		2.3.1 Choices for a Laptop Machine	14
		2.3.2 Choices for a Desktop Machine	15
		2.3.3 The Other Alternatives	18
		2.3.4 The choice of a display	18
		2.3.5 The use of a projector for display	20
	2.4	RAID Computer Systems	20
	2.5	The Linux Operating Systems	20
	2.6	Having the Best of Both Worlds - Windows and Linux	20
		2.6.1 Two or More Machines	20
		2.6.2 Dual Boot Machines	21
		2.6.3 The Windows Subsystem for UNIX-based-Applications	21
		2.6.4 Using Cygwin	21
		2.6.5 Virtual Machines	23
		2.6.6 Clusters of Computers	24
	2.7	Networking	24
	2.8	Security	24
	2.9	Compatibility of 32-Bit and 64-Bit operating systems	24
	2.10	Further Reading	25
		References and Endnotes: Chapter 2	26
Chapter 3		Software for Molecular Modelling and Computational Chemistry	31
	3.1	Chemical Structure and Molecular Modelling	31
		3.1.1 Structures obtained from experimental methods	31
		3.1.2 The Born-Oppenheimer principle	31
	3.2	Molecular Mechanics	32
		3.2.1 Molecular mechanics - the basics	32

	3.2.2	Force field parameters	33
	3.2.3	Molecular mechanics - how it works	34
	3.2.4	Molecular mechanics - limitations	34
	3.2.5	The choice of force fields	35
	3.2.6	The extended use of molecular mechanics	36
3.3		Method based on quantum mechanics	36
	3.3.1	Semi-empirical molecular orbital programs	37
	3.3.2	*Ab initio* molecular orbital methods	38
	3.3.3	Methods based on density functional theory	38
	3.3.4	Basis sets	39
	3.3.5	Internal coordinates and Z matrices	39
	3.3.6	Atomic units	42
	3.3.7	Scaling in computational programs	42
3.4		Graphical display software	42
	3.4.1	The requirements of molecular modelling	43
	3.4.2	How OpenGL satisfies these requirements	43
3.7		Molecular modelling software suppliers	45
		References and Endnotes: Chapter 3	46

Chapter 4		Using INTERCHEM for Molecular Modelling	51
4.1		Some words of advice	51
4.2		Building structures	51
	4.2.1.	Using INTERCHEM Sketch - basics	53
	4.2.2	INTERCHEM Sketch Modifying your drawing	55
	4.2.3	Using INTERCHEM Sketch - aliphatic structures	55
	4.2.4	Generating structures using SMILES strings	57
	4.2.5	Accessing the SMILES facility	57
	4.2.6	Building structures from fragments	62
	4.2.7	Using the INTERCHEM merge tool for building	65
4.3		The reliability of structures obtained by building	68
	4.3.1	Refining a structure using quantum mechanics (quinine)	70
	4.3.2	Getting the energies of structures using quantum mechanics	72
4.4		Getting structures from the published literature	73
	4.4.1	Getting structures from X-ray crystallographic data	74
	4.4.2	Information from X-ray crystallographic data files	74
4.5		Analyzing Structures	79
	4.5.1	Bond lengths	79
	4.5.2	Inter atomic distances	79
	4.5.3	Bond angles	79
	4.5.4	Torsion angles	80
	4.5.5	Pseudo torsion angles	80
	4.5.6	Inter planar angles	80
	4.5.7	Molecular mechanic calculations - geometric measurement	81
	4.5.8	Molecular volume	81
	4.5.9	Molecular formula and molecular weight	81
	4.5.10	Hydrophobicity	83

	4.6	Stereochemistry	83
	4.7	Geometric isomerism (cis/trans or E/Z isomerism)	84
	4.8	The use of random numbers in INTERCHEM	84
	4.9	Problems for you to solve, and questions for you to answer	84
		References and Endnotes: Chapter 4	86
Chapter 5		Molecular Modelling of Proteins and Nucleic acids	89
	5.1	Introduction	89
	5.2	The nature of proteins	89
		5.2.1 The structure of proteins	89
		5.2.2 The structures of nucleic acids	100
		5.2.3 Further reading	100
	5.3	Obtaining Structures for Proteins and Nucleic Acids	104
		5.3.1 Accessing the Protein Data Bank	104
		5.3.2 The options provided by the program PROTEINS	106
		5.3.3 How the extra data is stored in INTERCHEM 'D' Files	106
		5.3.4 Displaying proteins and nucleic acids structures	107
		5.3.5 Editing Protein and Nucleic Acid Structures	108
		5.3.6 Analyzing Protein Structures	109
	5.4	Protein Sequences	116
		5.4.1 Some Definitions	117
		5.4.2 Sequence Matching	117
		5.4.3 Background of Aligning Protein Sequences	118
	5.5	The program PRESTO	118
		5.5.1 Introductory exercise	118
		5.5.2 Aligning sequence Sets Globally and Locally	122
		5.5.3 Questions arising from the alignment experiments	123
		5.5.4 Making inferences from alignments	123
		5.5.5 Storing images from the screen in PRESTO	124
	5.6	Racemic protein crystals as sources of protein structures	124
	5.7	Problems for you to solve and questions for you to answer	125
		5.7.1 General instructions applicable to most of these problems	125
		References and Endnotes: Chapter 5	128
Chapter 6		Essentials of Stereochemistry and Conformational Analysis	131
	6.1	Chirality	131
	6.2	Conformation and conformational analysis	135
	6.3	Isomerism involving double bonds and rings	138
		6.3.1 Chirality in biphenyl derivatives and allenes	139
		References and Endnotes: Chapter 6	140
Chapter 7		Molecular Modelling and the Solid State of Materials	141
	7.1	Introduction	141
	7.2	Classification of solids	142
	7.3	Crystallography and the specification of crystalline structures	143
		7.3.1 The lattice concept	143
		7.3.2 The crystal lattice in two dimensions	144
		7.3.3 The crystal lattice in three dimensions	151
		7.3.4 Examining crystal structures with INTERCHEM	156

		7.3.5 Chirality and crystallography	162
	7.4	Origin of cohesive forces in solids	164
	7.5	Thermodynamics of crystalline solids and molecular modelling	168
	7.6	Lattice energy calculations	169
		7.6.1 Worked example---calculation for sodium sulfate	170
		7.6.2 Lattice energy calculations for organic molecular materials	175
	7.7	The shape of crystals	177
		7.7.1 BFDH approach for crystal habit prediction	178
		7.7.2 Potential deficiencies in the BFDH approach	180
		7.7.3 Attachment energy approach for habit prediction	181
	7.8	Envoi	182
References and Endnotes: Chapter 7			184
Chapter 8		The Source of Archived 3D Chemical Structure Information	189
	8.1	Introduction	189
	8.2	Structures of small organic molecules from X-ray crystallography	189
	8.3	Structures of inorganic compounds and metals	190
	8.4	The Protein Databank	190
	8.5	ZINC	190
	8.6	Interprobe Chemical Services 3D Database	191
	8.7	File formats	191
		8.7.1 XR Format	192
		8.7.2 TRIPOS MOL2 Format	192
		8.7.3 PDB format	195
		8.7.4 MDL format	198
		8.7.5 Crystallographic Information Files (CIF format)	202
		8.7.6 INTERCHEM D format	202
		8.7.7 Multi-structure files	204
	8.8	General comments about files of 3D coordinates	204
		8.8.1 Right handed coordinate systems	204
		8.8.2 The rules of engagement	205
	References and Endnotes: Chapter 8		206
Chapter 9		Molecular Modelling and Medicinal Chemistry	209
	9.1	The need for new (legal) drugs	209
		9.1.1 Recent history	209
		9.1.2 The economics of drug development	210
		9.1.3 The plight of pharma	210
	9.2	What makes a compound a drug	210
		9.2.1 The way drugs work - a basic classification	211
		9.2.2 How modern drugs are designed	211
		9.2.3 Some definitions	211
		9.2.4 Natural products as leads	213
		9.2.5 How molecular modelling helps drug design	213
		9.2.6 The drug design pathway	213
		9.2.7 High throughput screening	213
	9.3	Molecular modelling in drug design	214
		9.3.1 Virtual high throughput screening	214
		9.3.2 Docking	215

Contents

		9.3.3 Additional filtering processes	218
		9.3.4 Isosteres - Variations on a theme	219
References and Endnotes: Chapter 9			222

Chapter 10	Using Interprobe Software for Drug Discovery	227
10.1	Overview	227
10.2	Setting up the database using Cygwin	227
	10.2.1 The importance of the full stop (and some other hints)	228
	10.2.2 How the databases are organised	229
	10.2.3 Setting up the databases for use by QUICKSCAN	231
	10.2.4 Using QUICKSCAN	232
	10.2.5 Viewing the structures extracted using INTERCHEM	236
10.3	Docking experiments using INTERCHEM	236
	10.3.1 Getting the necessary files	236
	10.3.2 Docking of the ligand into the protein	240
	10.3.3 Docking of analogues of tamoxifen	242
	10.3.4 Generating series of ligands for docking experiments	242
	10.3.5 Other targets for tackling malaria	245
10.4	Drugs from natural products	248
10.5	Problems for you to solve	251
10.6	Envoi	251
References and Endnotes: Chapter 10		252

Appendices		253
Appendix A1	Mathematical Background	254
Appendix A2	Data Tables	261
A2.1	Table of Standard Bond lengths	261
A2.2	Crystallographic Space Groups	262
Appendix A3	Numbering of Steroid Structures	265
Appendix A4	Essential Information for mounting Interprobe software	267
A4.1	Software that is included on the compact disc	267
A4.2	Copying the contents of the CD onto your computer	267
A4.3	Setting up the program INTERCHEM	267
	A4.3.1 Testing the program INTERCHEM	267
	A4.3.2 Problems due to mismatched dynamic linked libraries	268
	A4.3.3 Problems due to the file STEER.DAT	269
	A4.3.4 Errors due to the settings on your display screen	270
	A4.3.5 Errors in other display parameters	270
A4.4	Setting up the program PRESTO	271
	A4.4.1 Testing the program PRESTO	271
	A4.4.2 Problems in running PRESTO	271
	A4.4.3 Effects of screen resolution on PRESTO	271
	A4.4.4 Effects of other screen settings when using PRESTO	271
A4.5	Setting up the other Interprobe programs	273
A4.6	Setting up other programs	274
A4.6.1	Setting up MOPAC	274

	A4.6.2 Miscellaneous programs and scripts	275
	A4.7 Other problems - errors in the programs	275
	A4.8 Removing and updating Interprobe software on your computer	276
	References and Endnotes: Appendix A4	276
Appendix A5 File Compression and Transfer of Files between Computers		277
	A5.1 File compression	277
	A5.2 Transfer of files between computers and operating systems	278

Answers to 'Problems for you to solve and questions for you to answer'	279
Chapter 4	279
Chapter 5	280
Chapter 10	280
Subject Index	287

Typographical conventions used in this book

(1) The fount that we use **generally** in the text is

        Times New Roman 12 point

(2) The fount that we use in the **endnote**s is

        Times New Roman 11 point

(3) When we wish to indicate that text is printed by a computer we show it in

```
Courier New 12 point
```

This fount is also used when we wish to indicate that you should type something in response to a computer.

(4) When we want to show a complete line of text produced by a computer, without wrapping, we will use a fount as small as `Courier New 8 point`.

(5) We use the convention of enclosing the **type** of response needed by the computer by enclosing the name describing the entity in chevron brackets thus:

```
<file name>
```

In this case your response should be an actual file name (**without** the brackets!).

(6) In displaying SMILES strings in the text, we would have preferred to have used an equal spaced fount like `Courier New`. Unfortunately this had the effect of seeming to introduce spaces before and after brackets thus:

```
c2cc(-c1ccccc1)ccc2CCC
12345678901234567890123
```

So we settled for Arial

        c2cc(-c1ccccc1)ccc2CCC
        12345678901234567890123

This gets rid of the spaces but characters do not have equal widths.

(7) In the instructions for using the Interprobe programs INTERCHEM and PRESTO that have graphical user interfaces (GUIs), we have adopted a convention of referring to the Menu Buttons by their Row and Column coordinates, usually following the wording on the button. The Menu Name may also be included, thus:

        Dual Mode Menu, Invert Structure (R1/C6)

(8) In displaying line drawings of 6-membered aromatic rings the presence of the six pi-electrons is shown by a circle inscribed within the hexagon. We prefer this notation to that showing alternating double and single bonds (the Kekulé notation). We do not extend this notation to the 5-membered rings of heterocyclic compounds such as pyrrole, furan and thiophen, preferring in these cases to show the double bonds.

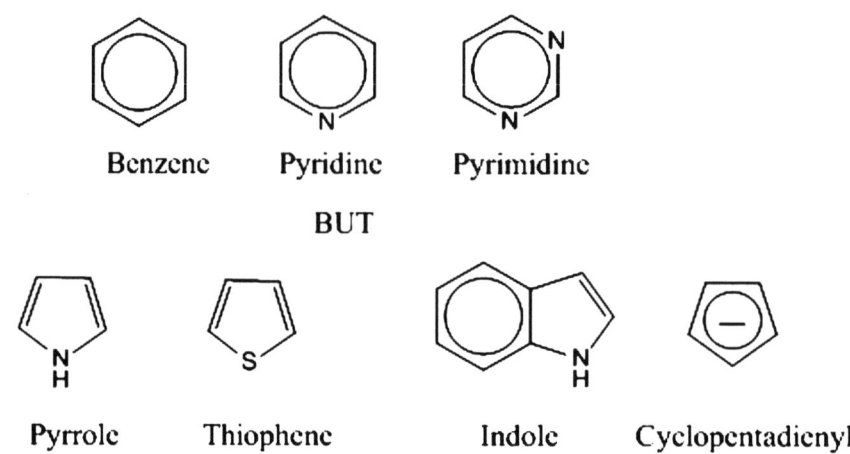

Chapter 1
Introduction

1.1 The beginnings – Some history

In 1929, in a paper entitled "Quantum Mechanics of Many-Electron Systems", the physicist P. A. M. Dirac made the following provocative statement:[1]

> "The underlying physical laws necessary for the mathematical theory of a large part of physics and the whole of chemistry are thus completely known, and the difficulty is only that the exact application of these laws leads to equations much too difficult to be soluble. It therefore becomes desirable that approximate practical methods of applying quantum mechanics should be developed, which can lead to an explanation of the main features of complex atomic systems without too much computation."

Presumably, the physicists who read this paper were mollified to some extent because some of their number would still be employed, but it seems that the paper was (fortunately) ignored by chemists, who carried on as usual. Indeed, in the eighty two years that have followed the appearance of Dirac's paper, it is likely that more new chemical compounds have been recorded than in all the previous years. What also happened was the "approximate practical methods", envisaged by Dirac, appeared, because of something that he only hinted at: "computation". Dirac shared the Nobel Prize for Physics in 1933 with Erwin Schrödinger.

While the majority of chemists were doing their own things ("Stamp Collecting" according to Ernest Rutherford – another physicist, who ironically got a Nobel Prize for Chemistry!) some physical chemists and chemical physicists did take Dirac's words to heart, and so 'Theoretical Chemistry' and then (when computers were invented) 'Computational Chemistry' came about.

1.2 About the book

This book is about "Molecular Modelling". We have given it a subtitle "Computational Chemistry Demystified" because we believe that *modelling* provides an easy entry point into computational chemistry, a branch of chemistry that has hitherto been misunderstood and undervalued, and thence into the rest of chemistry. While chemists have used many sorts of models for a long time, we concentrate our attention on *computer based modelling* because now this is the method of choice, and the necessary equipment is, if not already available, easily affordable.

We follow the present introductory chapter with one that deals with the choice of computer equipment. We do this because we believe that success in modelling and computation can be critically dependent on making the right choice. However, if you have already purchased your computer for other purposes, do not despair, what we have written may still help you in improving its performance, and to adapt it for molecular modelling.

While Chapter 2 deals primaily with the business of acquiring a computer suitable for Molecular Modelling, it also deals with wider issues of operating systems and handling more than one computer. If you already have a computer, we show how with very simple modifications (or none at all) it can do the job better. This chapter will also help those responsible for organizing a course in Molecular Modelling for the first time.

In the later chapters we show how to get the necessary chemical structural information, in forms that can be used by a computer. There is a vast amount of this that is freely available; free in the sense that it is easy to acquire, and also that it is in many cases available without payment.

Chapter 3 deals with principles of *molecular mechanics*. This is the easiest form of molecular modelling calculation to understand. The basic *ideas* about structure optimisation by minimizing the energy associated with a structure, that are at the heart of molecular mechanics, can be transferred to *molecular orbital methods*. We do not deal with these methods in detail, preferring to point you in the direction of the many excellent books that do. We do discuss the ways in which you may gain access to the programs that perform molecular orbital calculation.

Chapter 4 introduces the tools provided by INTERCHEM that allow you to build organic chemical structures of gradually increasing complexity. You are shown how to optimize the structures and then compare them with structures determined experimentally. The methods and examples should be handled equally well by other molecular modelling packages.

In Chapter 5 attention is turned to the larger structures of proteins and nucleic acids. There is a brief discussion of the salient features that distinguish these structures from those of smaller organics structures. We show how to gain access to the vast resources of the Protein Data Bank, and how to best display these structures to gain insight of their functions, and analyse them in various ways. The program PRESTO is introduced to allow the analysis and comparison of protein *sequences*.

Since stereochemistry and conformational analysis play important parts in molecular modelling, and indeed are underlying themes in our book, we offer brief summaries of these subjects in Chapter 6.

While most methods of computational chemistry deal with molecules in some sort of ideal (gaseous) state, and have difficulty dealing with molecules in solution, Chapter 7 deals with molecular modelling applied to the crystalline state. This is an area where progress is beginning to be made. The methods of molecular mechanics allow the many of the properties of crystals to be understood, and progress has been made in predicting the space group in which a chemical entity may crystallize.

In Chapter 8 we return to the basics of data handling. The information that is needed to define a structure, and the way that it is stored, has often leads to problems for people starting out in molecular modelling. It is not possible to discuss all the data formats that have been used, so we concentrate on those that have been mentioned in this book, or are commonly used elsewhere. We show those parts of a data file that are essential, and those that are not.

Perhaps the most compelling reason for writing this book comes from the use of molecular modelling in connection with Drug Design and Medicinal Chemistry, and this, and other related applications are dealt with in Chapters 9 and 10. The high cost of bringing a new drug to market has always made the companies involved turn to all available tools that might minimize this cost. Pharmaceutical firms have been held to ransom by the high prices of modelling software, and as our token contribution to the reduction of this cost, the Interprobe software is *offered without charge to all* under the GNU General Public License.

Two optical discs holding this software are provided with this book. You may copy and distribute the contents of these, and mount the software on any number of computers, provided you abide by the terms and spirit of the license.[2] The discs also have complete manuals for the programs INTERCHEM and PRESTO.

1.2.1 Some Caveats

It is perhaps necessary to define what we mean by a model in the context of chemistry. This is best done by showing what a model is not. Almost every molecular modelling software package will have an option (usually the default option) to show sulfur atoms as *yellow* balls. This does not imply that if we could actually see the atoms they would be coloured *yellow*. At the atomic scale the idea of colour is not meaningful. If you think carefully about it, every diagrammatic representation of a chemical structure is, to some extent, a model. The crystallographer who has, with great skill, determined the structure of (say) a protein is looking at a model. He or she is perhaps justified in being sceptical of a modeller who attempts to imbue a modification of such a structure with the same veracity as the original.[3] Provided that these limits of modelling are kept in mind, molecular modelling has a vital role to play in chemical research *and* education.

Molecular modelling is frequently dismissed as simply providing 'pretty pictures'. But many prestigious scientific journals make a feature of artwork that can rely in one way or another on molecular modelling.[4] To emphasize that modelling is respectable we would remind you, the reader, of the way that the structure of deoxyribonucleic acid was revealed by the extensive use of models.[5] Figure 1.1 shows how molecular modelling underpins computational chemistry, and how some of the other topics covered in this book are related.

1.3 Getting started

The usual advice given to anyone using a piece of equipment or software for the first time is *'Read the Manual'*. Experience has taught us that most people will ignore this. We acquiesce in this because we want to give you a taste of what molecular modelling can do. But we will assume that you have your computer set up, that the program INTERCHEM has been loaded on the machine, that you have started it by clicking on the appropriate icon on the desktop, and have pressed the escape key to get past the Welcome screen.[6]

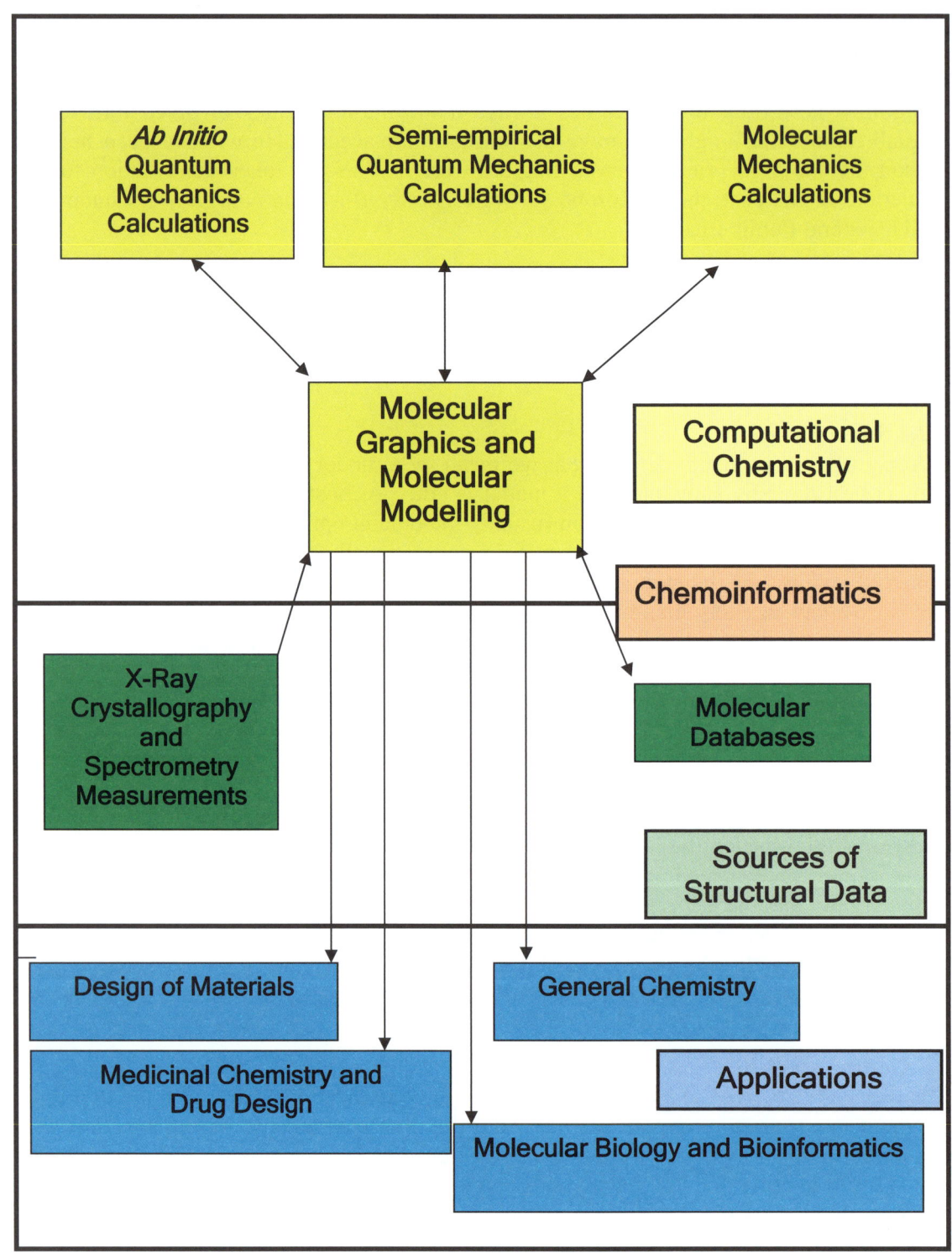

**Figure 1.1
Molecular Modelling and its
Role in Chemistry**

Introduction

If you move the mouse sprite around the buttons that are at the bottom of the screen, relevant help sentences will appear that will explain what each button does. You can click on the buttons to gain access to various tools. Not all of these will be appropriate for our first example (the molecule *valium)*; some are only relevant if you have two structures present; some are only relevant if you are viewing a large biomolecule or a crystal structure.

Using the mouse, click first on the button *Get from Catalogue* (Row 2/Column 8 in the Dual Display menu; henceforth we abbreviate the button coordinates to R2/C8), and then on buttons in the succession of menus that are revealed: *Aromatics and Heterocyclics, Valium, Confirm.* In the left hand window there will appear a simple 'wire-frame' structure of *Valium* (alias *diazepam*) (1).

What you have seen thus far is an entry point into *molecular graphics, molecular modelling,* and *computational chemistry.* In the remainder of this chapter, we will make use of INTERCHEM to explain the way these three subjects (and others) are connected, how you may gain entry to them, and profitably use them.

If you have succeeded so far, then you can load another structure into the right hand viewing area. Click on *Structure B* and repeat the steps used before but choose the structure *Biotin* (2).

The *wire-frame* depictions of *valium* and *biotin* show the bonds but gives no indication of the atoms that constitute the molecules. The identities of the atoms can be revealed in several ways; press each of the following (yellow) buttons in turn:

(1) *Ball and Cylinder* (R3/C9). This will show atoms as coloured balls (carbon and hydrogen - white, oxygen - red, nitrogen - blue, chlorine - green, sulfur - yellow), and the bonds as coloured cylinders (single - white, double - green, triple - blue, aromatic -cyan).

(2) *CPK Atoms* (R3/C10) This will show the atoms only as larger balls (approximately 1.5 times the van-der-Walls radii of the atoms).[7]

(3) *Cylindrical Bonds* (R4/C6). This will show the bonds only as coloured cylinders.

(4) *Wire Frame* (R3/C6). This reverts to the original display type, but more information can be revealed by pressing, in succession, the buttons: (a) *Bond-type Bonds* (R5/C1), (b) *Atom Numbers* (R5/C2), (c) *Chirality* (R5/C4). The bonds are first coloured as before, then a numbering of the atoms is shown (with colours appropriate to the atom types), and then (in the case of biotin) those atoms that are chiral are labelled with the letters *R* or *S* according to the Cahn - Ingold - Prelog (CIP) convention.[8] The numbering of the atoms is arbitrary in that it does not necessarily follow any prescribed numbering system. However the atoms can be numbered in a more logical manner, by techniques that are described later in the book.

We should emphasize that the computer is in fact modelling molecular models! We wrote earlier that the concept of colour as applied to objects on the atomic scale is not meaningful; the colours that are used here are arbitrary, though there is consensus among computer programs for some of the colours.

Perhaps the most important feature of the images on your computer screen is their three-dimensionality; that is their solidness. For the most part, chemists can happily work with two-dimensional diagrams of chemical structures, their training will allow them to do this, making use of clues provided in the diagrams (thick lines, dotted lines, arrows etc). However, there are situations where a three dimensional picture is essential. These occur especially when we are dealing with crystals or with large biomolecules such as proteins.

Molecular modelling uses a computer program to show you the shape of an object (molecule) that possibly has no resemblance to any familiar object. Contrast this with the tasks faced by computer programs for use by engineers or architects; we can easily recognize the shapes of cars and houses! We shall see later that molecular modeling programs use several special techniques to evoke three-dimensionality.

To finish this preliminary view of the molecular world, you should experiment by clicking on the green buttons. These allow you to rotate the models or make them larger or smaller.

1.4 Basics

For you to understand what is going on inside your computer that generates these pictures we have to define some basic chemical and mathematical concepts. We describe concepts here in the simplest way, but later we will need to qualify these definitions.

(1) We describe a molecule as being composed of (the appropriate numbers of) atoms of the elements joined (bonded) in specified ways.

(2) To give the molecule its three-dimensional structure we need to define the positions of the atoms in a suitable coordinate frame. Mostly we will use an orthogonal (*i.e.* rectilinear) Cartesian frame, with the measurements being in Ångstrom units.[9] However, for some purposes (*e.g.* crystallography) other coordinate frames will be used.

(3) Two types of bonds are commonly dealt with. Mostly we will be dealing with covalent bonds. We will further classify these as single (two electron), double (four electron), triple (six electron), and aromatic (a hybrid between single and double bonds, conveniently referred to as three electron bonds). The other type of bonding (ionic bonds) is found in ionic crystals. It was not until the middle of the twentieth century that current ideas of chemical bonding were developed. Pre-eminent in this area was the author of an influential book[10] (and subsequent Nobel Laureate), Linus Pauling.

1.5 Atoms and bonds

We have implied that we need to define the positions of the atoms in a chemical structure *and* the bonds that join them. Do we need to have both these items of information? It turns out that, in favourable circumstances, provided we know the identities of the atoms, and how they are bonded together, we can predict their positions. In other cases, if we know the positions of the atoms, we can predict the bonding. However, it certainly helps if both positional and bonding information is to hand.

1.6 Isomerism, stereoisomerism, configuration, and conformation

We will return to a discuss stereoisomerism in detail in Chapter 6 of this book, but it is appropriate to define these terms here. *Isomerism* is used to describe a situation where two or more different chemical substances can have the same molecular formula. In some

Introduction

cases the substances are very different, for example benzene (3) and dipropargyl (4) (both C_6H_6). Others may differ only in the placement of functional groups, for example butan-1-ol (5), butan-2-ol (6), 2-methylpropan-2-ol (7).

In their book *Stereochemistry of Organic Compounds,* Eliel and Willen[11] define *stereochemistry* as "chemistry in three dimensions". Stereoisomers are therefore isomers that differ in the disposition of the (otherwise identical) bonding of the constituent atoms or groups of atoms in space. It will become evident that stereochemistry is at the heart of molecular modelling.

The term *configuration* is used to characterize an arrangement of atoms or groups in one stereoisomer. The term *conformation* is used to characterize one arrangement of atoms or groups in a structure that may have several such arrangements existing in equilibrium.

1.7 Low dimensional structural information
Although a three dimensional structure can provide the most information about a compound, and for many purposes a two dimensional structures mentioned above can often suffice, a need has arisen for simple ways of characterizing structures. The most popular of these ways is SMILES (Simplified Molecular Interpretive Line Entry System).[12] Its popularity is due to its simplicity. It is the descendant of more complicated systems (Dyson[13] and Wiswesser[14] systems) that required extensive training to use. All of these systems aim to reduce the specification of a structure to a string (*i.e.* a line) of alphanumeric symbols. They arose because of a need to catalogue and classify (mainly organic) compounds. The Dyson system was adopted by the Chemical Abstracts Service, and the Wiswesser system had a brief popularity in the pharamaceutical and chemical industries.

SMILES strings have two uses; firstly to characterize and index already existing structures, and secondly to specify, as yet, unrealized structures. We call these two uses *descriptive* and *prescriptive* respectively. What makes SMILES attractive, is the fact that while for descriptive purposes for a particular structure there is a unique (*i.e. canonical*) SMILES string, when a need arises to specify a new structure *any* syntactically correct SMILES string will do. A version of SMILES, modified to handle stereochemistry in a prescriptive mode, is part of INTERCHEM (see paragraph 1.8).

Difficulties have arisen in the use of SMILES for both descriptive and prescriptive purposes, and this has given rise to alternative systems. One that seems to offer advantages, when used prescriptively, is SYBYL Line Notation (SLN) promulgated by the firm Tripos.[15] For descriptive purposes, and particularly for indexing, the InChi[16] system that is owned and supported by IUPAC will probably become the system of choice. The Open Source software for InChi from IUPAC is included in the software accompanying this book.

In classifying these low dimensional data, most chemists would accept that the structures that they draw (on the backs of envelopes, or beer mats) are two dimensional (2D), some would contend that when thickened or dotted lines are used (to introduce stereochemistry) we have 2.5D drawings. Whether you classify SMILES strings as 1D or 0D is a matter of choice. However we can leave such philosophical niceties to one side, and conclude this chapter by going back to INTERCHEM.

1.8 SMILES in INTERCHEM
Click on the button labelled *Build via SMILES* (R9/C2 in the *Dual Display* menu), and then enter the following SMILES string in the long type-in box: `C1CCCCC1`. This SMILES string is one that represents *cyclohexane*. Note that the letters must be in upper case; lower case letters would respresent *benzene*; the two occurrences of the number 1 show that the structure is cyclic. Then press the button *CONFIRM – Accept*. Dispose of the small window that shows the energy of the structure, and then press *Store as Structure C*.

Introduction

The usual result is (usually) the *twist-boat conformation* of cyclohexane. Now press *Lowest Energy Structure*. Then choose to generate sixteen trial structures (by pressing the appropriate button). When the structures have been generated, press *Store as Structure D*.

Dispose of the subsidiary menu by pressing *Abandon Structure EXIT*, and dispose of the extra windows by using *Escape*. The lowest energy structure will be the *Chair conformation* of cyclohexane. Cyclohexane at room temperature is an equilibrium mixture of this chair conformation (99.6%) plus a small amount of the twist-boat conformation (0.4%).[18]

You should now have models of these two conformers of cyclohexane in the display areas *C* and *D*, and you can study them in the various display modes, and find their molecular properties.

1.9 Where now?
You may already be in possession of other molecular modelling software, and wonder how INTERCHEM can help you. We believe that INTERCHEM has features that are absent from other software, so try it out. If you have had no previous practical experience of molecular modelling using a computer, this simple introduction and the INTERCHEM program should get you started.

If you are thinking that the computer you are using could do with improvement, or that you really ought to get a new one for the job, read the next Chapter on Computers for Molecular Modelling.

If you want to get on with doing some experiments, then go to Chapters 4, but refer to Chapter 3 when you need explanations.

If you have had problems getting the program INTERCHEM to work satisfactorily, read Appendix 4. This goes through the steps for getting the program INTERCHEM working.

You could usefully get the program PRESTO working. You will encounter PRESTO in Chapter 5 of this book.

If you still have a choice about purchasing a suitable computer, then you should first read Chapter 2 that deals with this.

The full INTERCHEM manual, and also that for PRESTO which we deal with later, are to be found as Word Documents and as PDF files on the compact disc that accompanies this book.

If our assumptions about having your computer set up are not correct, or if things do not work in the expected way set out above, then you will have to *read the manual*, or the crucial instructions for setting up the program that are contained in Appendix 4 of this book. Alternatively, you could ask a colleague for help!

References and Endnotes. Chapter 1

[1] P. A. M. Dirac. *Proc. Roy. Soc. A,* 1929, **123**, 714. This third sentence in this paper is often quoted at the beginning of books on Molecular Modelling.[18, 19] The paper as a whole is not so easy to understand! The statements have been analyzed in detail by W. Kutzelnigg, (*Theor. Chem. Acc.,* 2000, **103**, 182), and this paper should be consulted by those wishing to understand the philosophical implications. Chemists should not be too critical of Dirac; Graham Farmelo's biography (*The Strangest Man, The Hidden Life of Paul Dirac, Quantum Genius*, Faber and Faber, London, 2009) reveals that he was a man whose life was not straightforward.

[2] Web address *http://www.gnu.org*

[3] Until recently, the Protein Data Bank (see chapter 8) accepted structures based on molecular modelling experiments (i.e. without experimental X-ray or NMR evidence). Such models have now been relegated to a separate part of the databank and no new data of this type are being accepted.

[4] *The Journal of the American Chemical Society* succumbed and began featuring cover artwork in 2009.

[5] J. D. Watson, *The Double Helix*, Weidenfeld and Nicolson, London, 1968,; Paper back edition, Penguin Books, Harmondsworth 1970.

[6] Twenty five years ago we would have needed to explain what 'clicking', 'icon', and 'screen' meant in the context of using a computer; now it seems that the current generation has innate computer skills!

[7] CPK is an abbreviation for **C**orey-**P**auling-**K**olton; a type of (hardware) space-filling molecular modelling system.

[8] The convention for describing the type of stereochemistry typified by the tetrahedral valency of carbon (and other similar elements) was proposed by R. S. **C**ahn, C. K. **I**ngold, and V. **P**relog ; *Angew. Chem. Int. Ed. Engl.*, 1966, **5**, 385. Proposals for revisions were made by V. Prelog and G. Helmchen, *Angew. Chem. Int. Ed. Engl.*, 1982, **21**, 567.

[9] The Ångstrom unit (10^{-10} metre), abbreviated as Å, despite not being an SI unit, is widely used. It is a convenient since most chemical bonds have lengths between 0.8 and 3.0 Å.

[10] Linus Pauling. *The Nature of the Chemical Bond.* Third edition, Cornell University Press, Ithaca, 1960.

[11] E. L. Eliel and S. H. Wilen. *Stereochemistry of Organic Compounds.* John Wiley and Sons, New York, 1993.

[12] D. Weininger, *J. Chem. Inf. Comput. Sci.*, 1988, **28**, 31; D. Weininger, A. Weininger, and J. L. Weininger, *J. Chem. Inf. Comput. Sci.*, 1989, **29**, 97.

[13] G. M. Dyson, *A New Notation and Enumeration System for Organic Compounds.* Second Edition, Longmans, Green, and Co. Ltd., London, 1949.

[14] W. J. Wiswesser, *A Line-Formula Chemical Notation.* New York, Crowell, 1954; E. G. Smith and P. A. Baker, *The Wiswesser Line-Formula Chemical Notation (WLN) Third Edition.* Chemical Information Management Inc., Cherry Hill, 1975.

[15] S. Ash, M. A. Cline, R. W. Homer, T. Hurst, and G. B. Smith, *J. Chem. Inf. Comput. Sci.,* 1997, **37**, 71; S. W. Homer, J. Swanson, R. J. Jilek, T. Hurst, and R. D. Clark, *J. Chem. Inf. Mod.,* 2008, **48**, 2294.

[16] InChi™ http://www.iupac.org/inchi/index.html (Accessed 27 July 2011)

[17] Reference[11] p 690.

[18] A. Hinchliffe, *Molecular Modelling for Beginners..* Wiley, Chichester, 2008.

[19] S. M. Bachrach, *Computational Organic Chemistry*, Wiley, Hoboken, 2007.

Chapter 2
Computers for Molecular Modelling.

2.1 Caveats
Any discussion of computer hardware runs the risk of obsolescence in a very short time because the development and release of new equipment is so very rapid. We have to thank those games-playing enthusiasts and the online communications devotees for their eagerness in purchasing the latest equipment, which has prompted manufacturers to make ever more capable machines.

There are, however, many aspects of component choice that do not become obsolete rapidly, but can ensure a prolonged working life and help to maintain a competitive edge to a machine's performance. Outright speed of calculation, whilst important, is only one consideration in the design of machines for molecular modelling. Flexibility, security, and reliability are also important.

In this chapter we show how to acquire a computer for molecular modelling by posing and answering a series of questions, and making various choices. This process may not cover all possible answers and choices, but we hope that it will help you.

2.1.1 Is special computer equipment needed for Molecular Modelling?
The answer to this question is a qualified NO. All of the Interprobe software that is provided with this book, and that is designed to run under the Microsoft Windows operating systems, will run on an ordinary personal computer, either desktop or laptop. The only requirement is adequate screen resolution; this is discussed later in this chapter. Similar requirements are likely to apply when other molecular modelling software is used.

2.1.2 Are there choices to be made and what are they?
Despite the definitive statement in the last paragraph, there are choices that can be made that can affect the performance and convenience of the equipment. They fall into several categories:

(1) Choice of operating system
(2) Choice of hardware
(3) Choices that are involved if you want to modify existing hardware
(4) Choices based on costs

The choices in these categories are not completely independent, and so what follows in our explanations will sometimes involve linked solutions.

2.2 Choice of operating system
While strictly speaking, the operating system is just another piece of software, its choice does have an impact on choice of hardware, and so we deal with this first.

The current versions of some of the Interprobe software is only available for the Microsoft Windows operating systems, (but this need not be the most recent version of Windows).[1]

There are alternative operating systems, and, for various reasons you may want to use one. We will show you later in this chapter how to set up a computer system that can have more than one operating system present.

The alternative operating systems and associated hardware are:
- (1)　UNIX[2]
- (2)　Linux[3]
- (3)　Apple's Operating system OS-X[4]

UNIX is a proprietary operating system that was licensed by many workstation manufacturers (for example Sun Microsystems, Hewlett-Packard, and Silicon Graphics Inc), who modified it to there own requirements (and usually renamed it). UNIX is also the basis of Apple's current OS-X operating system.[4]

Linux is an open-source operating system that adheres to most of the standards (POSIX)[5] set by UNIX. Because of the success of Linux, it has largely displaced UXIX as the system of choice for workstation manufacturers (other than Apple) who want to offer an alternative to Windows.

Linux is important for molecular modeling since many of the *ab initio* molecular orbital calculating programs work best under Linux. Because of this, we will show you later in this chapter how to set up a computer system that can have more than one operating system present.

While there are many attractive features in Linux, (particularly the fact that it is FREE), the dominance of Microsoft[6] in business, academic, and personal computer sectors, means that most people would make Windows a first choice. It is because of this that we (reluctantly) bend to the prevailing wind. Having decided (or having been persuaded) that Microsoft Windows® 7 is the operating system that you want to install, there are (unfortunately) a bewildering set of choices to be made between the various versions. For reasons that we discuss later (paragraph 2.6.3), our recommendation is that you should choose Windows® 7 professional.

2.3 Choice of hardware - desktop or laptop?
This particular choice is put first; because it comes to this; are you a sedentary or a mobile worker? If your lifestyle demands that you move around a lot, then a Laptop computer might be your choice.

2.3.1 Choices for a laptop machine
Because of the way Laptop computers are put together, their hardware is not easily reconfigured after purchase. They usually have a single hard disc, and a motherboard that includes the components of a graphics card. Their inner workings are generally best left well alone! The important choice is the resolution of the screen because the screen is built into the machine. In Table 2.1 you will see that there are resolutions that are available on laptops that are marked as suitable for use by INTERCHEM. You should choose a laptop that has one of these resolutions. Unfortunately, the recommended resolution (1280 × 1024) can mean an expensive laptop.

If the laptop manufacturer is offering choices in the configuration of the machine, our recommendation is that you should choose to have the maximum amount of memory that the machine can handle, and the largest disc possible.

Laptop Machines have built-in touch-pads for moving the sprite and selecting from menus etc. For those people have difficulty using these devices, a mouse that plugs into a USB port is a useful addition. A carrying bag is almost a necessity. Depending on yours lifestyle, choose one that is large enough, but does not advertise the fact that its contents *are* valuable equipment by featuring the laptop manufacturer's name. If you are regularly carrying the laptop between home and work, consider buying a second power supply adaptor.

2.3.2 Choices for a desktop machine

If you are considering using a desktop machine (rather than a laptop) for molecular modelling, then the choices that you have to make are different. There are four options to begin with; these are listed roughly in order of increasing cost:

(1) To use and possibly modify an existing machine.
(2) To buy a new standard machine, and possibly modify it.
(3) To build one yourself.
(4) To buy a machine built to your specification from a specialist retailer.

The last two options have basically the same constraints, and the individual choices that they raise have impact in the other options; so we will consider them first together.

The hardware choices are considered in the following order (since the first choices have some impact on the later ones):

(a) Case and Power Supply (PSU)
(b) **C**entral **P**rocessing **U**nit (CPU)
(c) Graphics Card
(d) Motherboard
(e) Memory
(f) Hard Discs
(g) Optical Discs
(h) Other components and Accessories
(i) Putting it all together

(a) Choice of case and power supply

Computer cases frequently have a power supply already installed. To allow for expansion the case should be a 'full height' tower, and should have at least four 5¼ inch slots accessible from the front panel to accommodate discs. A case of this size should have at least a 500 Watt power supply. If that is not so, buy the case and power supply separately. Avoid cases that have spurious internal lights which only serve to attract insects. On the other hand, a case that has blower fans and air filters has an advantage.

The two largest consumers of power in a computer are the CPU and the graphics card(s). If there are a large number of discs they will also contribute. It is common for graphics cards to have extra power supply connections, besides those on the edge connector. Make sure these are provided. Unless you have a power hungry graphics card (or multiple CPUs) then a 500 Watt power supply will be adequate, otherwise choose a 600 Watt or even larger one.

(b) Choice of central processing unit

There are two main manufacturers of CPUs; Intel[7] and Advanced Micro Devices (AMD).[8] They both offer series of processor, nowadays featuring devices that have several *cores* (2 and 4 cores are common, 6 and 8 also available). Each core functions as a separate CPU, and these are used by a Windows operating system to allow background activity (housekeeping, communications etc.) to run in parallel with your programs. Our advice is *not* to choose the most recent, fastest, offering, but one that has become established, and has good reviews.

(c) Choice of Graphics Card

It may be that you are content to choose a motherboard that includes on-board graphics. If so you may skip this section; the graphics demands made by molecular modelling are not severe. A separate graphics board will usually offer the ability to drive two screens, and is essential if you want to use 3D stereo graphics. In that case an NVIDIA Geforce FX370 card is needed, as a minimum specification.[9]

(d) Choice of motherboard

Since the choice of CPU model and manufacturer determines the CPU *socket* type, the choice of motherboard is constrained to be one that has this socket. Make sure that the board that you choose has all the necessary additional features. A modern board will come equipped with an external network connection (RJ45 socket; 10/100/1000 Hz), internal connection for SATA discs (look for at least 6 connections), and several pairs (or triplets) of sockets for memory. There are usually PATA connections for up to two optical discs, (if these are absent look for 2 more SATA connections). It seems that connections for floppy discs are being dispensed with on modern boards, but if one is present it would be useful, provided that the front panel of the case has the necessary slot. If you have chosen to use a separate graphics card, make sure that the motherboard has the necessary PCI-express-16 connector.

(e) Choice of memory

The only real choice that you have for the memory is the *quantity*, since the type (number of pins) and speed is largely determined by the motherboard. If you can afford it, populate all the sockets with the largest sized chips. If economy is needed, there are constraints (rules) that pairs of sockets have chips of the same size, and sometimes that the pairs of sockets of lowest addresses shall have the chips of largest size. It is usually recognized that for efficient use of the CPU cores, 2 Gbyte of memory per core is needed, but this is not an absolute requirement. It is important to remember that if you subsequently decide to increase the memory size of a machine that is already using all of the sockets, you will end up with some redundant memory chips.

(f) Choice of hard discs

First there are some acronyms to define. ATA stands for **A**dvanced **T**echnology **A**ttachment; (a finer piece of obscurity is hard to imagine!). This term was applied to discs that had parallel data connections, usually in the form of ribbon cables, and now referred to as PATA discs to distinguish them from newer serially connected discs termed SATA discs. It is usual to think of serial transmission of data as being slower than parallel transmission; this disadvantage has been overcome, and SATA discs are now to be preferred, not least because the cables are smaller, and so interfere less with cooling airflow inside a computer.

If you look at a listing of discs in a catalogue of computer components, and (mentally) plot the price per Mbyte against disc capacity, you will find that there is minimum . Currently (March 2011) this minimum stands at discs of ~1TByte (= 1000 Mbyte) capacity. Choosing discs of this size would give you the most economical way of getting a large total disc memory, (ignoring the total cost). However, we would suggest that you use smaller discs to hold the operating system(s). Even the most greedy and bloated operating system will fit on a 160 Mbyte disc (currently the smallest size listed).

The discs that are used in desktop machines are 3½ inch wide and are designed to fit in the internal slots of the machine. The alternative sized 2½ inch discs are designed for laptops but can be mounted in desktops using adaptor brackets.

What we suggest is that using 3½ inch hard discs, you should have two of them mounted in caddies (also known as cartridges or drawers),[10] that can be mounted in slider devices in two of the front slots of the machine. If the motherboard allows for four SATA connections, two more discs can then be mounted internally in the normal way. The caddies can be unplugged and replaced *but only when the machine is powered off !!*

Having the discs in demountable caddies offers these advantages:
    (1)   The ability to change operating systems
    (2)   The ability to make backup copies of operating systems and data.
    (3)   The ability to have different projects on separate cartridges
    (4)   The ability to swap large amounts of data between different machines

(g) Optical discs drives

The computer will need at least one optical drive capable of reading and writing CDs and DVDs so that you can load software (including an operating system). This will occupy a third slot on the front panel of the case. The fourth slot could be used for a second optical disc drive. With this arrangement you would then be able to make copies of CDs and DVDs easily. Alternatively the fourth slot could be used for a third hard drive in a caddy.

(h) Miscellaneous components

If you are intending to build the machine yourself you should purchase any extra cables that you might need. (The motherboard kit is likely to include some, but you could need extra SATA cables). Other useful accessories include a floppy disc drive using a USB connection (if there is no provision for an internal floppy drive system).

Depending on the circumstances at your site, it might be prudent to invest in an *Uninterruptible Power Supply (UPS)*. This would certainly protect your equipment from an unnecessary shutdown due to a momentary glitch in the power supply; it would also cope with outages of a few second duration caused by lightning strikes on the power transmission network, and, if the necessary software is loaded, would cleanly shutdown a single computer system in there is a sustained loss of power. UPSs are separate boxes that incorporate stand-by batteries that are kept fully charged. When mains power is interrupted, the batteries are brought into use to provide alternating current supply at the correct voltage.

(i) Putting it all together
It is probably best, particularly if you have had no experience of assembling a computer, to entrust the building to the supplier of the equipment. The supplier will normally be prepared for a modest extra outlay to do this, and make sure that the whole assembly works. There would then be some liability on the supplier to rectify serious faults. The supplier would normally install the version of Windows that you select, but tell the supplier that you wish to *activate* the operating system yourself and insist that you are supplied with the all the discs used to mount the operating system. This is important so that you can re-establish the system in the event of, for example, disc failure.

2.3.3 The other alternatives
While the discussion in the preceding sections is concerned with two alternative ways of getting a bespoke machine, the points made also cover the enhancement of a machine purchased off-the-shelf. Adaptation of an older machine has special problems. If it will run the software that you require, it would be best not to upgrade it. You would thereby gain experience, and save money, both of which would be handy when you come to buy a new system.

2.3.4 The choice of a display
When a desktop computer is purchased, the necessary keyboard and mouse are usually in the bundle. However, the display is usually not included in the price, and it is left to the purchaser to make a choice. For molecular modelling it has been traditional to choose a screen with resolution of 1280×1024 pixels. (Acronym SXGA). This is the size of screens traditionally supplied with graphics workstations, and was adopted as standard when INTERCHEM was being designed. Now there is a profusion of different screen sizes; (see Table 2.1 where other possible choices are listed). The physical size of the display is also important; too small and it will be difficult to read small characters; too large and it will occupy a lot of space on your desk. However, a wide screen does have the advantage of being able to display two pages of text side-by-side.

Nowadays it is certain that any display that you purchase for use with a computer will be a liquid crystal display (LCD). In those that are described as LED (light emitting diode) displays, the LED designation only refers to the way the *backlighting* of the screen is produced. To use an LCD screen for modelling, it is best to run it at its *native resolution*, and this is usually the same as its maximum resolution. You *can* run such a screen at lower resolution, but this will alter the aspect ratio and distort the pictures.

The older *Cathode Ray Tube* (CRT) devices have gone out of fashion, because they are not energy-efficient and are very heavy. They do have an advantage in that the aspect ratio of the display can be adjusted within limits. If you acquire an old computer system that has a screen of this type, and has adequate resolution, do not reject it; it still will do the job.

Table 2.1
Standard Screen Resolutions

Resolution X × Y	Aspect Ratio	Pixel Count	Acronym*	Availability on Desktop	Availability on Laptop	Availability on Projector	Suitable for Molecular Modelling INTERCHEM + PRESTO	Suitable for Molecular Modelling General
640 × 480	4:3 = 1.3333	307,200	VGA	Legacy	Legacy	Legacy	No	No
800 × 600	4:3 = 1.3333	480,000	SVGA	Legacy	Legacy	Common	No	No
1024 × 768	4:3 = 1.3333	786,432	XGA	Common	Common	Common	Possible	Yes
1280 × 768	5:3 = 1.6666	983,040	Organic LED				No	Yes
1280 × 800	8:5 = 1.6	1,024,000	WXGA	Common	Common		Possible	Yes
1366 × 768	1.7786	1,049,088	WXGA TV format		Common		No	Yes
1440 × 900	8:5 = 1.6	1,296,000	WXGA+	Common	Common		Possible	Yes
1280 × 1024	5:4 = 1.25	1,310,720	SXGA	Common	Common	Rare	Yes, Recommended	Yes
1400 × 1050	4:3 = 1.3333	1,470,000	SXGA+	Common	Common		Yes, Recommended	
1680 × 1050	8:5 = 1.6	1,764,000	WSXGA+	Common	Rare	Rare	Yes. Recommended	Yes
1600 × 1200	4:3 = 1.3333	1,920,000	UXGA	Rare	Rare	Rare	Yes	Yes
1920 × 1080	16:9 = 1.7778	2,073,600	HD1080 †	Common	Costly	Costly	Yes, could be Useful	Yes
1920 × 1200	8:5 = 1.6	2,304,000	WUXGA	Common	Costly		Yes	Yes
2048 × 1536	4:3 = 1.3333	3,145,728	QSGA	‡				
2560 × 2048	5:4 = 1.25	5,242,880	QSXGA	‡				
3200 × 2048	25:16 = 1.5625	6,553,600	WQSXGA	‡				
3840 × 2400	8:5 = 1.6	9,216,000	WQUEGA	‡				

Notes:

The screen sizes are arranged in order of the total number of pixels, which should be reflected as the order of increasing costs.

* In the Acronyms the individual letters have the following meanings:

V = Video; G = Graphics; A = Array; W = Wide; X = Extended; U = Ultra; S = Super; Q = Quad; + = Plus.

† The resolution of 1080 pixel rows in the Y coordinate is that used in High Definition Television.

‡ The sizes that are characterised as 'Quad', are often realised as 'walls' of screens, driven from multiple graphics cards

2.3.5 The use of a projector for display

Frequently there is a need to connect a computer (usually a laptop) to a projector, so that a group of people may, for example, interactively view the results of an experiment. Projectors of native resolution 1280 × 1024 (or better) are expensive (typically over £2000). Using a projector of lower resolution than that of the host can have either of two undesirable effects; the resolution of the host laptop may be degraded, or the projector display may be cropped. For group viewing, an alternative to a projector that is worth considering is the use of one or more large screen television sets. Sets with 32 inch screens and 1920 × 1080 resolution cost around £250.

2.4 RAID disc systems

There is some uncertainty as to what the acronym RAID[11] stands for; **R**edundant **A**rray of **I**ndependent **D**iscs, or **R**edundant **A**rray of **I**nexpensive **D**iscs. The latter term was the original one but either description will do! The scheme was devised at a time when discs were not as reliable (individually) as they are now, to increase the reliability of the system as a whole, by using two or more discs together, or to increase the speed of access to discs. Use of RAID can be justified in certain molecular modelling systems, but not generally.

2.5 The Linux operating systems

Linux[3] has evolved from a kernel system, written by Linus Torvalds, into a family of complete operating systems. These are 'Open Source' systems that are available at *no cost* usually by downloading from a website, and also *for payment* from commercial companies. Linux systems are important for molecular modelling, since many *ab initio* molecular orbital programs are best run under Linux. There are also complete modelling packages that can be run on *either* Windows, Linux, or Apple systems. More details are given in Chapter 3.

2.6 Having the best of both worlds - Windows and Linux

There are some features of the Linux operating system that makes it attractive for molecular modelling. We have suggested one solution to the problem of using both Windows and Linux together, by having separate caddies that contain the operating systems. This approach has the disadvantage that only one operating system may be used at a time; the machine must be powered down before changing from one system to another.

There are also problems concerning the different formats for text files adopted by the different operating systems. When only a single machine is available, transfer of data between operating systems is best handled by using a USB 'memory stick' as intermediary.

2.6.1 Two or more machines

If cost is not a problem, having two machines is the best way of having Windows and Linux. The two machines are connected using an internal network. The configurations of the machines need not be identical; if only one were to be used for graphics displays, then the other need not have an expensive graphics card. Linux has a software tool called *Samba* that allows discs to be shared between the operating systems.

If your desk space is limited and you decide you need more than one machine, consider purchasing a KVM switch; this would allow sharing of a single **K**eyboard + **V**ideo screen + **M**ouse between 2 or 4 machines.

2.6.2 Dual and multi-boot systems

If a version of Linux is running on a computer, it is possible to arrange for the Linux loader (called GRUB) to control the operation of several operating systems on the machine. Each operating system must be mounted in a separate memory partition, and if a Windows system is included, it must be the first system to have been loaded. Only one operating system may be used at a time, and while it is possible to have different versions of Linux share the use of facilities, it is not easy to arrange to have Linux and Windows systems communicating with each other. (See section 2.6.4 for a solution to this problem by use of Cygwin).

2.6.3 The Windows Subsystem for UNIX-based-Applications

The *Ultimate* version of Windows 7 has the capability (not available in other retail versions) of using the Windows **S**ubsystem for **U**nix **A**pplications (SUA). The extra code must be downloaded from the Microsoft Server as it is not on the installation disc. The subsystem provides a *Command Line Interface* to UNIX, in the form of two *shells*; *C-Shell* and *Korn-Shell*. Using this subsystem (preferably with a wide screen, or two screens) it is possible to run Windows jobs and Linux jobs simultaneously.

At first sight, the *Ultimate* version of Windows 7 is attractive. (The difference in price between the *Ultimate* and *Professional* versions of Windows 7 is only 5%). However, we have found two reasons for not recommending the *Ultimate* version:

(1) It is rarely offered as the operating system installed on a new computer, and thus needs to be purchased separately.

(2) Evaluation of the SUA found it wanting for some of our applications. Crucially it does not provide a satisfactory way of handling data in compressed formats produced on Linux systems. This is probably due to its being a **UXIX** system, and Microsoft's unwillingness to make any concessions to **LINUX**. Specifically it does not provide the ***gzip*** package and its component commands: *gzip, gunzip,* and *zcat*. These are necessary for uncompressing files that have the '.gz' suffix. This form of file compression is used by the Protein Data Bank.

2.6.4 Using Cygwin

Cygwin is a free version of LINUX that is maintained by the firm Red Hat, as an operation separate from both its *Enterprise* version, and the free version called *Fedora*. From within a running Windows system, it is downloaded and setup very simply. It differs from other versions of Linux in that it is specifically designed to run in a Windows environment.

In order to get a taste of what it can do, and assuming that you have a working Windows system, using your default browser go to:

```
http://www.cygwin.com/setup.exe
```

This will connect you to a central server, and will provide a question and answer session that is concerned (mostly) with protecting your system; you can trust it by answering 'yes' to most questions. The penultimate question will ask 'Do you want to load from the Internet', answer 'yes'. This will provide a drop-down menu of 'Mirror Sites'. Choose one appropriate to your location. In the United Kingdom choose:

```
http://www.mirrorserver.org
```

This will result in the display of a hierarchically expandable menu system. The top entry is labelled 'All' and is preceded by a +/- checkbox, and followed by a 'Change' icon and the word 'Default'.

Clicking 'Next' at the bottom of the screen will load a minimal system as the default. However, you can expand each level of the menu, to reveal the various packages, and the individual contents of each package. You can then change each entry by clicking the change icon. This will go through the sequence:

```
Default>> Install>> Reinstall>> Uninstall>> Default
```

If you select 'Install' for the highest level 'All', you will install everything! This can take some time (typically 90 minutes for the download stage, and a further 30 minutes for setting up the system on your computer). However, what you will have is all the Cygwin offering, and this is probably the easiest option

The final act of the setting up procedure will provide an icon on your Windows desktop labelled **Cygwin**. Clicking on this will give you a command-line window with a line:

<Username>@<Machine Name> followed by a line with a $ prompt.

You now have a **Linux** system working alongside your **Windows** system! .

If you type: 'pwd' (print working directory) you will get the response:
/home/<Username> (the normal home directory in Linux)

This directory will normally be empty to start with, but you can begin to build up directories and files here, independently of your Windows files. You will notice that LINUX uses the forward slash character (/) to separate levels in it file structure, whereas Windows uses the backslash character (\ ).

To see how the LINUX and Windows file systems are related type the following commands at the $ prompt:

```
cd ../../          Descend two levels in the file structure
ls -l              List the files and directories
cd cygdrive        Change the directory to cygdrive
ls -l              List the files and directories
```

You should now see a single letter file name corresponding to each drive and partition of a drive in your Windows system. For a single hard-drive you should see only one entry; the lower case letter c. A reference to a CD/DVD drive (typically d) will only appear if there is a disc in the drive,

Now type:

```
cd  c              Change directory to   c
ls -l              List directories and files
```

You should now see a listing of the directories on the Windows system of the computer. The listing will be in LINUX format, and some of the columns will not have much information. However, this should reassure you that the two file systems are connected. With some care it is possible to transfer information between the Windows and LINUX file stores in either direction.

If you are not familiar with LINUX (or UNIX) remember that LINUX commands are nearly always entered using **lowercase** letters, and that in LINUX filenames and directory names are **case sensitive**. Thus the filenames, `rubbish`, `Rubbish`, `RuBbiSh`, and `RUBBISH` all refer to different files or directories.

Table 2.2 shows corresponding commands in Windows and LINUX. More information on Linux commands can be found in the useful reference books.[22, 23, 24, 25]

Table 2.2
Corresponding Commands in Windows and LINUX/UNIX

Purpose	Windows Command Line	LINUX/UNIX
Directory listing	`Dir`	`ls`
Change directory	`cd <directory name>`	`cd <directory name>`
File listing from beginning	`Type <file name>`	`more <filename>`
List lines at end of file		`tail <filename>`
Print name of current directory	`Carriage return`	`pwd`
Execute command	`.\<command>`	`./<command>`
Compress (in place)		`gzip`
Uncompress (in place)		`gunzip`
Uncompress and list		`zcat`
Uncompress to file		`zcat > <filename>`
Find string in file		`grep <string> <filename>`
Search Manual page (*i.e.* help)	`Help 'subject'`	`man <command name>`
Search the manual pages		`apropos <string>`
Concatenate files and display them		`cat <filename>, <filename>, ...`
Copy file	`Copy <file name>`	`cp <file name> <path>`

In Table 2.2 'chevron' brackets '<xxx>' are used to delimit strings that represent file names or commands *etc.* that should be typed as part of the command. The brackets themselves should not be typed.

2.6.5 Virtual machines

The firm *Sun Microsystems* had developed software called *VirtualBox*, before being acquired by the firm *Oracle*. Sun had offered the software for no payment, and this arrangement has been continued by Oracle. The package comes in separate 32 bit and 64 bit versions for Windows, Linux, and Apple systems, and is easy to download from the Oracle site. The package is straightforward to load and there is good documentation as well. What it allows you to do for example is, working on a Linux system to load a version of Windows. This will run *inside* the host system (Linux) and provide you with a (nearly) full Windows system. The word *nearly* needs to be emphasized; some facilities might be missing. There can also be problems in communicating between the host and slave systems.

The other versions of VirtualBox allow Windows to be run inside an Apple machine, and Linux to be run inside a Windows machine. You are not limited to one slave machine; you could have (for example) several incarnations of a single licensed copy of Windows running inside a Linux box.

2.6.6 Clusters of computers

Linux has come to the fore as the operating system of choice if you wish to run *Clusters* of computers. These clusters have become known as *Beowulf* clusters.[26] All the hardware that is needed is the computers themselves, plus networking equipment, and a KVM switch so that you can used one keyboard + screen + mouse to control them all. The computers do not need to be identical, but it helps if they are. Cluster computing is often thought of as difficult to master. This is because one way to implement such systems needs special co-called message passing software. However in some applications a simple controlling program will be adequate. We provide an example on the software disc that accompanies this book.

2.7 Networking

It is certain that you will require equipment to connect your computer(s) to the internet, and to each other. In a commercial or educational environment it is likely that facilities for this will be provided for you, and may not be under your complete control. In a private situation for internet connection you will need an **A**symmetric **D**igital **S**ubscriber **L**ine (ADSL) service. This is a service running on a modified telephone line. The service provider may supply you with the necessary ADSL modem, but for some purposes it might be preferable to use a third party device. This applies if you wish, for example, to run a website or a file server. Windows 7 includes code that allows you to set up a local area network (LAN), and one of the books that we list tells you how to do this.[13]

For the local internal connections between your computers you will need a network switch. These devices are available to connect 4, 8, or more computers; they will also provide connections between computers and networked printers, scanners, and cameras.

2.8 Security

If your computers are connected to the Internet, it is necessary to provide security to prevent attacks on them from outside. There are free packages, notably from the firm AVG, but the same firm will, for a modest annual payment, give more comprehensive protection. This takes the form of anti-virus, anti-spyware, and anti-rootkit programs, and daily updates of their associated databases. You can also protect each computer by implementing a firewall (Windows 7 provides this). As a frontline defense there is usually a firewall in an ADSL modem.

When an operating system is being loaded, there is usually a requirement to provide the name of an administrator, and an associated password. This password should provide strong protection. This means that it should be at least eight characters in length, should include uppercase and lowercase letters, and digits.

Compared to Windows, Linux systems are usually considered to be more resistant to attack. However when Linux has been loaded for the first time, there are present a variety of unnecessary 'pseudo' users (*e.g.* 'printer', 'guest'). All of these should be removed.

2.9 Compatability of programs with 32-Bit and 64 bit operating systems.

Although all of the Interprobe software for Windows (INTERCHEM, PRESTO, QUICKSCAN, PROTEINS *etc.*) was originally developed for 32 bit systems, it works on 64 bit systems. This applies to most software from other suppliers. Because of this, when there is a choice, you should always choose to use or load a 64 bit system.

2.10 Further reading

For the following topics these books are recommended:
Windows Operating Systems,[12, 13, 14, 15] UNIX Operating Systems,[16, 17, 18, 19, 20, 21] Linux Operating Systems,[22, 23, 24, 25] Cluster Computing.[26, 27, 28]

Table 2.3
Download Sites

Firm/ Software	Site	Access Date
Linux		
Fedora	http://fedoraproject.org	28 April 2011
OpenSUSE	http://software.opensuse.org	28 April 2011
Ubuntu	http://www.ubuntu.com	28 April 2011
CERN / Scientific	http://linux.web.cern.ch/linux	28 April 2011
Cygwin	http://www.cygwin.com	28 April 2011
Microsoft		
Updates	http://www.update.microsoft.com	28 April 2011
SUA	http://www.microsoft.com/downloads/en/details.aspx	28/April 2011
Nvidia		
Drivers	http://www.nvidia.co.uk/Download/index.aspx?lang=en.uk	28 April 2011
ATI (graphics)		
Drivers	http://www.amd.com/us/Pages/AMDSupportHub.aspx	25 June 2011
Oracle		
Java	http://java.com/en/download/index.jsp	28 April 2011
OpenOffice	http://www.openoffice.org	28 April 2011
VirtualBox	http://virtualbox.org	28 April 2011
Independent		
Drivers, etc	http://xpdrivers.com	28 April 2011

References and Endnotes. Chapter 2

[1] Microsoft now brings out a new version of the Windows operating system roughly every four years. At that point there is frenzy among computer professionals, to acquire stock of the previous version of the system that they are familiar with, before a traumatic period ensues while the problems of the new system are dealt with. At the present time (March 2011) the current version is Windows 7, and Service Pack 1 has just been issued. So we have a few quiet years before the problems start again. The versions of Windows on which INTERCHEM, PRESTO and the other Interprobe programs run are:

	Issued	**Comments**
Windows NT 4.0	August 1996	No longer supported by Microsoft.
Windows 98	June 1998	No longer supported by Microsoft.
Windows 2000	February 2000	No longer supported by Microsoft.
Windows XP	October 2001	Considered most satisfactory recent version.
Windows Vista	January 2007	Vista has many disadvantages.
Windows 7	October 2009	Better than Vista.

The later versions of Windows (XP onwards) are available in various editions and for both 32 bit and 64 bit systems. The Microsoft Store has the current prices in the UK (excluding VAT) and USA (which should be considered for comparison only):

Home Premium	£120	$120	(for either 32 or 64 bit versions)
Professional	£176	$200	(for either 32 or 64 bit versions)
Ultimate	£184	$220	(package includes 32 and 64 bit versions)

[2] Development of UNIX was started by AT&T (Bell Labs) in 1969, to provide a multi-user operating system for use 'in house'. Coincidentally the programming language C was developed as a aid to coding the system. AT&T offered the system (so-called System V) to academic institutions for no charge. From this, a separate version of UNIX was developed by the University of California at Berkeley, known as Berkley Standard Distribution (BSD). This was also widely distributed. The differences between the two versions, and the fact that the ownership of the original version was acquired by Santa Cruz Operation (SCO), (which was less tolerant of free or unlicensed use of (what it preceived to be) its property), caused problems, and was one stimulus for the development of Linux. Many workstation manufactures (Sun Microsystems, Silicon Graphics Inc., Hewlett Packard Ltd.) has licenced UNIX (and often renamed it), but most of these companies have now replaced it by Linux.

[3] Linux is an Open Source operating system that owes its origins to a modest project started by Linus Torvalds at Helsinki University in 1991. Torvalds is still responsible for overseeing the *kernel* of the system. The remainder of the operating system's tools and facilities have been contributed by a large number of people acting in cooperation. All of the code is contributed under *Open Source* principles. This means that anyone is free to use it without payment of royalties. The only conditions are that any modifications and improvements, and any software into which it is incorporated is also made available under the same Open Source principles.

Versions of Linux are issued both by commercial firms and not-for-profit organisations. The commercial organisations offer what are known as *Enterprise Editions,* for which a charge is made, and for which maintenance and help facilities are provided. The conditions of the Open Source license require that Linux must also be made available without charge (except for the cost of media). The commercial firms fulfil this obligation by offering free versions of Linux in parallel to the commercial versions. Since most people will *download* such a package, (rather than *purchase* a disc), there is essentially no cost involved. In addition, not-for-profit organisations have sprung up as *downstream* suppliers, who offer versions that may be only slightly changed from the parent version. The taxonomy of Linux versions is very complex; there are over 250 versions! The principal suppliers of Enterprise versions and the free versions derived from them are shown in Figure 2.1.

Computers for Molecular Modelling 27

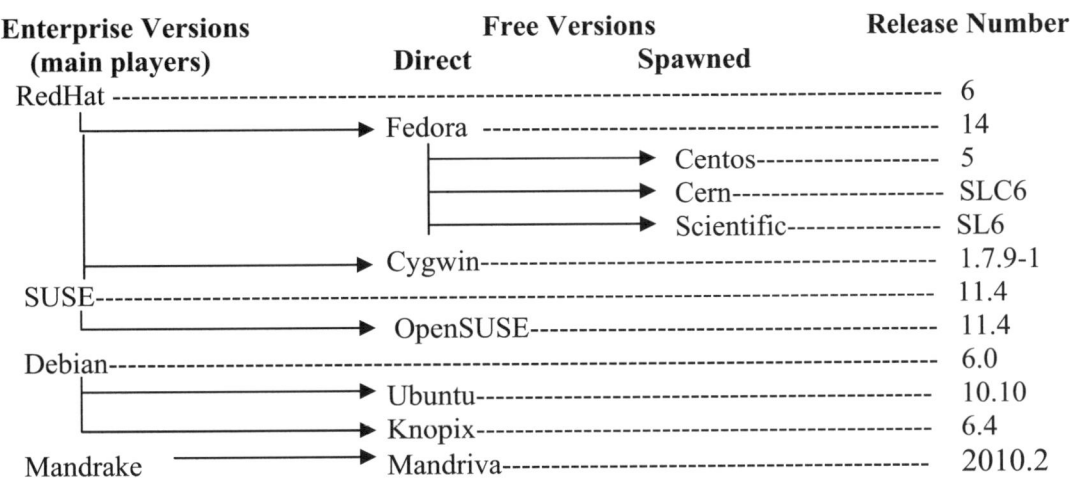

Figure 2.1
A Limited Taxonomy of Linux Distributions

Only a limited number of distributions are listed. They are mostly ones that have been tested and used by the authors.

Scientific Linux (also known as Fermi-Lab Linux) and CERN-Linux are very similar and result from collaboration between the two laboratories.
Ubuntu is unique in having two planned distributions annually in April and October. The release numbers listed are not necessarily the latest ones. When choosing any of the releases, go for the latest regular version (not a beta release).

The main 'problem' with Linux is its success! All the distribution will contain a large number of applications and utilities, with strange names that give no hint as to their purpose. While these will have been checked for errors, there has probably been no selection based upon their utility. Linux inherits the idea of shells from UNIX. Common ones are *C-Shell, Bourne-Shell, Bourne-again Shell,* and *Korn-Shell*, all with slightly different syntax. This variety tends to deter the faint hearted. Nevertheless with patience and perseverance mastery will reap its rewards.

[4] Computers made by Apple Inc. used Motorola 68000 series chips as CPUs until around 2006. At that time the firm switched to using the Intel chips (Core Duo), and thus came into line with hardware used in most personal computers. The traditional Apple operating system was also changed to one based on UNIX. It now has the generic name 'Darwin', and is also known as OS-X. Each new version has a name based on names animals of the cat family: thus Tiger -> Leopard -> Snow Leopard (current version) -> Lion (projected Summer 2011). This Intel architecture facilitates running programs designed for Windows, using either Apple's emulator (Boot Camp) or Oracle's VirtualBox.

[5] POSIX stands for **P**ortable **O**perating **S**ystem for UN**IX**. It serves as a standard for both UNIX and Linux. You will see systems classified as 'POSIX compliant'.

[6] Windows runs on 90% of personal computers, Apple machines account for 8%, the remainder use Linux. This statistic minimizes the importance of Linux by ignoring its use on many large server systems, notably those used by Google.

[7] http//www.intel.com

[8] http://www.amd.com. AMD have acquired the graphics card manufacturer ATI.

[9] http://www.nvidia.com
NVIDIA have specialised in the production of high performance graphics cards, which they sell under two brand names: GeForce for the consumer market; Quadro for the professional market. They have also pioneered the use of the parallel architectures in graphics cards for parallel computing. The brand names Tesla and Fermi are used for these activities. Cards designed specifically for parallel computing are also produced. In this connection the firm has devised a programming language called CUDA.

[10] The caddy systems are made by StarTech. (http://www.startech.com) and have part numbers as follows:
Serial ATA Removable Hard Drive Enclosure including Drawer (White): DRW110DAT
Serial ATA Removable Hard Drive Enclosure including Drawer (Black): DRW110DATBK
Removable Hard Drive Drawer (White): DRW110CAD
Removable Hard Drive Drawer (Black): DRW110CADBK
In the United Kingdom they are available from CPC (http://www.cpc.co.uk)

[11] RAID or Redundant Arrays of Independent Disks is a scheme (a range of protocols) for using sets of discs to provide systems that are less prone to disruption due to disc failure, or to improve speed of accessing discs. The scheme is applicable to all operating systems, and was first used on UNIX systems. It works best when all of the discs in the array are identical in type and size (or at least have the same capacity.

There are several protocols with different aims; the three most commonly encountered are:

RAID 0. In this, two discs are arranged so that alternate blocks of data are stored on alternate discs. This results in faster access but if either of the discs fails then *all* the data becomes invalid. More than two discs may be used when the technique is known as *striping*.

RAID 1 In this, two discs are used in a so-called *mirror* arrangement; identical data is stored on both discs. If either disc is corrupted, then the second disk can be brought into use. Thus reliability is achieved at the expense of halving total capacity.

RAID 5 This protocol uses three or more discs to achieve recovery of data after a single error occurrence. A simplified explanation goes as follows. Suppose N discs are used, then striping is used to split successive bytes of data among $N-1$ discs. Then a bit-wise logical XOR summation of these $(N-1)$ bytes is stored on the extra disc. Errors are detected by reading the data back. If an error is found in one of the discs, then an XOR summation of the data on the error-free-disc *and* the checksum disc gives back the correct pattern of the incorrect byte.

There are limitations in what can be achieved by RAID; it will not correct for errors generated elsewhere in a computer (although similar techniques of check-summing are used to detect and *correct* errors in the CPU and memory). RAID will not prevent errors introduced by viruses *etc.*, nor errors caused by power supply problems. RAID can be provided by *software* techniques, but the only satisfactory way is to use special *hardware* and to have redundant power supplies.

Reference Books - Windows Operating Systems.

These four books are listed in order of increasing dissatisfaction and harshness of criticism of Microsoft Windows. By consulting all of them you will be able to improve your understanding of the way Microsoft operates, and of the way Windows could and should operate!

[12] W. R. Stanek, *Windows 7 Administrator's Pocket Consultant*, Microsoft Press, Redmond, 2010

[13] R. Cowart and B. Knittel, *Windows7 in Depth*, QUE Publishing, Indianapolis, 2010.

[14] P. Thurrott and R. Rivera, *Windows 7 Secrets*, Wiley Publishing Inc., Indianapolis, 2009.

[15] D. A. Karp, *Windows 7 Annoyances: Tips, Secrets, and Solutions: Tools and Techniques to Improve Your Windows 7 Experience*, O'Reilly, Sebastopol, CA, 2010.

Reference Books - UNIX Operating Systems

[16] Sams Development Team (Eds.), *UNIX Unleashed, Fourth Edition*, SAMS Publishing, Indianapolis, 2001.

[17] A. Robbins, *UNIX in a Nutshell, Fourth Edition*, O'Reilly, Sebastopol CA, 2005.

[18] S. R. Bourne, *The UNIX System*, Addison-Wesley, London, 1982.

[19] H. McGilton and R. Morgan, *Introducing the UNIX System*, McGraw-Hill, New York, 1983

[20] P. S. Wang, *An Introduction to Berkeley UNIX*, Wadsworth, Belmont CA, 1988

[21] M. J. Rochkind, *Advanced UNIX Programming, Second Edition*, Addison-Wesley, Boston, 2004

Reference Books - Linux Operating Systems

[22] E. Siever, A. Weber, S. Figgins, R. Love, and A. Robbins, *LINUX in a Nutshell, Fifth Edition*, O'Reilly, Sebastopol CA, 2005.

[23] M. K. Dalheimer and M. Welsh, *Running LINUX, Fifth Edition*, O'Reilly, Sebastopol CA, 2005

Note that many of the books entitled ' **** *Linux Unleashed* ' are available in new editions shortly after a new version of the featured operating system appears. The precise titles for references [24] and [25] may differ from those shown here.

[24] M. McCallister, *OpenSUSE Linux Unleashed*, SAMS Publishing, Indianapolis, 2008.

[25] A. Hudson and P. Hudson, *Fedora Unleashed, 2008 Edition*, SAMS Publishing, Indianapolis, 2008.

Reference Books - Cluster Computing

[26] T. L. Sterling, J. Salmon, D. J. Becker, and D. F. Savarese, *How to Build a Beowulf*, MIT Press, Cambridge MA, 1999.

[27] W. Gropp, E. Lusk, and T Sterling, *Beowulf Cluster Computing with Linux, Second Edition*, MIT Press, Cambridge MA, 2003.

[28] W. Gropp, E. Lusk, and A. Skjellum, *Using MPI, Portable Parallel Programming with Message-Passing Interface, Second Edition*, MIT Press, Cambridge MA, 1999.

Chapter 3
Software for Molecular Modelling and Computational Chemistry

3.1 Chemical structure and molecular modelling
Defining the molecular structure of a substance is frequently the aim of chemical investigation and, in the past, investigations of this type could extend to a personal lifetime of research for a single compound. Now, with the plethora of techniques available to the chemist, structure determination is often a routine matter.

3.1.1 Structures obtained from experimental methods
If the question is put to a knowledgeable experimental chemist today; 'What method would you choose to determine the structure of a new compound'? The answer is most likely to be (with some obvious provisos) 'X-Ray crystallography'. This technique is now the one considered to be the most convenient and reliable, since it can yield directly the three dimensional structure. However, in later chapters we will examine critically its perceived status. Other chemical and spectroscopic techniques that might be mentioned in the answer yield only the nature of atoms that are present, their connectivity, and hints about the distances between the atoms. In these circumstances modelling may become an essential tool. This is true in other cases where the aim is to establish whether a hypothetical structure is capable of being made, or to calculate the properties of known substances.

What do we mean by 'defining a molecular structure'? At the simplest level this requires specifying the chemical identity and relative positions of the atomic nuclei of which a molecule is comprised. This is what we do, maybe somewhat casually, when we sketch a regular hexagon on a piece of paper and declare this to be the structure of the molecule benzene. The positions of electrons associated with the atomic nuclei are implicit and not explicit in our simple model. However, it is possible to perform many useful molecular modelling tasks without calculating the positions of electrons from first principles. This is fortunate as it allows us to utilize the analogy of a mechanical system of masses connected by springs to formulate molecular models. If we limit ourselves to such models then some fairly simple concepts from classical physics are sufficient to explain how the models operate.

3.1.2 The Born-Oppenheimer principle.
The idea expressed in the preceding paragraph might seem arbitrary. What about the electrons – after all are they not what holds a molecule together? It is justified by what is variously known as the Born-Oppenheimer principle or Born-Oppenheimer approximation. The view taken is that, because the masses of atomic nuclei are so much greater than the mass of an electron, the motions of the nuclei can be considered separately from the motions of the electrons. In particular, it is assumed that the electrons are capable of moving faster, and that they follow the motions of the nuclei instantaneously. This forms the basis of the branch of molecular modelling known as molecular mechanics. It has limitations; in particular when we need information concerning electronic structure, we have to turn to calculations that involve quantum mechanics.

3.2 Molecular mechanics

In a classical formulation, a molecular mechanics program is capable of refining an initial structure (Cartesian coordinates of the atoms plus their connectivity). It does this by considering the forces acting on the atoms (hence the alternative name 'Force Field Method'), and minimizing the strain energy of the structure. The most developed versions of molecular mechanics deal with organic compounds, and what follows next is applicable chiefly to this class of compounds (the 'organic set' of elements are usually defined as: C, H, N, O, the halogens, S, P, and sometimes B, and Se).

3.2.1 Molecular mechanics - the basics

In order of decreasing importance the components of the strain energy are:

(1) Bond stretching energy ($E_{Stretch}$)

$$E_{Stretch} = \sum_{1}^{nBonds}(k_n/2)(r_{ij} - r_0)^2 \qquad (3.1)$$

The summation is over the energy of all of the bonds; k_n is the force constant for the particular type of bond being considered (the division by 2 is conventional since the formula is based on Hooke's Law), r_{ij} is the distance between the pair of atoms involved in the bond, and r_0 is the standard bond length expected for the particular type of bond.

(2) Energy from deformation of bond angles (E_{Bend})

$$E_{Bend} = \sum_{1}^{nBondpairs}(k_n/2)(\theta_n - \theta_0)^2 \qquad (3.2)$$

Here the summation is over all pairs of bonds meeting at a common atom, k_n is the force constant, θ_n is the actual angle, θ_0 is the standard bond angle.

(3) Energy arising by deformation of 'equilibrium' bond torsion angles ($E_{Torsion}$)

$$E_{Torsion} = \sum_{1}^{nTorsions}(k_n/2)(1 + \cos(m\vartheta_n - \vartheta_0)) \qquad (3.3)$$

The summation is over each *definable* torsion angle for each bond in the structure. *Definable* in this context means that the bond shall be either a single, double, or aromatic bond (***not*** a triple bond), and shall have atoms attached at both ends that allow a torsion angle to be defined. The factor m is the number of minimum energy points in a full 360° rotation about the bond (3 for single bonds; 2 for double bonds). The calculation ultimately involves the positions of four atoms, two that define the bond itself, and two that allow the definition of the torsion angle. Triple bonds (acetylenic bonds) are not included in the calculation since all the four atoms involved lie in a straight line (or nearly so), and the torsion angle is therefore not definable. For single carbon-carbon bonds, each bond has nine torsions to be considered; for carbon-carbon double (or aromatic) bonds the number is four; for a carbonyl group there are no torsions.

(4) Energy from van-der-Waals forces between pairs of (non-bonded) atoms (E_{vdW})

$$E_{vdW} = \sum_{i=1}^{nvdW} \sum_{j=i-1}^{nvdW} [A_{ij}/r_{ij}^{12} - B_{ij}/r_{ij}^{6}] \qquad (3.4)$$

The formula shown here is based on the Lennard-Jones potential.[1] It is one of several alternative ways of defining this term. The twelfth power term represents the repulsive forces between atoms when they are close together; the sixth power term models the attractive forces at larger distances. The parameters A and B are specified for each combination of atom types, and r_{ij} is the distance between a pair of atoms

(5) Energy from interaction between electronic charges on pairs of (non-bonded) atoms (E_{Elec})

$$E_{Elec} = \sum_{i=1}^{nElec} \sum_{j=i-1}^{nElec} (C/\varepsilon)[Q_i Q_j / r_{ij}] \qquad (3.5)$$

The summation is done over all pairs of atoms that are neither bonded together nor bonded to a common atom; r_{ij} is the distance between a pair of atoms, Q_i and Q_j are the (partial) charges on the two atoms (measured as fractions of an electron charge), and C is a constant that converts the energy into kcal/mole (or kJ/mole). ε is the dielectric constant of the medium.

The total *strain* energy for the structure is then given by equation 3.6

$$E_{Total} = E_{Stretch} + E_{Bend} + E_{Torsion} + E_{vdW} + E_{Elec} \qquad (3.6)$$

3.2.2 The force field parameters
Where do we get the numbers that are used in these equations?
The parameters (force constants, A, B, C) in equations 3.1 to 3.5 are dimensioned and scaled so that the energies of each of the terms are in consistent units (kcal/mole or kJ/mole).

The atomic charges Q_i and Q_j are measured in fractions of the charge of an electron, and if the structure bears no net charge, they should sum to zero. In the absence of charges obtained from (say) quantum mechanical calculations, the charges are usually calculated by the method of Gasteiger and Marsali.[2]

In equation 3.5 the divisor ε is the dielectric constant of the medium; whether this should really be treated as a *constant* is a matter of debate. (See section 3.2.4).

The force constants in equations 3.1, 3.2, and 3.3 can be related to the force constants of vibrational (infrared) spectroscopy, although it is unusual to take the numbers directly from that source. The environments of the atoms concerned are examined in detail; thus for carbon atoms, not only will the type of the bond itself (single, double, etc.) be considered, the nature of adjacent bonds and the general environment (*e.g.* presence in a small ring) will be taken into account.

The inter-atomic distances (r_{ij}) can be calculated from the atomic coordinates by the use of a 3D version of Pythagoras's theorem (Formula 3.7).

$$r_{ij} = [(x_i - x_j)^2 + (y_i - y_j)^2 + (z_i - z_j)^2]^{1/2} \qquad (3.7)$$

What is not so obvious is that the inter bond angles and torsion angles are also readily available from these distances by the methods of vector algebra (see Appendix 1).

3.2.3 Molecular mechanics – how it works

When a structure is submitted to a computer for molecular mechanics calculations, the standard parameters (force constants for each type of bond, standard distances for each atom pair, etc.) are abstracted from a database and fed into the program. The expanded form of equation 3.6 allows the partial differential coefficients of the total energy with respect to each of the Cartesian atom coordinates to be abstracted. The refinement of the structure is then a matter of minimizing the strain energy.

To do this, the program will invoke special software for the minimization. In practice different software might be used in separate stages of the process. These methods go by various names, (*Method of steepest descent, conjugate gradient method, Broyden-Fletcher-Goldfarb-Shanno algorithm*). Because this whole field of minimization is specialized, we make reference to a source book on numerical methods.[3]

All of the processes work by first selecting that atom coordinate that has the largest differential coefficient. The sign and magnitude of this will indicate the way in which the coordinate should be changed, and by how much. After making the necessary change a new set of coefficients is calculated, and the process is repeated. This *iterative* process will stop when changes in the energy become too small to matter.

Whether the final refined structure will correspond to a unique *global* energy minimum will depend, to some extent, on the starting structure. In situations where there are no hints as to what the global minimum energy structure might be, it is possible to keep repeating the refinements using randomized starting coordinates until an acceptable structure is obtained. The process for a structure having N atoms is an attempt to find the minimum in a hypersurface in $3N$ dimensions. It is generally agreed that this process is not guaranteed to succeed even for one dimension![4]

If a molecular mechanics calculation is being used as an adjunct to an experiment, the experiment itself might provide additional information. An example of this a 2D NMR spectrum; if the distances between pairs of atoms, or torsion angles can be inferred, it is then possible to add extra distance and torsion constraints, with high force constants, at the start of the calculation.

3.2.4 Molecular mechanics - limitations

Molecular mechanics calculations are most frequently applied to organic structures. The variety of, and the difficulty of characterizing, bonds in organometallic and inorganic structures have made use in these areas difficult. To illustrate this point, in ferrocene one way of formulating this molecule is to have five bonds from the iron atom to each of the carbon atoms in the two cyclopentadienyl rings (ten bonds in all) (1) in Figure 3.1. All evidence shows that in ferrocene derivatives with a single substituent in each of the cyclopentadienyl rings there is only one isomer. This implies that the connections between the iron atom and the rings are better represented by (2), where the pseudo bonds between the iron atom and the centres of the rings are freely rotatable. The example, ferrocene, is an archetype for a whole class of organometallic compounds, but many more individual cases would need to be included in any practical framework for molecular mechanics to work in these circumstances.

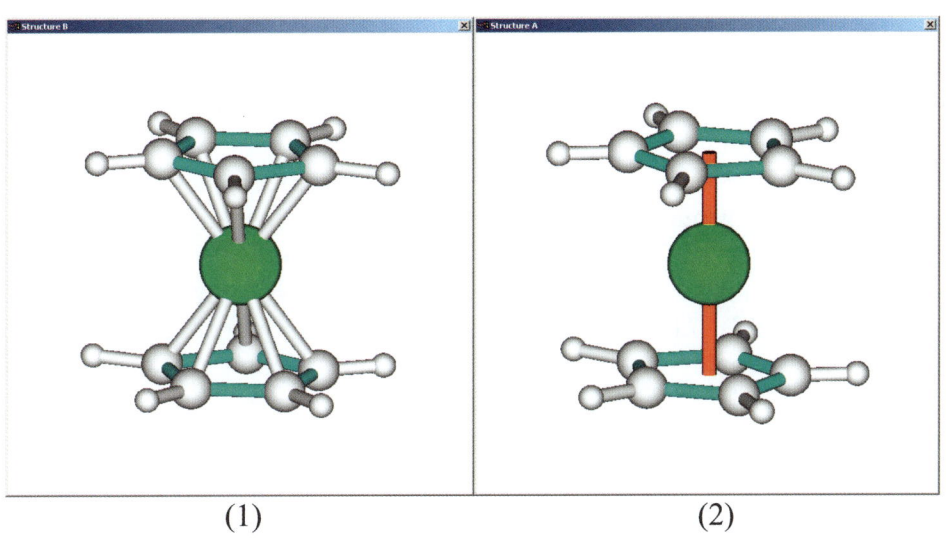

(1)   (2)

Figure 3.1
Alternative representations of Ferrocene
(Carbon and hydrogen atoms shown in white, iron atoms in green)

A difficulty of specifying the dielectric constant in the Coulombic energy term (equation 3.5) has already been alluded to. In calculating structures of small molecules in the gas phase it is usual (and reasonable) to use a value of unity (the dielectric constant of free space) for the space between the atoms. With large biomolecules, such as proteins, the problem arises in choosing a single value. Over short distances it is still reasonable to use a low value for the 'oily drop' model of closely packed hydrophobic amino acid residues. However, to model the situation where water is occluded in the structure, a value approaching that of liquid water (78.0) is more appropriate.

3.2.5 The choice of force fields

The pioneering work of Allinger led to the series of force fields MM1, MM2, MM3, and MMP1. The development of these was discussed in a book published in 1982.[5] Allinger has applied hindsight to review this early work,[6] and put it in perspective. This article also discusses the alternative formulations of the constituent energy terms.

By surveying the publications on molecular mechanics it soon becomes evident that there are a large number of different force fields. If molecular mechanics works at all, why are there so many choices? In section 3.2.4 the difficulty in handling transition metal π-complexes was mentioned. The fact that a quite different type of bonding is involved, and therefore different types of parameters and functional form, provides some explanation.

In other applications, notably in handling proteins and nucleic acid structures,[7, 8] the use of special force fields has been essential. The reasons for this are due to the need for handling hydrogen bonds; the fact that these molecules are polymers is also relevant. While a standard force field will handle the structure of a dipeptide *in the gas phase* with ease, when a structure with many peptide bonds is modeled *in solution* it is necessary to adopt a different approach.

In the past when dealing with proteins, the 'united atom' approach has sometimes been used. This was necessary when computers had limited power. In this approach, for example, a methyl group is treated as a single entity, instead of a separate carbon and three hydrogen atoms. There have been attempts to devise force fields that would be more generally applicable; notably the DREIDING force field,[9] and UFF (Universal Force Field). There are three papers that announced this force field; the first[10] discusses the basics of the method, while the second shows applications to purely organic molecules,[11] and the third applications to compounds containing main group elements.[12] With parameters for covalent bonds between elements in the whole periodic table, transition metal coordination compounds are modelled well, but π-bonding is not covered.

3.2.6 The extended use of molecular mechanics

In later chapters in this book we show how the ideas and techniques of molecular mechanics are incorporated in other applications, for example, where crystal forces need to be taken into account, and where we need to model the interaction of small molecules and enzymes. In both these cases it is frequently assumed that only the 'long range' forces, modelled by equation 3.4, are involved.

3.3 Methods based on quantum mechanics

In the restricted area of defining structures, the aims of quantum mechanics calculations and those of molecular mechanics are similar. Thus a typical quantum mechanics calculation would seek to minimize the energy given by an expression summarized as equation 3.8.

$$E = E_E + E_N \tag{3.8}$$

This equation puts in symbolic form the idea of the Born-Oppenheimer principle, in separating the molecular structure's energy dependencies on the atomic nuclei (E_N) and the electrons (E_E). Because an electron has a mass that is so small (compared to that of typical atomic nuclei), Heisenberg's uncertainty principle tells us that its position and motion cannot be handled in the same ways that those of atoms can. The way forward, in handling electronic energy in molecules, has been to invoke the ideas of electron density (the probability of finding an electron at some volume in space) and electronic orbitals (the shapes and dispositions or these volumes). The calculations are frequently referred to as *Molecular Orbital* (MO) *calculations*.

These calculations have been traditionally been divided into two kinds: (a) semi-empirical, and (b) *ab initio* quantum mechanical methods, but there is really no sharp dividing line between the rigour of the two. The largest molecule for which the structure can be determined analytically (*i.e.* without iterative calculations) is the hydrogen molecule (H_2). For anything more complex than this, the methods have to be tuned in some way. The iterative programs for minimizing energies use the same algorithms (steepest descent *etc.*) that are used in molecular mechanics calculations; the differences are in the functions that are being handled.

We do not wish to minimize the achievements of MO methods; their value when compared with molecular mechanics, is in their ability to provide methods of calculating a whole range of properties of structures (infra red and Raman spectral frequencies, dipole and quadrupole moments). In this book, devoted to the *practicalities* of molecular modelling, we recognize that there are many recently published books that deal with the *theory* of quantum mechanics, and refer you to them.[13,14,15,16,17,18,19,20,21,22,23,24,25] We will concentrate on pointing to the originators of the programs, to explain how to set up calculations, and also to provide guides through the maze of acronyms.

3.3.1 Semi-empirical molecular orbital programs.

This field of research owes much to the work of M. J. S. Dewar and his co-workers. The outcome was two programs MOPAC and AMPAC, that at first were almost identical, but have subsequently diverged. The first of these was freely available as the version MOPAC-93. This came about because the author Dr. J. J. P. Stewart, was at the time employed at the United States Air Force Academy, and the program was deemed to be a publication of the United States Government, and therefore not capable of being copyrighted. This has resulted in MOPAC being probably the most widely used quantum chemistry code, being made available in numerous commercial modelling packages. Unfortunately these distributions often differ from the original code. Later versions of MOPAC *are* copyrighted and distributed under license by Stewart Computational Chemistry.[26]

The second program AMPAC-9 is the direct descendant of the Dewar group's work. It is distributed under license by Semichem.[27]

The belief that (in the case of *ab initio* methods at least) MO calculations work without recourse to experimental information is not true. Both types of program are *parameterised* by reference to experimental heats of formation of a range of compounds. The range has to be wide enough to cover all the elements and bonding types for the intended use of the program. For semi-empirical MO programs the parameterisations have names that belie the philosophy behind the method. They are listed here with their abbreviations:

(1) INDO   Intermediate Neglect of Differential Overlap
(2) MINDO   Modified Intermediate Neglect of Differential Overlap
(3) MNDO   Modified Neglect of Differential Overlap.

The key word is 'Neglect' implying 'How simple do we need to make the program so that we can achieve reasonable accuracy in a reasonable time?'
Then in a more positive move 'How can we accommodate more elements?'

(4) AM1   **A**ustin **M**odel **O**ne   Celebrating Dewar's move to Austin, Texas
(5) PM3   Stewart's **3**rd **P**arameter **M**odel (extension of AM1)
(6) PM6   Further extension to include transition metals

Of these 'methods' numbers (2) to (6) are currently available in MOPAC and AMPAC.

3.3.2 *Ab initio* molecular orbital methods

The most widely quoted program is GAUSSIAN.[28] Other programs are NWCHEM,[29] and GAMESS; this program exists in two versions GAMESS-US,[30] and GAMESS-UK.[31]

In contrast to semi-empirical methods, the aim in *ab initio* programs is to make the programs as accurate as possible, if necessary by using more computing power. This is shown by the way the methods of increasing power and complexity are characterized. Here the jargon word is *basis set*, and the various basis sets have acronyms that will indicate (to the cognoscenti) just what parameters and variables are being used.

Recall that in molecular mechanics methods, a minimum energy structure is being sought by varying the positions of the atomics nuclei. In quantum mechanics methods (both semi-empirical and *ab initio*), it is the minimum in the electronic energy that is being sought. This is done by varying the participation and characters of the *molecular orbitals*. In semi-empirical methods the orbitals are (in general) restricted to those that are recognised as being involved in bonding. Thus with molecules containing only elements of the first three periods of the periodic table (the 'organic set') only s and p bonding orbitals are invoked. When *ab initio* methods are used, choice of the basis set can control the participation of d, f, and higher orbitals (even when these higher orbitals are not usually associated with the elements concerned).

3.3.3 Methods based on density functional theory (DFT)

To some extent the *ab initio* methods mentioned in the last section have been eclipsed by new techniques based on DFT because they are faster. Fortunately all of the *ab initio* programs mentioned in section 3.3.2 have code for DTP applications. A Density Functional method is chosen simply by selecting an appropriate basis set.

What does 'Density Functional Theory' mean? Firstly ignore the word 'Theory'; in this book which aims at practicalities we have deliberately excluded any equations that deal with *quantum mechanics*. Secondly the word 'Functional' is a mathematical term, and we suggest that you refer to a mathematics text to see what it means. The important word is 'Density' meaning *electron density*. What distinguishes DFT methods is the concentration on the *overall electron density* surrounding the atomic nuclei. In contrast, semi-empirical methods consider the electron density in the *bonding orbitals* and *ab initio* methods are distinguished by their inclusion of the higher orbitals.

The beginnings of DFT can be traced to early days of quantum mechanics, but while the development or molecular orbital methods forged ahead, it had to wait until 1964 and 1965 when two seminal papers were published; Hohenberg and Kohn, Kohn and Sham.[32] These are recognised as laying the foundations for DFT. The practical applications have only appeared in the last decade of the 20th Century. For another review see Nagy.[33] The books mentioned earlier[13, 14, 16, 17, 18, 19, 21, 22, 23] have sections that review Density Functional *Theory*.

3.3.4 Basis sets

Both *ab initio* and DFT programs use *basis sets* as the way of choosing the level of rigour. In practical terms these are files of numerical parameters. Each of the programs[28, 29, 30, 31] mentioned in section 3.3.2 has representative basis sets built in, and one of these needs to be chosen at the start of a calculation. In fact there are very many more basis sets that can be called upon. There is a web site called *Basis Set Exchange (BSE)*;[34] the site allows a user to download a basis set (out of a total of 415) tailored for just the elements needed for a particular problem, and in the correct format for the user's program. Elements are selected from a graphic of the periodic table. The paper[35] describing the facility is mostly concerned with the way the database is organized rather than the details of the basis sets..

3.3.5 Internal coordinates and Z-matrices

When calculations are started in a modelling program, it is usual (whatever the ultimate aim of the calculations) for there to be steps where the structure is altered so that the energy associated with the structure is minimized. We have introduced this idea before and called it 'optimisation' or 'energy minimization'. The various calculation methods do this in their own ways, but the basis is a stepwise alteration of the coordinates of the atoms. With a molecule containing N atoms, this means that there are $3N$ atomic coordinate variables to be considered. Leaving aside for the moment the details of how a calculation depends on this number ($3N$), it is clear that the larger the number the longer the calculation will take, and the more it will cost. It was realized early on that the number could be made smaller. The scheme is as follows:

(1) The first atom (i) is defined and placed at the origin of the coordinate system;
    No parameter is assigned

(2) The second atom (j) is assigned a single coordinate (distance vector ij)

(3) The third atom (k) is assigned two variables:
    (a) The distance from the second atom (vector jk)
    (b) The angle that is subtended by the vectors kj and ij at atom.

(4) The fourth atom (l) and all subsequent atoms are assigned three variables
    (a) The distance form from the third atom (vector lk)
    (b) The angle that is subtended by the vectors lk and kj at atom k
    (c) The torsion angle defined by the atoms $l\ k\ j\ i$

The torsion angles are defined, for example in the case of the fourth and subsequent atoms, as being that angle through which vector *ij* would need to be turned to make it coincide with vector *kl*, when looking down vector *jk*. The angle has positive sign when the turn is made clockwise; negative sign when the turn is anti-clockwise.

With this scheme it is easy to see that in a molecule with four or more atoms (*N*) only *3N-6* variables are needed, instead of *3N* when Cartesian coordinates are used. A problem arises when *N=4* and the atoms lie in a straight line, since in these circumstances it is not possible to define a torsion angle. For the same reason it is necessary to ensure that the situation of four consecutive atoms in a line, is not encountered elsewhere in defining the starting coordinates.

It is natural to assume that the atoms need to be specified in an order that allows them to be bonded: *i* to *j*, *j* to *k*, *k* to *l*, and so on, but this is not so. This scheme for specifying a starting geometry, has some merit when the number of atoms is small, but thankfully it is no longer a requirement for most programs, and is usually available as an option. The table of numbers for this scheme is known as a *Z-Matrix* because the triangular shape is reminiscent of the letter *Z*. The method of specifying coordinates as *distance, inter bond angle, signed torsion angle*, is also known as *Internal Coordinates*.

In most programs, although it is no longer obligatory for a user to specify internal coordinates when defining a structure, the program will often convert a set of Cartesian coordinates into internal coordinates in the course of calculations, because the calculations can be done more efficiently. Be aware that, because of this, the order of the atoms in the output file may be different from that in the starting file.

Using input in the form of internal coordinates allows the use of any symmetry that a structure may have to reduce the number of variables in the optimisation process. In this scheme a single line in the input file would take the modified form:

NS, AN, NS1, P1, NS2, P2, NS3, P3

Here *NS* is the serial number of the new atom being defined, and *AN* is its atomic number (or symbol). *NS1, NS2,* and *NS3* are serial number of atoms already defined, and *P1, P2,* and *P3* are character strings that point to separate tables where the numerical quantities are defined.

An example of such a scheme featuring the structure of *benzene* (which has six-fold symmetry) is shown in Table 3.1 together with the corresponding data that does not invoke symmetry. When in optimising a structure the program needs to refine variables, it is the *target variables* that are changed, not the *pointers*.

The reasons for the move away from Z-matrix plus internal coordinate usage is a consequence of the nature of computations now being carried out; structures of interest in the biomedical and medicinal chemistry areas frequently lack extensive symmetry. Moreover, if what symmetry there is in a simple structure is invoked, doing so is in effect restricting the outcome of calculation where the object is to model its interaction with a larger structure, lacking any symmetry, such as a protein.

Table 3.1
Z-Matrix inputs for benzene

NS	AN	NS1	Distance	NS2	Angle	NS3	Torsion
1	6						
2	6	1	1.4200				
3	6	2	1.4200	1	120.0		
4	6	3	1.4200	2	120.0	1	±0.0
5	6	4	1.4200	3	120.0	2	±0.0
6	6	5	1.4200	4	120.0	3	±0.0
7	1	1	1.0800	2	120.0	3	+180.0
8	1	2	1.0800	3	120.0	4	+180.0
9	1	3	1.0800	4	120.0	5	+180.0
10	1	4	1.0800	5	120.0	6	+180.0
11	1	5	1.0800	6	120.0	1	+180.0
12	1	6	1.0800	1	120.0	2	+180.0

(a)

NS	AN	NS1	P1	NS2	P2	NS3	P3
1	6						
2	6	1	D1				
3	6	2	D1	1	A1		
4	6	3	D1	2	A1	1	T1
5	6	4	D1	3	A1	2	T1
6	6	5	D1	4	A1	3	T1
7	1	1	D2	2	A2	3	T2
8	1	2	D2	3	A2	4	T2
9	1	3	D2	4	A2	5	T2
10	1	4	D2	5	A2	6	T2
11	1	5	D2	6	A2	1	T2
12	1	6	D2	1	A2	2	T2

(b)

D1 = 1.42	D2 = 1.08	(Angstroms)
A1 = 120.0	A2 = 120.0	(Degrees)
T1 = 0.0	T2 = 180.0	(Degrees)

(c)

In a simple Z-Matrix there is only one table having eight columns, corresponding to the top table shown here (**a**). The columns hold: the atom serial number (NS), its atomic number (or symbol) (AN), and three pairs of columns. These hold serial numbers of atoms already defined (NS1 etc), and actual numbers for the distance, angles, and torsions, relating the atom NS to them,

In the second form of the Z-Matrix, designed to exploit the symmetry of the molecule, there are two parts. The numerical parameters (NS1 etc) are replaced by pointers in the table (**b**) that point to entries in the third table (**c**), and it is these that are optimized.

In this case of benzene, if symmetry is taken into account, only six parameters need be optimized, compared to 30, if symmetry is not used.

There *is* an advantage in using internal coordinates for simple symmetrical molecules, such as benzene, but this is completely lost on dealing with molecules that are devoid of any symmetry. These are the majority of structures in, for example, drug design application of modelling. Hinchliffe (page 78 in his book *Molecular Modelling for Beginners*[16]) predicts the end of the Z-Matrix input method. Anyone who has tried to generate input coordinates from a floppy physical model using a ruler and protractor can only rejoice.

3.3.6 Atomic units

In their innermost workings, most *ab initio* and DFT programs will use what are known as *Atomic Units* rather than SI units. This is to avoid multiplications and divisions by large numbers. There is a further advantage in that many of the constants (*e.g.* Planck's constant, π, the mass and charge of the electron) are automatically taken care of. Sometimes output coordinates are delivered back to the user in these units, and this can cause confusion. This usually shows up by structures having apparently longer bonds than they should have, due to confusion between the size of the *Bohr Radius* and *Angstrom units*. Atomic units are defined in Table 3.2.

Table 3.2
Atomic Units and Other Units

Name of Unit	Measuring	Symbol	Definition in SI units	Remarks
Bohr Radius	Length	a_0	$0.52917706 \times 10^{-10}$ m	* Unit = Bohr
Ångstrom Unit	Length	Å	1.0×10^{-10} m	
Mass of electron	Mass	m_e	$0.9109534 \times 10^{-30}$ kg	*
Mass of proton		m_p	$1.6726485 \times 10^{-27}$ kg	
Mass of neutron		m_n	$1.6749543 \times 10^{-27}$ kg	
Energy	Energy	E_h	4.3598×10^{-18} J	* Unit = Hartree
Charge on electron	Charge	Q, e	1.6022×10^{-19} C	*
Rydberg constant		R_∞	1.097373177×10^7 m^{-1}	*

The *atomic units* are marked *
Source: *Handbook of Chemistry and Physics 56th Edition,*
Cleveland, CRC Press, 1975

3.3.7 Scaling in computational programs

We have hinted already that the demands that different computational programs place on computer resources is not the same. We would expect that all programs would require *time* (*t*) and *memory* (*m*) that would depend, in some way, on the number of atoms (*N*) in a molecule. But would the dependence be linear ($t \propto N$), or quadratic ($t \propto N^2$), or based on some other function? To find the answer it is usual to look for the most demanding part of the calculations. Thus, in the case of molecular mechanics programs, equations 3.4 and 3.5 show that the number of summations involved in finding the van-der-Waals and Coulombic energies varies as the square of the number of atoms. Computer scientists refer to this using the 'Big – Oh' notation; thus: 'the program involves $O(N^2)$ steps' is to be read as 'the program involves on the order of N^2 steps'. Similar analysis (looking for nested summations) of *ab initio* programs yields an $O(N^4)$ dependency, while DFT programs have $O(N^2)$ dependency.

3.4 Graphical display software

The history of computing in the last thirty years has seen the appearance, waxing, waning, and (in some cases) oblivion of manufacturers, some of which have left us valuable legacies. Digital Equipment Corporation (DEC) showed, in the form of VAX computers, how disc memory could serve when core memory was short, Sun Microsystems Inc. gave us Java, Javascript, and Virtual Box before being acquired by Oracle. But Silicon Graphics Inc. (SGI), still surviving, after bankruptcy in 2009, has given us the gem in the form of OpenGL. This is a tool that arose out of earlier proprietary software. It is prefixed 'open' since SGI has spun off a separate organisation that has participation from other, potentially rival, firms. The aim of this consortium is to make graphical display software more portable. Although the software is not Open Source, there are acknowledged freeware versions (for example Mesa). In principle, the source code of programs that incorporate OpenGL should be transferable between computers running different operating systems: Linux, Unix, Mac OS X, and Microsoft Windows. In practice this goal is rarely completely achievable.

Microsoft was accepted as a member of the OpenGL consortium, but this has not prevented this firm from devising alternative software called DirectX.

3.4.1 The requirements of molecular modelling

The demands that molecular modelling places on a computer display system are, when compared to the requirements of the entertainment industries, relatively modest. The chief requirement is to be able to give to a viewer insight into the shape and other characteristics of a molecular structure. Given that the very idea of shape in this context is hard to grasp, it is easy for molecular modelling to be misused. Hence we start with some facts and caveats.

(1) Molecular modelling on a computer screen is concerned with showing pictures of molecular models!

(2) In scientific publications *colour* can be used in arbitrary ways to make particular points clear, but to do this effectively, *colours* need to be chosen carefully.

(3) In modern computers the *red*, *green*, and *blue* components of the display are capable of being set with 8-bit precision. This means that for each component there are 256 levels of intensity, and when combined there are 256^3 (= 16,777,216) shades.

(4) The human eye is not capable of distinguishing all these shades (many of them are effectively *black*). Nevertheless this diversity of shades is essential for techniques such as lighting and shadowing.

(5) When their true shape is not easily discernable, the ability to show the three-dimensional nature of molecules is a key requirement.

3.4.2 How OpenGL satisfies these requirements

The features of OpenGL[36, 37, 38, 39, 40] that distinguish it from earlier attempts at the provision of cross-platform graphics languages are *efficient handling of 3D geometry, lighting models, surface modelling, texture mapping, animation, font handling, and 3D stereo viewing*. Not all of these features are of interest in molecular modelling. So here we concentrate on those that are.

3.4.2.1 Loop structure

Each graphical object is created by a loop structure that is part of the OpenGL language. These objects can in turn be assembled into larger objects.

3.4.2.2 Three dimensional viewing - Z buffering
This relies on the provision of a Z-buffer. This means that in addition to the 2D coordinates (horizontal X, vertical Y) a third (Z) coordinate is provided for. The numerical values of this coordinate are interpreted, by displaying those objects that are nearer to the viewer more brightly than those that are further away. This technique is frequently called *Depth Cueing*.

3.4.2.3 Colour
The ability to set the background colour, and also the colour of each component of a graphical object is crucial. OpenGL treats the three components (*red*, *green*, and *blue*) of colour itself as *colour coordinates*. There is a fourth coordinate (called *alpha*) that is used to control how two or more colours are blended together, how Z-buffering is controlled, and how lighting is interpreted.

3.4.2.4 Primitive shapes
Besides the ability to draw lines, OpenGL has a library of primitive 2D shapes (triangles, quadrilaterals, polygons *etc.* but not circles). There is no provision in OpenGL itself for 3D objects, but these are catered for in readily available libraries. The OpenGL Utility Toolkit (GLUT) library has spheres, cones, all of the platonic solids, and, (what is the icon of OpenGL), a teapot! The real power of the language lies in the ability for the programmer to construct libraries of other objects.

3.4.2.5 Lighting
This is the essential feature that allows 3D objects to be shown properly. Without it a white sphere appears as a uniformly white circle. A clue as to how the three dimensionality may be realized, is given if an icosahedron is drawn with the faces in different colours. If a regular icosahedron is taken and each triangular face is divided into three smaller triangles, and this process is repeated sufficiently, the result is a reasonably good approximation to a sphere. Lighting effects are achieved by calculating the perceived colours of the faces when they are illuminated by light with specified colour and direction. There is the ability to involve the specular and diffuse reflectances and also the texture of the surfaces. A key requirement is the calculation of the normal vectors (normals) to each of the faces in the object. Lighting is demanding in computer time, and the calculations are usually performed on the dedicated hardware of a graphics card.

3.4.2.6 Animation
Animation is done by repeatedly drawing a picture as the scene changes. For this to be successful, each frame has to be complete before it is displayed. To do this a technique known as *double buffering* is used; each new scene is assembled and stored in the *back* buffer, while the current scene is displayed from the *front* buffer. When the complete scene has been stored the two buffers are *swapped*, and the next scene is sent to the back buffer. Good animation requires that the buffers can be swapped sufficiently fast (≥ 50 frames/sec), and that the changes in the scene are small. In molecular modelling animation it is mainly used to move or rotate the picture of a molecule. In this way it is possible to reveal parts that are hidden, or for which the detail is not clear.

3.4.2.7 Three dimensional stereo viewing

The simplest and least expensive way to use stereoscopy is to draw two images in contrasting colours (red + green) into each frame. The two images differ in such a way that they simulate the images perceived by the left and right eyes. If the calculations are done correctly then the effects of both perspective and parallax need to be taken into account, but it is usual to ignore the perspective. Parallax just involves rotation by 3 to 5 degrees on the vertical axis. When viewed with spectacles, with appropriately coloured lenses, the combined images (red + green -> yellow) are interpreted by the human brain to give the 3D effect. An alternative to the red + green system is to use red + blue yielding a magenta image. The use of spectacles with coloured lenses is known as *anaglyph stereoviewing*. This sort of stereo viewing, as well as that which uses spectacles with orthogonally polarised lenses and a twin polarized projection display, is frequently known as *passive stereo viewing*.

A better 3D effect is achieved if what is known as *quad buffering* is used. In this, for every scene, two angularly displaced images are stored in the *right-back-buffer* and *left-back-buffer*. The swap command now exchanges these buffers with the corresponding front buffers. For a static image the two front buffers are rapidly displayed alternately. The screen image is viewed with special spectacles that have lenses for the left and right eyes, which are alternately and synchronously rendered opaque or transparent. To achieve a satisfactory effect, the system is required to have a refresh rate is ≥ 50 frame-pairs/sec. This means that the display unit must have a refresh rate of ≥ 100 frames/sec. Until recently only cathode ray tube (CRT) display units were able to achieve this speed. However liquid crystal display (LCD) units capable of this speed have now become available. These units together with the viewing spectacles, signaling hardware and appropriate graphics boards are expensive. The real advantage of this type of system is that colour information is not used in creating the 3D stereo effect, and can therefore be used in the conventional way. This type of stereo- viewing where the spectacles are switched by an electrical or electro-optical signal is known as *active stereo viewing*.[41]

3.5 Molecular modelling software suppliers

This book features the software from Interprobe Chemical Services, but there are many more programs that are available. The reference section of this chapter lists some of the agencies from which these may be obtained.[42, 43, 44, 45]

3.6 Further reading and keeping up to date

Two serial book series that are worthy of your attention are: *Reviews in Computational Chemistry*[46] and *Annual Reports in Computational Chemistry*.[47]

References and Endnotes. Chapter 3

[1] J. E. Lennard-Jones, *Proc. Roy. Soc. Ser. A.,* 1924, **106**, 463

[2] J. Gasteiger and M. Marsali, *Tetrahedron,* 1980, **36,** 3219; *Croatia Chem. Acta,* 1980, **53**, 601.

[3] W. H. Press, S. A. Teukolsky, W. T. Vettering, and Brian P. Flannery. *Numerical Recipes in Fortran. The Art of Scientific Computing, Second Edition,* Cambridge University Press, Cambridge, 1992. pp. 387-448.

[4] Reference 3, p. 387

[5] U. Burkert and N. L. Allinger, *Molecular Mechanics*, ACS Monograph 177. American Chemical Society, Washington, 1982.

[6] J. P. Bowen and N. L. Allinger, in *Reviews in Computational Chemistry*, ed. K. B. Lipkowitz and D. B. Boyd. VCH Publishers, New York, 1991, Volume 2, pp. 81-95.

[7] B. R. Brooks, R. E. Bruccoleri, B. D. Olafson, D. J. States, S. Swaminathan, and M. Karplus, *J. Comput. Chem.,* 1983, **4**, 187.

[8] A.D. MacKerell Jr., D. Bashford, M. Bellott, R. L. Dunbrack Jr., J. D. Evenseck, M. J. Field, S. Fischer, J. Gao, H. Guo, S. Ha, D. Joseph-McCarthy, L. Kuchnir, F. Kuczera, F. T. K. Lau, C. Mattos, S. Michnick, T. Ngo, D. T. Nguyen, B. Prodhom, W. E. Reiher III, B. Roux, M. Schlenkrich, J. S. Smith, R. Stote, J. Straub, M. Watanabe, J. Wiórkiewicz-Kuczera, D. Yin, and M. Karplus, *J. Phys. Chem. B,* 1998, **102**, 3586.

[9] S. L. Mayo, B. D. Olafson, and W. A. Goddard III, *J. Phys. Chem.,* 1990, **94**, 8897.

[10] A. K. Rappé, C. J. Casewit, K. S. Colwell, W. A. Goddard III, and W. M. Skiff, *J. Am. Chem. Soc.,* 1992, **114**, 10024.

[11] C. J. Casewit, K. S. Colwell, and A. K. Rappé, *J. Am. Chem. Soc.,* 1992, **114**, 10035.

[12] C. J. Casewit, K. S. Colwell, and A. K. Rappé, *J. Am. Chem. Soc.,* 1992, **114,** 10046.

Further reading for Quantum Mechanical Methods

The books are listed in reverse order of publishing year. Many of the books deal with other aspect of computational chemistry besides quantum mechanics.

[13] J. H. Jensen. *Molecular Modeling Basics*, CRC Press, Boca Raton, 2010.

[14] T. Heine, J.-O. Joswig, and A. Gelessus. *Computational Chemistry Workbook Learning Through Examples*, Wiley-VCH, Weinheim, 2009.

[15] H.-D. Höltje, W. Sippl, D. Rognan, and G. Folkers. *Molecular Modeling, Third Edition*, Wiley-VCH, Weinheim, 2008.

[16] A. Hinchliffe. *Molecular Modelling for Beginners, Second Edition,* . Wiley, Chichester, 2008.

[17] S. M. Bachrach. *Computational Organic Chemistry*, Wiley, Hoboken, 2007

[18] J. Gasteiger and T. Engel (eds). *Chemoinformatics, A Textbook*, Wiley-VCH, Weinheim, 2003.

[19] D. W. Rogers. *Computational Chemistry Using the PC, Third Edition*, Wiley-Interscience, Hoboken, 2003.

[20] T. Schlick, *Molecular Modeling and Simulation, An Interdisciplinary Guide,* Springer-Verlag, New York, 2002

[21] F. Jensen. *Introduction to Computational Chemistry*, Wiley, Chichester, 1999.

[22] A. R. Leach. *Molecular Modelling, Second Edition*, Addison Wesley Longman, Harlow, 1996. (Second edition 2001).

[23] J.-P. Doucet and J. Weber. *Computer-Aided Molecular Design: Theory and Applications*, Academic Press, London, 1996.

[24] G. H. Grant and W. G. Richards. *Computational Chemistry*, Oxford University Press, Oxford, 1995.

[25] R. E. Christoffersen. *Basic Principles and Techniques of Molecular Quantum Mechanics*. Springer-Verlag, New York, 1989.

Computer Programs

[26] MOPAC™. Stewart Computational Chemistry, 15210 Paddington Circle, Colarado Springs, CO 80921, USA. Skype: Jimmy.Stewart2 (Between 1500 and 2200 hrs GMT); email: MrMOPAC@OpenMOPAC.net. See J. J. P. Stewart, *J. Comp. Chem.*, 1989, **10**, 221; *J. Comp.-Aided Mol. Design.*, 1990, **4**, 1. *J. Mol. Mod.*, 2007, **13**, 1173.

[27] Semichem Inc. 12456W, 62nd Terrace, Suite D, Shawnee, KS 66216, U.S.A. Phone: +913-268-2271; Fax: +913-268-3445; URL: www.semichem.com

[28] GAUSSIAN is founded on the work of John Pople; a mathematics graduate of Cambridge University and Nobel Laureate, who, having settled in the United States, worked at Carnegie Mellon University, Pittsburg and at then Northwestern University, Evanston, Illinois. The program GAUSSIAN was for some time freely available, but in recent years (since Pople's death) it has been made available under license commercially by the firm Gaussian Inc. 340 Qinnipiac St. Bldg 40, Wallingford CT 06492 USA. Phone: +1-203-284-2501; Fax: +1-203-284-2421; URL: www.gaussian.com.
The latest version of the program is GAUSSIAN 09. The firm Gaussian Inc has been criticised for its licensing policies, in particular for its refusal to license the software to any research group that produces competing computer code, and for its discouraging any publication of comparisons of the performance of GAUSSIAN with that of other programs. See: *Chem. Eng. News*, 1999, **77**, (28), 27; 2004, **82**, (10), 29; J. Giles, *Nature*, 2004, **429**, 6989. For a rebuttal see: M. Fritsch, www.gaussian.com/g_misc/silly.htm. Another requirement is that when the use of GAUSSIAN is acknowledged in a publication, a reference containing a long list of persons who have worked for Gaussian Inc. is required. If the merit of a program were to be measured by the number of column-centimetres devoted to references to it, then GAUSSIAN would win hands down!

[29] NWCHEM. This program, developed at Pacific Northwest National Laboratory, Richland, WA 99352, USA, is open source. It is distributed by CPC Program Library, Queen's University, Belfast, N. Ireland. It is described in: M.Valiev, E. J. Bylaska, N. Govind, K. Kowalski, T. P. Straatsma, H. J. J. Van Dam, D. Wang, J. Nieplocha, E. Apra, T. L. Windus, and W. A. de Jong, *Comput. Phys. Commun.*, 2010, **181**, 1477.

[30] GAMESS-US. This has been developed by Mark Gordon's group at Ames Laboratory, Iowa State University. It is available at no cost to both academic and industrial users, under a license that does not allow further distribution outside the recipient's group. The source code is included. M. W. Schmidt, K. K. Baldridge, J. A. Boatz, S. T. Elbert, M. S. Gordon, J. H. Jensen, S. Koseki, N. Matsunaga, K. A. Nguyen, S. J. Su, T. L. Windus, M. Dupuis, J. A. Montgomery, *J. Comput. Chem.*, 1993, **14**, 1347. http://www.msg.ameslab.gov.

[31] GAMESS-UK. This program has been developed by Martin Guest's group at Daresbury Laboratory, Warrington, UK. It is derived from GAMESS-US, but it is now sufficiently different to be regarded as a separate program. It is available to workers at UK academic institutions under a no cost license from Dr. Jens Thomas at STFC, Daresbury Laboratory, Warrington, WA4 4AD, UK. Phone +44-(0)1925-603849; Fax: +44-(0)1925-603240; URL: www.cse.stfc.ac.uk; email: jens.thomas@stfc.ac.uk. The program is described in: M. F. Guest, I. J. Bush, H. J. J. van Dam, P. Sherwood, J. M. H. Thomas, J. H. van Lenthe, R. W. A. Havenith, and J. Kendrick, *Mol. Phys.*, 2005, **103**, 719.

DFT Methodology

[32] P. Hohenberg and W. Kohn, *Phys. Rev. B.*, 1964, **136**, 864; W. Kohn and L. J. Sham, *Phys. Rev.A.*, 1965, **140**, 1133.

[33] Á. Nagy, *Physics Reports*, 1998, **298**, 1.

Basis Sets

[34] https://bse.pnl.gov/bse/portal (Accessed 8th September 2010)

[35] K. L. Schuchardt, B. T. Didier, T. Elsethagen, L. Su., V. Gurumoorthi, J. Chase, J. Li, and T. L. Windus, *J. Chem. Inf. Model.*, 2007, **47**, 1045.

OpenGL

[36] D. Shreiner, M. Woo, J. Neider, and T Davis, *OpenGL® Programming Guide, Fifth Edition, The official Guide to Learning OpenGL® Version2*, Addison-Wesley, Upper Saddle River, NJ., 2006. (The Red Book).

[37] D. Shreiner (editor), *OpenGL® Reference Manual, Fourth Edition, (The Official Reference ocument to OpenGL Version1.4)*, Addison-Wesley, Boston, 2004. (The Blue Book).

[38] M. J. Kilgard, *OpenGL™ Programming for the X Window System.* Addison-Wesley, Reading, Mass., 1996. (The Green Book).

[39] R. J. Rost, *OpenGL® Shading Language, Second Edition,* Addison-Wesley, Upper Saddle River, NJ., 2006. (The Orange Book).

[40] R. S. Wright Jr. and M. Sweet, *OpenGL Super Bible, Second Edition,* Waite Group Press, Indianapolis, 2000.

[41] The salient difference between the two types of stereoviewing is that in *active stereo* the two images, (one for each eye) are presented to the brain sequentially, whereas in *passive stereo* the images are presented simultaneously. The terminology is not really appropriate. A better term for *active stereo* is *frame-sequential stereo*.

Molecular Modelling Software

[42] Tripos International, 1699 South Hanley Road, St. Louis MO 63144-2319, USA; Telephone +1314 647 1099;
Tripos, Martin-Kollar-Strasse 17, Munich D-81829, Germany,
Telephone +49 89 45 10300.
URL: http://tripos.com (Accessed 2 August 2011)
Products: Sybyl.
Commercial

[43] Accelrys Inc. 10188, Telesis Court, Suite 100, San Diego, CA 92121, USA,
Telephone +1-858-799-5000.
URL: http://accelrys.com (Accessed 2 August 2011)
Products: Pipeline Pilot, Drug Discovery Studio, Materials Studio.
Commercial

[44] Wavefunction Inc. 18401 Von Karman Avenue, Suite 370, Irvine, CA 92612, USA,
Telephone +1 949-955-2120.
URL: http://www.wavefun.com (Accessed 2 August 2011)
Products: Spartan
Commercial

[45] Mark A. Thompson, Planaria Software LLC, Seattle, WA, USA
URL: http://www.arguslab.com (Accessed 2 August 2011)
Products: ArgusLab,
Commercial, free to academic institutions.

[46] The latest issue is *Reviews in Computational Chemistry*, Volume 27, (Editor K. B. Lipkowitz, Editor Emeritus D. B. Boyd), Wiley, Hoboken, N.J., 2011

[47] The latest issue is *Annual Reports in Computational Chemistry*, Volume 6, (Ed. R. A. Wheeler), Elsevier, Amsterdam, 2010.

Chapter 4
Using INTERCHEM for Molecular Modelling

In this chapter you will learn how to use INTERCHEM to acquire 3-dimensional structures of both small and medium sized molecules. We deal with large macromolecular entities in the next chapter. Small molecules can be easily built up from atoms or from pre-built component parts. Both small and large molecules can be obtained from databases that are (more or less) freely available on the Internet. You will be introduced to the various ways for building structures, and you should be able to judge what is the best method to use in particular circumstances. One object of modelling is to allow you to examine the predicted properties of these structures as a prequel to studying them in the laboratory.

4.1 Some words of advice
First plan what you want to do before you start. Treat modelling like any other form of experimental work. Fortunately you will not be required to undertake risk assessment (unless you are involved in installing a large and heavy computer!).

A major consideration is planning where you store the results of your modelling. For each project create a new directory (*folder* in Windows terminology) in the user area of your computer, and create subdirectories as you go along. Aim to have relatively few files (less than twenty) in each directory; there are exceptions to this when you are generating large numbers of structures, or abstracting them from a database.

Keep the lengths of the names of directories and files short (the standard '8.3' convention for files[1] has advantages) but do not use single character names. Windows allows for names with embedded spaces, but this can be confusing, and is best avoided.

Keep separate written notes of what you have done, or if you are fortunate to have a computer with two screens, use one screen as an Electronic Laboratory Notebook for recording crucial steps.

Finally in this list of exhortations make regular backups of your work on a portable file store (*e.g.* a USB Stick).

Figure 4.1 shows the Entry - (Dual Display mode) screen of INTERCHEM; this is the default display, and starting from it you can access all the facilities of the program, through other display modes

4.2 Building structures
INTERCHEM provides five methods of building structures:
(1) Sketching.
(2) Using SMILES strings.
(3) From pre-built fragments.
(4) By merging of structures.
(5) AUTOBUILD; a separate program for assembling large numbers of structures.
What follows now are worked examples illustrating the first four of these methods.

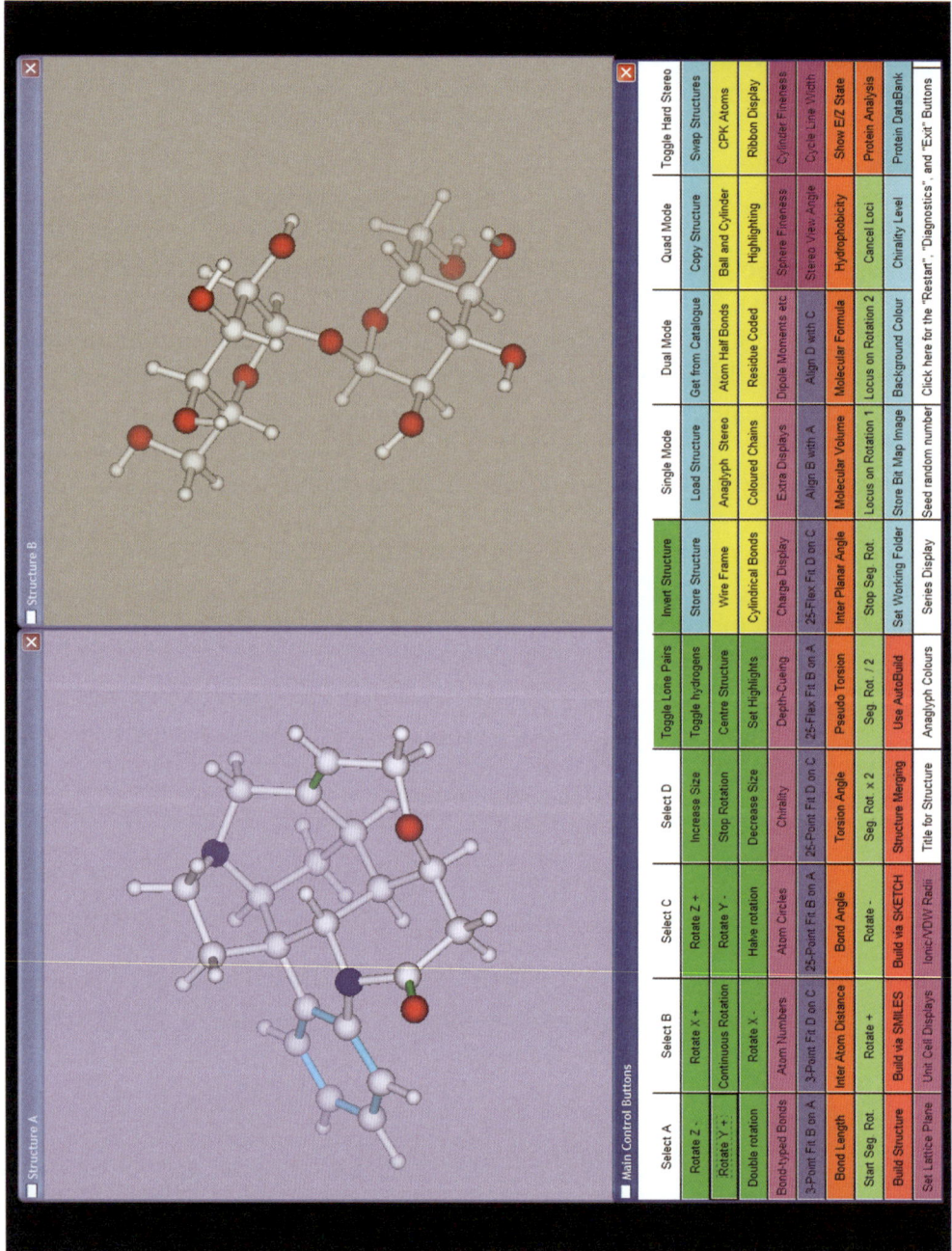

Figure 4.1
INTERCHEM Dual Display Mode
Structure A is Strychnine, Structure B is Trehalose

4.2.1 Using INTERCHEM sketch - basics

Start up INTERCHEM in the usual way, and from the *DUAL* display mode click on *Build via SKETCH* (R9/C3). The display will change so that the right hand window now shows a hexagonal array of green circles. You will use these circles as guides (a template) in your sketching. Moving the mouse sprite to anywhere within a circle and clicking the left button will tell the computer to define a carbon atom to be at the centre of the circle, and also define that atom as the starting point for a bond. The spacing of the circles is 1.5 (in arbitrary units). These are interpreted by the program as Ångstrom Units, since 1.5Å is a good compromise for the inter-atomic distance of most bonded atoms, (except hydrogen). You can choose to draw structures with either *single bonds* (Button R3/C8) or *aromatic bonds* (Button R2/C8). Start the drawing by clicking Button R3/C6. Then click on a series of circles with the ***LEFT*** button to trace out the structure you want.

As a general rule use adjacent circles, and if you are tracing out a chain structure (an acyclic structure not involving rings) make a zigzag pattern. (If you were to click on three circles in a straight line it is not easy to tell afterwards how many atoms and bonds are present). If you are making a ring structure the grid of circles lends itself to the formation of the ubiquitous 6-membered rings

To complete the drawing of the basic skeleton of atoms, (which are to start with all carbon atoms by default), click on the ***RIGHT*** button for the last circle. This takes the program out of the continuous drawing state. This last step is very important, since if you do not stop the drawing, you are liable to create unwanted atoms and bonds in your structure. You can always check on the ***Status*** of the drawing process; this is displayed in a separate window at the lower left hand side of the screen. There is a button (*End Bond Drawing* R3/C7) that, if necessary, will force the drawing process to stop.

When you click on a circle that has already been used (*i.e.* already has an atom assigned to it), a new atom is *not* created. However, if you retrace a path between circles, *new bonds are temporarily created.* This feature is not an error, and is purposely allowed so that you can draw complicated structures in one continuous sequence. *Note that making a bond twice is not a method for making a double bond!*

The *Status Window* tells you whether the drawing process is ON or OFF, and whether single or aromatic bonds have been chosen. It also tells you the total number of atoms and bonds that you have created. There is a button *Tidy Sketch* (R2/C7) that will remove duplicated bonds (and any unconnected atoms). The revised numbers of atoms and bonds is shown immediately in the *Status Window*.

4.2.1.1 Example - ethylbenzene

In this exercise you will draw two joined sequences of atoms in separate operations. First, using *Aromatic Bonds* (*Sketch Mode* R2/C8) draw a simple hexagon on the screen. Next switch to *Single Bonds* (*Sketch Mode* R3/C8) and, starting from one of the carbon atoms of the hexagon, attach a two-atom chain. The aromatic ring will be shown coloured in *cyan*, and the two-carbon side chain will be coloured *white*. The structure that you have sketched will look like Figure 4.2(a).

Now click on the button *Convert 2D->3D* (R4/C6). An optimised 3D structure will now appear in the left hand window. Although there were no hydrogen atoms specified in the drawing, the appropriate hydrogen atoms are added in the 3D structure. You can copy this structure to one of the four main display areas (*A, B, C,* or *D*) of INTERCHEM, (buttons R1/C2, R1/C3, R1/C4, R1/C5) or directly save the structure in a file (R7/C4).

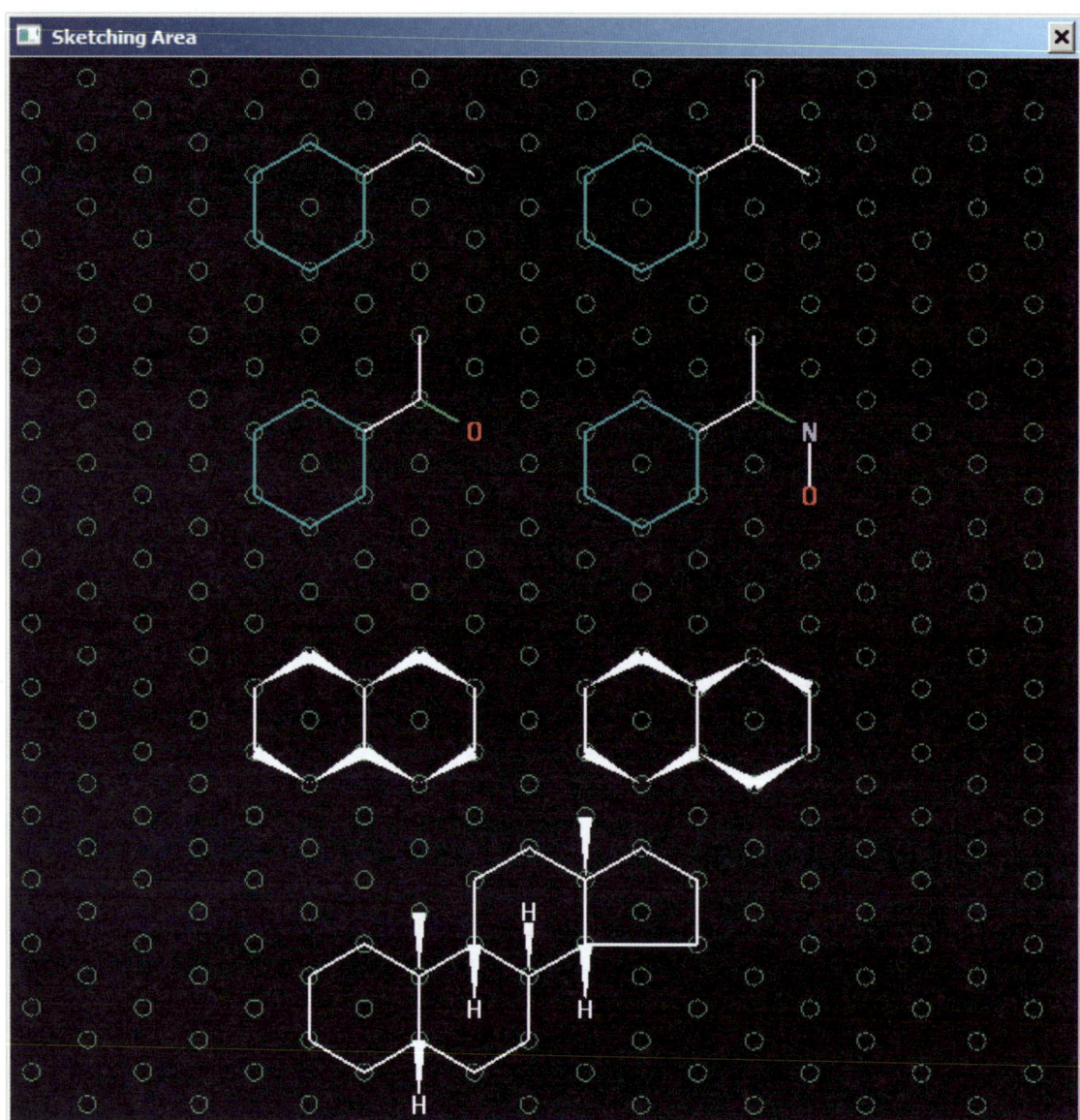

Figure 4.2

Input for Sketching
In order: left to right; top to bottom
(a) Ethylbenzene, (b) Isopropylbenzene,
(c) Acetophenone, (d) Acetophenone oxime
(e) *trans*-Decalin, (f) *cis*-Decalin,
(g) 5α,8β,9α10β,13β,14α-Androstane

4.2.2 INTERCHEM sketch – modifying your drawing

Your original sketch persists in the right hand window. When you drew the sketch, all of the *atoms* were carbons, and you were limited to drawing *single* or *aromatic* bonds. To be able to draw other types of structure, buttons are provided to *Change Atoms* (R6/C6), *Change Bonds* (R6/C7), *Delete Atoms* (R6/C8), and *Delete Bonds* (R6/C9). To define a bond, in order to alter (change) or delete it, you need to click on both of the two *Atoms* that are joined by the bond (the order in which you pick them does not matter). When you change an atom or bond you are offered choices from menus for the new atom or bond types. If you delete an atom, *all* of the bonds connected to the atom are also deleted.

4.2.2.1 Example - acetophenone (methyl phenyl ketone)

Modify the ethyl benzene structure that you created in the following way: using single bonds, add a further carbon atom to the side chain to make isopropyl benzene (Figure 4.2 (b)). Next click on *Change Atoms* (R6/C6), and then pick the new atom and change it to an *oxygen* atom. Then choose the bond that joins this atom to the rest of the structure, and change it to a *double* bond (Figure 4.2(c)). Now if you click on *Convert 2D->3D* (R4/C6), a 3D structure of acetophenone will be generated in the left hand window.

4.2.2.2 Example - acetophenone oxime

You can modify further the forgoing 2D structure (acetophenone). Click on *Change Atoms* (R6/C6), and pick the *oxygen* atom and change it to *nitrogen*. Next make a new (single) bond starting from this nitrogen atom. Finally change the terminal (carbon) atom of this bond to *oxygen* (Figure 4.2(d)). Clicking on *Convert 2D->3D* (R4/C6) will now give a 3D structure of acetophenone oxime. It is likely that this will not be the best conformation for this molecule; this could be improved by use of the program MOPAC (see later in this chapter).

4.2.3 Using INTERCHEM sketch with aliphatic and alicyclic structures

The 2D structures that you have drawn so far, and the 3D structures that have been derived from them, have been based largely on aromatic ring structures. These rings are essentially planar (although there have been side chains that are not necessarily planar). Because of this, the conversion of the 2D *drawing* to a 3D *structure* is a (relatively) straightforward task for the computer. In contrast generating structures of aliphatic (open chain) compounds and alicyclic compounds presents more problems. These problems are related to the possibility of having multiple isomers or conformers.

4.2.3.1 Example - cyclohexane

Choose *Use Single Bonds* (R3/C8). As before, draw a hexagon on the screen by picking adjacent circles using the LEFT mouse button; finishing the closure with a RIGHT button click. The bonds will now appear as white lines signifying that they are in fact single bonds. Now click on *Convert 2D->3D* (R4/C6). The 3D structure that appears in the left hand window will be (most likely) the *chair* form of cyclohexane.

However occasionally you will get the *twisted boat* form. This ambiguity is caused by the way the molecular mechanics program is constructed; it incorporates a degree of randomness in its workings. To cope with this, the *Sketch* facility has buttons that can modify the 2D drawing by providing 'hints' to the 3D structure, and the use of these is illustrated by two examples. When naphthalene is fully hydrogenated, a mixture of two isomeric products is formed: trans-decalin and cis-decalin. These are distinct stable compounds that, in the real world, can be separated (for example by gas chromatography).

4.2.3.2 Example – trans-decalin
Using single bonds, draw two fused cyclohexane rings. Draw the two rings side by side; this is not strictly necessary, but we assume in what follows that you draw them in the same way. Next modify the structure by clicking on *Pull Atom Up* (R2/C9) and the click on one of the top atoms of the structure. Repeat these steps for the other top atom. Now use *Push Atom Down* (R3/C9) with the two bottom atoms. The drawing should now be as shown in 4.2(e). The thick ends of the wedge shaped bonds are connected to the atoms that are nearest to you, the thin ends of the bonds are connected to the atoms that are furthest away. Using *Convert 2 ->3D* (R4/C6) will produce 3D structure of trans-decalin with the two fused cyclohexane rings both in the chair form; the *trans* designation indicates that the *hydrogen* atoms attached to the bridge head atoms are on *opposite* sides of the rings.

4.2.3.3 Example – cis-decalin
Take the 2D drawing made for the preceding example, and click on *All Atoms in Plane* (R3/C10). This will remove the hints that were applied previously. Instead, apply *Pull Atom Up* (R2/C9) to the top left and bottom right atoms, and *Push Atom Down* (R3/C9) to the top right and bottom left atoms; Figure 4.2(f) shows the result. Using *Convert 2 ->3D* (R4/C6) will produce the 3D structure of cis-decalin with two fused cyclohexane rings both in the chair form (as before); the *cis* designation indicates that the *hydrogen* atoms attached to the bridge head atoms are on *the same side* of the rings.

4.2.3.4 Example – androstane
Steroids are an important class of natural products. They are characterized by having a central skeleton of four fused rings of carbon atoms; three six membered rings and one with five atoms. In nature these rings are functionalized by substitution and the presence of double bonds, and in some cases by rearrangement of the skeleton. Androstane is a hydrocarbon and can be regarded as the archetype steroid. It has six chiral centres, and thus can exist as 2^6 (=64) isomers. For this example we build just one of these. Construct the skeleton shown in figure 4.2(g). Note that two of the six so-called *bridgehead substituent atoms* are carbon atoms, the other four are hydrogen atoms. It is necessary to explicitly draw *all* these atoms, and convert four of them to hydrogen atoms. The stereochemistry is now directed by pulling both substituent bridgehead carbon atoms *UP,* and pulling or pushing the other bridge-head substituent (hydrogen) atoms, so that the sequence of these atoms is alternately *DOWN* and *UP.* When the conversion to a 3D structure is done, you will get that for 5α, 8β, 9α, 10β, 13β, 14α–androstane. The Greek letter α denotes that the substituent (hydrogen or carbon atom) is below the (average) plane of the ring system; the letter β means that it is above the plane.[2] (See appendix A3 for more information on the nomenclature of steroids).

Using INTERCHEM for Molecular Modelling

4.2.4 Generating structures using SMILES strings

The sketching method that has been described in the preceding sections in this chapter has required that we have some idea about the stereochemistry of the target structure. In favourable cases, the one refined structure that is obtained will be the one you want! The second method uses SMILES[3] strings as starting points, and while it will initially give you a single structure, it also is capable of generating a series of structures. This allows you to investigate what structures (*i.e* conformers) might be possible for a specified set of connected atoms. The basic rules needed for constructing SMILES strings (for use by INTERCHEM) are given in Table 4.1

4.2.5 Accessing the SMILES facility

Since some features of this facility can generate automatically large numbers of files, it is more than ever important to have a folder (subfolder – subdirectory) assigned to each job that you tackle.

SMILES is started from the *DUAL* display by clicking *Build via SMILES* (R9/C2). This will cause a new horizontal window to appear. This has a section for inserting the SMILES string, and several buttons. The next step is to enter into the typing space the appropriate string, and then to click on CONFIRM—ACCEPT. After a time, (that depends on the size of the structure), a window will appear showing the *excess* energy associated with this structure. Clicking ESCAPE will clear this window, and a subsidiary menu will appear giving you the option (amongst others) of storing and displaying the structure in one of the four standard display areas. The SMILES string that you have entered is retained and will be displayed the next time you invoke SMILES.

This is useful since it is quite easy to make errors in entering SMILES strings. The program will detect if the string has *syntactical* errors. If you have made a mistake, the redisplayed string can be corrected in the usual way, using backspacing to erase characters; CONFIRM-ACCEPT will then process the corrected string.

The first molecules that we deal with purposely go over ground that was covered when using the SKETCH tool. This is so that you can assess the merits and demerits of the two building methods.

4.2.5.1. Aliphatic hydrocarbons

Refer to Table 4.1 for the basic rules for writing SMILES strings. Here you will find that to describe simple aliphatic structures involving the common 'organic set' you use the normal atomic symbols of the elements. Thus the three strings:

$$CC \quad CCC \quad CCCC$$

will generate the three homologous aliphatic hydrocarbons ethane, propane, and n-butane. The use of upper case letters for the symbols implies that the bonds are single bonds that do not need to be specified.
When double bonds are needed, use the 'equals' symbol. Thus:

$$C=C \quad \text{and} \quad C=CC=C$$

will generate ethene (ethylene) and 1,3-butadiene respectively.

For triple bonds use the 'hash' symbol (#). Thus:

$$CC\#CC$$

gives but-2-yne.

Branching in a structure is indicated by the use of round brackets. Thus the strings:

$$CC(C)C \quad \text{and} \quad CC(C)(C)C$$

will give isobutane and 2,2,dimethylpropane (neopentane) respectively.

4.2.5.2 Cyclic compounds
To specify the SMILES strings for a compound containing one or more rings proceed as follows:
(1) Draw the structure using pencil and paper and then break each ring where there is a single (or an aromatic) bond.
(2) Label each of the two atoms that were originally joined with the *same single digit number after the atom symbol*.
Thus for cyclohexane use:

$$C1CCCCC1$$

For decalin use:

$$C12CCCCC1CCCC2$$

Note in this last example that the first carbon atom is labeled as the break point in the two rings and consequently has two adjacent digits.

At first sight it might seem that the use of single-digit labels limits the specification of only nine ring breaks. This in not so, since once one of the digit pairs has been 'used up', (*i.e.* completed) it is possible to use it again. Thus, to specify a SMILES string that will generate the androstane isomers, we could use either:

$$CC12CCCCC1CCC3C2CCC4(C)C3CCC4$$
$$\text{or}$$
$$CC12CCCCC1CCC1C2CCC2(C)C1CCC2$$

It is possible to have ring break numbers greater than nine (up to 99) if the digit pair is preceded by a percent symbol. Thus cyclohexane could be specified as:

$$C\%11CCCCC\%11 \quad \text{or all possibilities to} \quad C\%99CCCCC\%99$$

4.2.5.3 Benzene and other aromatic compounds
Referring to Table 4.1, you will realize that *Benzene* can be specified by the string:

$$c1ccccc1$$

If this string is entered into the typing area and CONFIRM—ACCEPT is pressed, you will be able to display the structure in a display window.

Using INTERCHEM for Molecular Modelling

This is not the only way that the benzene structure may be specified; an alternative using upper case letters and 'equals' symbols to indicate double bonds is:

$$C1=CC=CC=C1$$

Using this will give a Kekule structure that has alternating double and single bonds. If you display this you will see that the lengths of these alternating bonds are not the same.

Table 4.2 shows the SMILES strings for other representative aromatic hydrocarbons, and more general structures.

4.2.5.4 Chain branching and functional groups

In section 4.2.5.1 it was explained how, the SMILES strings of branched chain hydrocarbons are specified, by enclosing the branched part in parentheses. This technique can be applied to any reasonable depth using nested brackets and is not limited to carbon chains. Thus for 3-bromobenzaldehyde we can write:

$$c1c(C=O)c(Br)ccc1 \quad \text{or} \quad Brc1cc(C=O)ccc1$$

These are two equally valid strings. Again note that hydrogen atoms are not usually specified. When INTERCHEM generates a 3D structure hydrogen atoms are added to satisfy the normal valences of the atoms that are present in the string. However the need to add hydrogen atoms correctly requires that the bonding in some functional groups be specified in ways that are not immediately obvious. (See Table 4.2)

4.2.5.5 Specification of stereochemistry and conformation in SMILES

The way that stereochemistry is denoted when INTERCHEM uses SMILES strings is summarized in Table 4.1 as Rule 5. This is different from the way that other modelling software packages tackle the problem. This is because we require INTERCHEM to produce final structures of correct stereochemistry *and* conformation. These two requirements are distinct, but they both use an extra way of labeling atoms using characters enclosed in curly brackets.

The simple codes {R} or {S} (or the corresponding ones with lower case letters) force the *following* atom (carbon, nitrogen or sulfur) to respectively have the *R* or *S* stereochemistry according to the CIP convention.[4] The way the program handles this is to initially ignore these prefixes, but after the 3D structure is generated, it tests the configuration of the first labelled atom; if the configuration of the atom is wrong, the whole 3D structure is converted to its mirror image; otherwise it is left alone. This simple approach means that only the first atom labeled with {R} or {S} in the string can be treated in this way.

Conformation is handled by labelling *pairs* of atoms with strings, that specify the torsion (or pseudo torsion) angles that relate the atoms. This technique takes care of both E/Z (*cis/trans*) double bond isomerism and also rotational (conformational) isomerism about single bonds. Examples are shown in Table 4.2.

Table 4.1 Rules for Constructing SMILES Strings for use by INTERCHEM

(1) Atoms in the following symbol list are specified by these symbols:
H, B, C, N, O, F, Si, P, S, Cl, Br, I
(1a) Hydrogen does not need to be specified in most cases.
(1b) Atoms in 6-membered aromatic rings (*e.g.* benzene) are specified by lower case letters:
c, n, o, s
(1c) Aromatic 5-membered rings (*e.g.* furan) are specified with upper case letters and double bonds.
(1d) Other elements can be indicated by their symbol closed in square brackets:
[Au]

(2) Bonds are specified by the following symbols:

Single bonds: - (hyphen)
Double bonds: = (equals sign)
Triple bonds: # (hash symbol)
Aromatic bonds: : (colon)

(2a) Single bonds do not need to be specified between non-aromatic atoms
(2b) Aromatic bonds do not need to be specified between aromatic atoms

(3) Branching is indicated by enclosing the branched atoms within parentheses
(3a) Branching may be specified to depth 40.

(4) Ring structures are indicated by (formally) breaking the ring at any single or aromatic bond and appending to the two atoms, that defined the broken bond, identical single-digit numerical labels.
(4a) Atoms that are involved in more than one ring may be labeled by more than one single digit label.
(4b) Although not usually necessary, two-digit labels may be used for defining rings if the digits are preceded by the percent symbol: (*e.g.* %45)

(5) Stereochemistry can be indicated by preceding atom symbols with extra symbols enclosed in braces (curly brackets { and }).
(5a) The single letters R or S denote the configuration of the following atom according to the Cahn-Ingold-Prelog convention. While any number of such designations may be present, INTERCHEM takes account of the first one only.

(5b) Stereochemical relationships of pairs of atoms is denoted by preceding each of the atoms by identical groups of characters:

Xnn

Where X is one of the characters: T or t, C or c, G, or g, +, or – , and nn is a one- or two-digit number. Atom pairs labelled in this way and separated by two (or more) other atoms define a torsion (pseudo-torsion) angle subject to the following limits:

T or t	+170 to +190	(trans)
C or c	−10 to +10	(cis)
G or g	−70 to +70	(gauche)
Q or q	+75 to +105	(allenic)
+	+50 to +70	
−	−70 to −50	

(5c) An atom can have up to four stereochemical labels, (as defined by rules 5a or 5b), each enclosed in a separate pairs of braces.
(5d) In order to specify stereochemistry hydrogen atoms may be included.

(6) The end of the SMILES string is denoted by a space.

Table 4.2 Examples of SMILES Strings for use by INTERCHEM

Name	SMILES	Notes
n-Butane	CCCC	
2,2-Dimethyl propane	CC(C)(C)C	Rule 3
Benzene	c1ccccc1	Rule 4
Pyridine	n1ccccc1	
Ethyl cyanate	CCC#N=O	N1
Ethyl isocyanate	CCN#C=O	
Mercury fulminate	O=N#C[Hg]C#N=O	Rule 1c, N1
Nitrobenzene	c1ccccc1N(:O)(:O) or c1ccccc1N(:O):O	N2
Nitrosobenzene	c1ccccc1N=O	
n-Butyl nitrite	CCCCON=O	
n-Butyl nitrate	CCCCON(:O)(:O) or CCCCON(:O):O	
Benzaldoxime	c1ccccc1C=NO	
Tetramethylallene	CC(C)=C=C(C)C	N3
Methanesulfonic acid	CS(=O)(=O)O	
p-Nitrobenzenesulfonic acid	c1cc(N(:O):O)ccc1S(=O)(=O)O	Rule 3a
Cyclohexane	C1CCCCC1	
1,4-Dimethylcyclohexane	CC1CCC(C)CC1	
D-(+)Camphor	C{R}C12CC(=O)C(C2(C)(C))CC1	Rule 5a
L-(-)Camphor	C{S}C12CC(=O)C(C2(C)(C))CC1	N6
Hexahelicene	c1ccc2ccc3ccc4ccc5ccc6ccccc6c5c4c3c2c1	N4
Diazepam	c12ccc(Cl)ccc1C(c3ccccc3)=NCC(=O)N2C	Rule 4a
1R-trans-1,2-Dimethylcyclohexane	C{R}C1({T1}H)C({T1}H)(C)CCCC1	Rules 5b,5d
1S-cis-1,2-Dimethylcyclohexane	C{S}C1({G1}H)C({G1}H)(C)CCCC1	
L-Alanine	N{S}C(C)C(=O)O	
L-Alanine in zwitterion form	N(H)(H)(H){S}C(C)C(:O):O	N5
Benzo(b)furan	c12ccccc1C=CO2	Rule 1c
Cubane	C12C3C4C1C5C4C3C25	
Cycloheptene	C1CCC=CCC1	
Trans-cycloheptene	C1C{T1}CC=C{T1}CC1	Rule 5b

Notes

Many of the structures are discussed in the text.
The rules refer to those listed in Table 4.1.

N1 This string implies pentavalent nitrogen, but it is interpreted correctly.

N2 Note the way that the nitro group is coded.

N3 This will generate a structure that is planar. Use the build-from-fragments facility to rotate one half of the structure about one of the double bonds in order to place the methyl groups in a tetrahedral arrangement (see text).

N4 This interesting structure can be coded without any branching.

N5 Note how both the fully protonated nitrogen cation and the carboxylate anion are coded

N6 The strings for the camphor isomers show the use of nested parentheses to specify branched chains and loops.

4.2.5.6 Coping with isomers or conformers when things don't work as expected!

In the preceding paragraph, we have assumed that the process of converting the SMILES string into the desired 3D structure is successful. This is not always true where there is the possibility of there being more than one isomer or conformer. There are two options that help in these situations. After having run a trial that has given an incorrect structure, you have the option of doing multiple runs and either, (a) keeping the isomer that has the lowest energy, or (b) keeping a series of structures. (In both cases the number of runs can be chosen as a power of two from 4 to 512). Thus to be certain of getting the chair form of cyclohexane, choose to get the lowest energy form, (eight runs are usually sufficient). To get cis-decalin (see Table 4.2) run a series (16 or 32) and select from the resultant structures, one that has energy 2.03 kj/mole.

With large ring saturated hydrocarbons you can search for all possible conformers (*i.e.* investigate the conformation energy surface). Cyclododecane is a classic case.[5] (The SMILES is C1CCCCCCCCCCC1). Make sure you create a new directory to hold the results and (after the obligatory initial run) choose to save the results from 512 runs. The program will ask for a four-character filename prefix (choose something memorable). When the runs are finished, use the facility to display a series of structures (*DUAL* Display R10/C6), and from within that, the tool to sort the series according to the energy (R1/C10). Sorting the structure files also produces a text file that lists the energies of the structures, which you can print. The series-display tool has many uses. In this case, you need only to use the facility to display the structures in a right-hand window. Click on *Setup B for SERIES* (R9/C3). Then enter the name of the folder that stores the structures, and click on *Interchem D format*. The display will respond with the statement of the number of structures of this type that are in the folder. Click on *Sort Series* (R1/C10). Then click on *Start AUTO Load* (R9/C7) to start the automatic display of the structures; these will be in ascending order of (excess *i.e.* strain) energy. To stop the automatic display click on *Stop AUTO Load* (R10/C7) with the **right** button. If you wish to see the energies click on *Toggle Text Display* (R9/C10), and manually step through the structures using the set of buttons in rows 9 and 10, columns 4, 5, and 6. If you have access to Burkert and Allinger's book,[6] try correlating the results of this experiment with the structures for cyclododecane that they discuss.

4.2.6 Building structures from fragments

A third way of generating structures is to assemble them from pre-built partial structures. To access this method, start from the *DUAL* display mode and press *Build Structure* (R9/C1). The *DUAL* display will change and show a different menu, the two windows are now labelled *Base Structure* (on the left), and *Fragment* (on the right).

The pre-built structures are contained in the *Fragment Catalogue*. This is a collection of some 400 structures, mostly of real compounds, that have been optimized (energy minimized) using the program MOPAC.[6] These provide the partial structure fragments from which larger structures can be built. It is also possible for structures generated in other ways (*sketching* or from SMILES strings) to be used. The process is illustrated by examples.

4.2.6.1 p-Aminobenzamide

Press button *Load Base* (R1/C5). From the menu that appears select (in this order) *Aromatics and Heterocycles , Confirm, Benzene, Confirm*. The wire-frame structure of benzene with the atom numbers will appear in the left-hand window. Next press button *Load Fragment* (R1/C6). From the menu select *Simple Aliphatic Structures, Confirm, Formic Acid, Confirm*. The line structure of formic acid will appear in the right-hand window.

Next press the button *Join* (R2/C3). You will be presented with a series of menus that invite the selection of the serial numbers of the atoms to be joined in the base structure and the fragment, and the atoms that are to be eliminated. For this example choose to join atom *3* in the benzene structure to atom *1* in formic acid. The (hydrogen) atoms to be eliminated are respectively *9* in benzene, and *2* in formic acid. The structure in the *Base* window changes to give the structure of benzoic acid. The structure in the *Fragment* window is not changed (and could be used for a second addition process).

Next, load the fragment window with ammonia. Press *Load Fragment* (R1/C5). Then select *Simple Inorganic Structure, Confirm, Ammonia (planar), Confirm*. Proceed to join the fragment to the base structure as before. Select the atoms *6* (in the base structure) and *1* (in ammonia) and the atoms for elimination *11* (base), and *2* (ammonia).

The base structure will be changed to p-aminobenzoic acid. It is necessary now to convert this to the corresponding amide. This is done most simply by changing the oxygen atom of the hydroxyl group into a nitrogen atom. Click on the button *Alter Atoms* (R5/C1). Then select atom *13*, then *Nitrogen*. To complete the structure press button *Add Hydrogens* (R4/C4). It is now necessary to correct some of the bond lengths and bond angles, so select *Optimize Base* (R4/C2). After this you can store the base structure in a normal INTERCHEM-D file (R2/C5), and/or copy the structure to one of the display areas (R2/C1 to C4).

4.2.6.2 Quinine

The second example has been chosen to illustrate some of the other features that are provided for building from fragments. Quinine (1) is of interest because of its long-standing use in as a prophylactic treatment of malaria (and because of W.H Perkin Senior's unsuccessful (!) attempted synthesis that led to the discovery of the dyestuff mauveine)[7]. Its structure is:

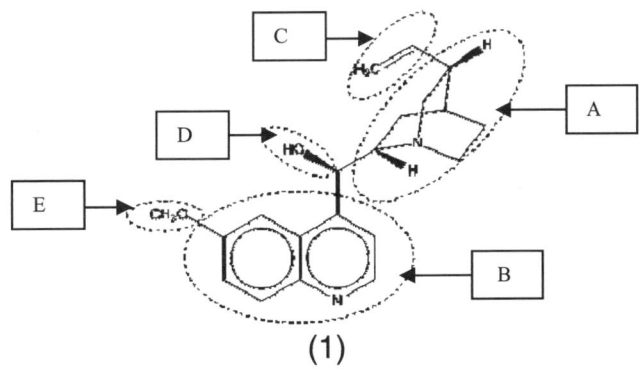

(1)

It comprises two (substituted) ring structures (A and B) joined by a single carbon atom bearing a hydroxyl group (D). We can recognize five stages in the construction process. When using this method of building a structure, it is beneficial to save temporarily the contents of the Base structure at the end of each stage (R2/C5). Use a file name like `temp001D.DAT` (say) for the first stage etc. When you have completed the building process successfully these temporary files can be deleted.

(1) We begin by accessing the top right hand ring system (A). The basic skeleton for this is in the fragment catalogue as *Quinuclidine* under the submenu *Miscellaneous Natural Produce Skeletons*. Find this and load it into the *Base Structure* (R1/C4). Next *Load Fragment* (R1/C6), choosing *Simple Aliphatic Structures –> Ethylene*. Join the two structures. Recognize that quinuclidine has a three-fold axis of symmetry. This means that there are three equally valid ways of joining this structure to the ethylene fragment. (Rather than specifying the choice of atoms to make the connection, we leave you to make all the decisions henceforth!)

(2) Next form the carbon atom in the substructure D. You can do this in two ways; either load the fragment area with *Methane* from *Simple Aliphatic Structures* and join this to the base structure (R3/C3). Alternatively you can change the appropriate hydrogen atom to a carbon atom (R5/C1) add hydrogens (R4/C4), and optimize the base structure (R4/C2).

(3) Now load the fragment area with *Quinoline.* This is in the *Aromatics and Heterocyclics* submenu of the catalogue. This will form the part of the structure of quinine labelled B. Now make the junction between the base structure and this new fragment. It is probably best to turn the pictures, of both the base structure and the fragment, on the computer screen, so that they correspond to their arrangements in the picture (above) of quinine. To do this, select *Display Base* (R3/C1) or *Display Fragment* (R1/C7), and use the green buttons on the right of the menu area.

(4) The last fragment to be added is *Methanol* from the *Simple Aliphatic Structures* submenu. This will form the part of the structure labeled E. Make the appropriate junction to the *oxygen* atom of methanol.

(5) The final step is to place the oxygen atom that is in the part of the structure labeled D. This is done most easily by altering one of the two *hydrogen* atoms to an *oxygen* atom. However it is crucial to get the stereochemistry correct, and it is easiest to adopt a trial and error approach! First Click on *Chirality* (R8/C4). Click on *Alter Atoms* (R5/C1) and pick the number of one of the hydrogen atoms. The chirality label of the carbon should appear as *R*.

If the label is *S*, you have picked the wrong hydrogen atom. In this case click on *Undo Last* (R3/C6); (this reverses the last alteration that you made to the base structure) then choose the other hydrogen atom. (*Note that only the last change made to the Base Structure can be undone*). Then click on *Add Hydrogens* (R4/C4) and *Optimize Base* (R4/C2).

Using INTERCHEM for Molecular Modelling

Finally and most importantly save the results of your efforts by clicking on *Store Base in File* (R2/C5). This will be stored in the your chosen folder; (but if you have forgotten to create one you can do that now). Choose to store it in INTERCHEM-D format, and give it the filename `quinineD.DAT`. We will make use of this structure later in this chapter.

Besides the chirality label (*R*) of the carbon atom bearing the hydroxyl group, there are three other atoms bearing labels all in the quinuclidine ring, the atom bearing the olefinic two-carbon chain is also *R*, the other two on the quinuclidine ring are *S*. Note that the configuration of these atoms cannot be independently controlled, If any of the four chirality labels does not follow this pattern, then look for an error in the way that the component structures have been joined,

4.2.7 Using the INTERCHEM merge tool for building

There are some compounds for which models are not easily built by the three methods that are outlined in the preceding sections. These compounds are characterized by having two or more parts not linked by covalent bonds. To allow modelling of these classes of compound, you can use the *Merge* tool that is provided in INTERCHEM. This tool was devised for other purposes to start with, and it is worthwhile listing all of its uses here:

(1) Modelling of *clathrate* compounds.

(2) Modelling the interactions of small molecules with the surfaces of crystals.

(3) Modelling of the interactions of proteins or nucleic acids on the one hand and small molecule ligands on the other, *i.e.* 'Docking'. We will return to this in Chapter 10, which deals with modeling in connection with Medicinal Chemistry.

(4) As a tool for assessing the similarity or differences of two or more structures by providing overlay pictures; such pictures do not of course represent real molecules.

(5) As a tool for building models of these 'difficult' molecules.

4.2.7.1. Ferrocene

We will deal with the first four of these uses elsewhere in this book, but show you here how to use it for building a model of *ferrocene*.[8] The structure consists of two parallel planar *cyclopentadienyl* rings (C_5H_5) that sandwich an iron atom. There is a five-fold axis of symmetry that goes from the centroid of one ring through the iron atom to the centroid of the second ring. The distances of the centroids of the two rings to the iron atom are 1.706Å. This distance is important because you will need to make sure that the two cyclopentadienyl rings are separated by exactly twice this distance (see below).

Begin by going to the *Build Structure* tool (R9/C1 in *DUAL* Display Mode). Next press *Load Base* (R1/C5 in the *Building Menu*). Choose the *Aromatics and Heterocyclics* menu, and from that *cyclopentadienyl*. The structure that appears in the Base window has the five membered ring of carbon atoms (plus hydrogen atoms), the carbon-carbon bonds are all aromatic (coloured *cyan*). There are two extra atom (numbered 11 and 12). Atom 11 is a pseudo atom that is at the centroid of the ring. Atom 12 is surplus to our requirements, so *Delete* it (R5/C3). Now rotate the structure so that the plane of the ring is at right angles to the plane of the computer screen - press *Rotate Y-* (R4/C9) twelve times. Next *Copy Base to A* (R2/C1) **and** *Copy Base to B* (R2/C2). Then press *Exit Building* (R8/C10).

The program will now have returned to the *DUAL* Display Mode menu. Next press *Structure Merging* (R9/C4), and from the new menu press *Merge B with A* (R1/C1). What will appear is a single image of the two (identical) structures that are in areas *A* and *B*, now superimposed. Within the Merge tool structure *A* is referred to as the *Host* structure (*cyan* coloured), and structure *B* as the *Guest* structure (*magenta*). The aim now is to move the two structures apart so that the distance between their centroids is 3.412 Å. Press *Add Good Contacts* (R20/C1). In the table that appears, enter the number 11 (the number of the centroid pseudo atoms) in both columns 1 and 2, and the number 3.140 as the *Low Limit* and 3.144 as the *High Limit* on the first line. Then press the button *Accept*. Now press *Show Good Contacts* (R20/C2). A line will appear, joining the two centroid atoms. This is coloured *Red* if the two ring centroids are too close, *Blue* if they are too far apart, and *Green* if the distance is within the limits. The aim now is to adjust the distance so that the line turns green. Move either or both of the structures *Left* or *Right* (Buttons R2/C1, R3/C1, R2/C2, R3/C2) until this happens. The default movement is initially 1.0 Å, but this can be made smaller by pressing *Halve Shift* (R18/C2). The halving can be applied repeatedly, so very fine adjustments can be made in the separation distance.

Next press *Complete Merging* (R30/C1). From the menu that appears, choose *Bind HOST and GUEST as a SINGLE STRUCTURE*. The program will return to the *DUAL* Display Mode, and the new combined structure will be in area *A*.

Now click on *Build Structure* (R9/C1) and *Copy A to Base* (R1/C1). To make the picture clearer, rotate the structure on the Y-axis (R4/C9). The next step is to create a further pseudo atom that is at the midpoint of atoms 11 and 22 (the two centroids). Press *Form Dummy Atom* (R4/C5), and then choose from the menu the numbers 11 and 22, pressing *Confirm* after each.

Pressing *Confirm* a third time will create the new (dummy) atom and this will be numbered 23. Now press *Alter Atom* (R5/C1). The menu that appears caters for elements outside the 'Organic Set' by requiring that *OTHER* be chosen followed by the atomic number (in this case 26 for iron). The next step is to create bonds between the iron atom and the centroids of the two rings. For this use *Form Bond* (R5/C5) and in each case choose *Ligand Bond*.

Now create *dummy bonds* between each of the *dummy atoms* (that are the centroids of the two cyclopentadienyl rings) and one of the carbon atoms in the appropriate ring (Atom11 to atom 1 and atom 22 to 12). These extra (invisible) bonds are necessary so that we can rotate the two ring structures with respect to each other.

The final step is to adjust the alignment of the two rings so that the hydrogen atoms of one ring do not eclipse those of the second ring. Rotate the structure so that the rings are parallel to the plane of the screen. Then press *Start Segment Rotation* (R6/C1). For the first atom choose 11 (one of the centroids) and the second atom 23 (the iron atom). It is not necessary to monitor a distance in this case, so press *Accept* for a third time without choosing an atom. Now press *Rotate+* (R6/C2) and/or *Rotate −* (R6/C3) until you are satisfied that the carbon-hydrogen bonds are not eclipsed and symmetrically disposed. Then press *Stop Segment Rotation* (R6/C5). Finally store the structure in an appropriately named file (R2/C5) (say `ferrocnD.DAT`).

It is important to note that when the INTERCHEM merging tools are used, lateral motions of the two structures can only be made along the principal axes of the display, and rotations can only be made around these axes. However the two monitoring facilities (*good contacts* and *bad contacts*) measure true distances between atoms.

4.2.7.2 Buckminsterfullerene

The announcement in 1985 of the discovery of a new allotropic form of carbon[9] was significant in many ways, not least because it was followed by articles in the popular press, and resulted in the awards of Nobel Prizes to three of the discoverers. The molecular form of carbon C_{60} has generated intense interest among scientists because of its icosahedral symmetry, and among non-scientists because of its resemblance to the shape of a soccer ball. It presents some problems in modelling. The method that is described here uses combinations of techniques that have been described in the forgoing sections. The screen displays at some of the key stages are shown in Figure 4.3.

The first step is to make a structure for half of the molecule using the *Sketch* tools. The result should be like that shown in Figure 4.3 (a). Draw the structure initially using single bonds and then change the bonds in the six-membered rings to be alternating single and double bonds *in the way shown*. (***DO NOT make the bonds aromatic, and make sure that the bonds shared by the five-membered and six-membered rings are single***). The structure does not look totally symmetrical, but this is because the grid of circles provided in *sketch* relies on a hexagonal pattern, and conflicts with the five fold symmetry of the structure being built. At this stage use *Convert 2D to 3D* (R3/C6) to get a structure in the left-hand window; (Figure 4.3(b)), if you rotate this structure (on either the X or Y axes) you will see that it is saucer shaped. It is wise (but not necessary) to store the structure in the areas C and/or D at this stage (in case of mistakes later). Then exit the *Sketch* facility (R7/C10).

The second step uses the tools that are present in the building-from-fragments facility. When you enter this from the *DUAL* Display menu, and then *Display Base*, the 3D structure from the sketching process is already there! (This is because the *Sketch* process uses the *Base* structure area for the results of 2D to 3D conversion). The structure does require modification since it contains hydrogen atoms. Get rid of these by pressing *Delete Hydrogens* (R5/C3). Note that this is not the same as *Toggle Hydrogens* (R2/C7), which only hides the atoms from the display. The result is shown in Figure 4.3(c). Copy the base structures to *both* areas *A* and *B* (buttons R2/C1 and R2/C2), and exit from the building process (R8/C10).

For the third step we use the *Structure Merging* (*DUAL* Display menu R9/C4). From the *Control for Merging Structures* menu choose *Merge B with A* (R1/C1). What appears to be a single structure will be displayed. Next press *Orthogonal View* (R15/C3) and move the two structures apart; *Move Host Right* (R2/C1) and *Move Guest Left* (R3C2). One of the two structures now needs to be rotated through 180° on the Y axis (*i.e.* clockwise). Press *Rotate Guest Y+* (R10/C2) repeatedly until the concave surfaces of the two saucer shaped structures are facing each other. The resulting display should look like Figure 4.3 (d). Then press *Orthogonal View* (R15/C3) again (Figure 4.3 (e)). To complete this step press *Complete Merging* (R30/C1) and the option *Bind HOST and GUEST as a SINGLE STRUCTURE*. The result of this action is to deposit the (as yet separate) parts of the structure in *Structure Display Area A*.

The final step remains to add ten new bonds to join the two halves together. For this use the building-from-fragments facility again. First *Copy A to Base* (R1/C1). The picture that you will see will be similar to Figure 4.3 (e). You must add the ten new *double* bonds to each of the gaps around the periphery of the circular display. When all the extra bonds have been made, there will be a series of connections alternately double and single, approximating to a circle, around the periphery of the drawing, as in Figure 4.3 (f). Having made the ten bonds, press *Optimize Base* (R3/C5). The display should now show the fullerene C_{60} as in Figure 4.3 (g). You can verify that the structure is symmetrical by rotating it about the three orthogonal display axes. You can also display it as a *Ball and Cylinder* view (Figure 4.3 (h)). **However, the most important thing to do is store the completed structure in a suitably named file (say `fullernD.DAT`) in a known location!**

4.3 The reliability of structures obtained by building

The preceding sections of this chapter show the basic methods of building chemical structures, together with instances where these methods have been used in combination. Questions remain as to how reliable such structures are:

(a) Is such a built structure unique?

(b) How well do structures built or optimized using different methods agree with one another?

(c) How well do built structures agree with those determined experimentally, for example by X-ray crystallography?

(d) Is there a best structure for a chemical substance?

To answer these questions we will make use of the structure of *Quinine* built in section 4.2.6.2.

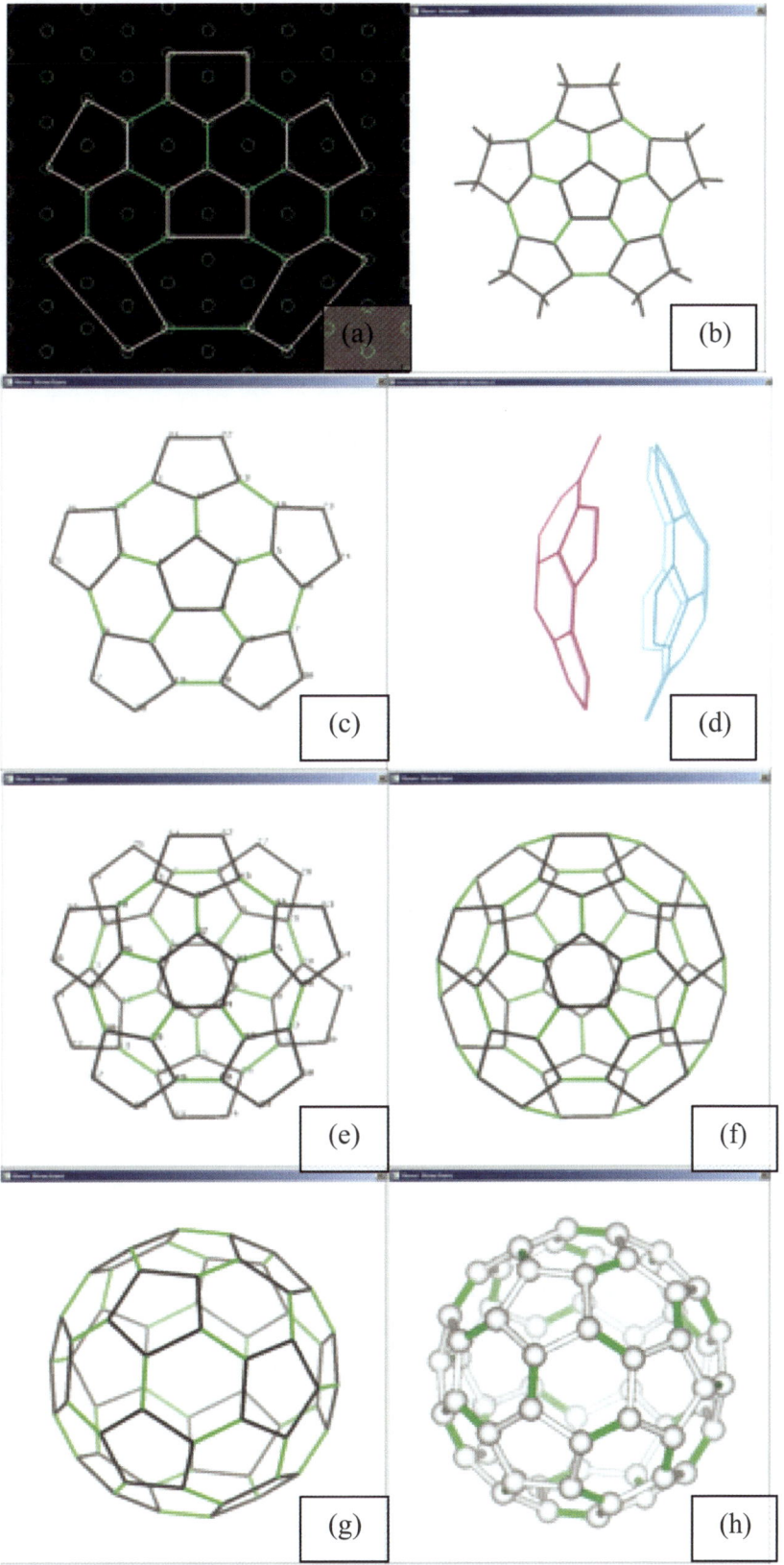

Figure 4.3
Stages in building Fullerene C_{60}
(a) Starting sketch, (b) After conversion to 3D, (c) After removal of hydrogen atoms,
(d) After duplication using merging, (e) Before adding extra bonds, (f) After adding extra bonds,
(g) After optimisation, (h) After optimisation –Ball and cylinder view. Double bonds shown green.

4.3.1 Refining a structure using quantum mechanics (quinine)

Using INTERCHEM load the structure of quinine that you stored previously (R2/C7) into structure area A. Then click on *Store Structure* (R2/C6). Choose to store as *Mopac'dat'Format*. This option automatically takes you through the steps for running a job using the MOPAC program.

You should name the file `quinineM.DAT`, and it should be stored in the directory that you have used for all your files for this project.

Choose to *Accept* (R4/C9) the default keyword parameters, and add a meaningful *Heading* when prompted.

Finally press *Start Mopac Calculation with the File just created*.

A blank text window will appear and persist while the calculation is being performed.

When it disappears press *Select B* (R1/C2), then *Load Structure* (R2/C7) and then *Mopac 'PRO' Format*.

The output file generated by MOPAC will be called `quinineM.pro`. Select this file. (To see this file listed in the directory you will need to use the filtering mechanism provided by the Windows operating system; this can be different in different versions of Windows).

From the final submenu at the bottom of the screen select *No Additional Features*. The structure of quinine as refined by MOPAC will appear in the right hand window.

It is usual for the structure output by MOPAC to differ from the starting structure by having been rotated. A simple way to align the two structures is to press *Align B with A* (R6/C7) and then *Direct Shapes*. The two structures will now be aligned in a similar way. But just how similar are they? Use *Structure Merging* (R9/C4), then *Merge B with A* (R1/C1). The two structures are now overlaid. Structure A (X-Ray structure) will be coloured *Cyan*; structure B (the MOPAC output structure) will be coloured *Magenta*. Pressing *Orthogonal View* (R14/C3) (repeatedly) will turn the overlaid structures through a right angle on the vertical axis (and back again). There is no scale provided for the display, but you can judge the differences between the structures by noting that typical carbon-carbon bond lengths are 1.35 to 1.5 Ångstrom Units. You can exit from the Merge display by pressing *Exit without Merging* (R30/C3). (Press *Escape* to clear the Welcome screen, and then *Dual* (R1/C8) to display the structures). To see in detail how the two structures differ you can choose to compare the individual bond lengths, bond angles, and torsion angles. (Buttons R7/C1 to C5).

We are now able to provide answers to some of the questions posed earlier. The comparisons that you have made show you that for question (a) the answer is NO. There is no unique structure. Question (b) is also answered; the structure of quinine, built from fragments (and optimized using Molecular Mechanics), differs from that where additional refinement using the program MOPAC (a semi-empirical Molecular Orbital program) has been used.

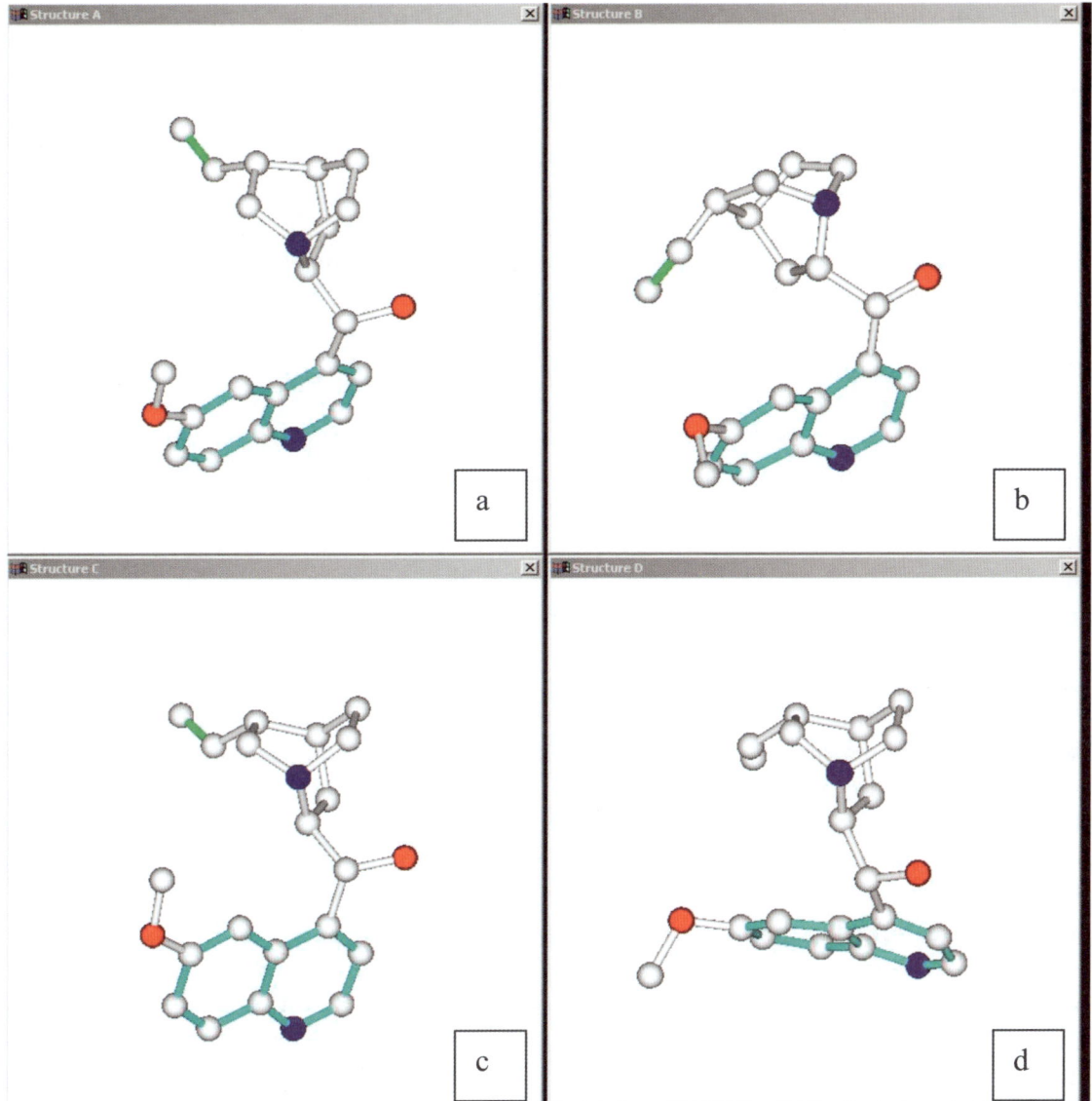

Figure 4.4
Comparison of Structures of Quinine.

The two pictures on the left are taken from the crystallographic data of B. Pniewska and A. S. Purzycka.[10] This structure was solvated with 0.5 molecule of toluene and 1 molecule of water for each quinine molecule. These solvent molecules have been removed from the pictures. For clarity hydrogen atoms are not shown.

The two structures on the right are different poses of the same structure obtained from multiple interpretations using INTERCHEM of the SMILES string:

C=CC(C1)C(C2)CCN1C2{R}C(O)c3ccnc4c3cc(OC)cc4

followed by selection of the lowest energy structure with correct stereochemistry. This structure was then subjected to further optimisation using AM1 in MOPAC.

The two comparisons (a with b and c with d) used overlaying of selected atoms in the quinoline, and the quinuclidine moeties respectively

Figure 4.4 shows comparisons of the structure of quinine determined by X-Ray crystallography and a structure obtained by refining one generated from the SMILES string using MOPAC. The two structures clearly differ.

Why structures derived by different methods differ is not unexpected. In the case of quinine, the reasons are to do with the fact that the two parts of the basic skeleton (quinoline and quinuclidine rings) are connected through the carbon bearing the hydroxyl group. The two single bonds involved allow rotation. The other appendages (the methoxyl group, the hydroxyl group, and the ethylene group) provide a further five bonds allowing rotation. If we allow for three minimum energy arrangements for each of these bonds, (as in ethane), there will be 3^7 (=2187) minima in the 8-dimensional energy surface.

Finding the global energy minimum by computational methods is known to be difficult; when the computation finds the edge of one of the (2187 in this case) local minima, the energy (and the structure) is driven to the bottom of a pit from which there no escape. To search for another minimum, the process needs to be started again from a new randomly chosen conformation.

Computational methods are usually set to work on isolated molecules (*i.e.* in a vacuum) and assume that the dielectric constant for the medium is 1.0. In these cases the charges on the atoms play a much larger role than when an assembly of molecules is in solution. In polar solvents, like water, the dielectric constant is much higher. In the crystalline state individual molecules affect one another, and usually fix the bonds, that with isolated molecules are capable of some rotation. Add to this the fact that many substances can crystallize in more than one form (polymorphism) and we are led to the view that there is generally no unique best structure for a substance.

4.3.2 Getting the energies of structures using quantum mechanics

Quantum mechanics calculations are used when we wish to get more details about structures than are provided by molecular mechanics calculations. We can illustrate this by looking at the simple compound benzoic acid (1).

We will first model the monomer form of this substance, to decide which of two possible structures (2) and (3) is the most likely. Start by using the SMILES string:

c1ccccc1C(=O)O

to generate a single structure, and then use the option to generate a set of 32 structures. (Make sure that you save all of the files that you create in suitably named directories!) Next use the method outlined in Section 4.2.5.6 to look at this series of structures. (*Dual Display Mode* - R10/C6). Sort the files first (R1/C10), and then display them one by one having switched the text display on (R9/C10). You should find that roughly half of the structures have energies of -1.24 kJ/mole, with a second set with energy of +9.82 kJ/mole. These two sets of structures correspond to the structures (3) and (2) shown above. The lower energy structures is stabilized by virtue of a hydrogen bond between the hydrogen atom of the OH group and the oxygen atom of the carbonyl group (shown as a dotted line).

We can also compare the energies of these two structures using the program MOPAC. Take a representative structure from each set and submit it to the program as outlined in section 4.3.1. The detailed results of the calculations are contained in the output files that are suffixed *M.out and (in summary form) in the files *M.pro. These can be summarized in Table 4.3.

Table 4.3
Thermodynamics of Benzoic Acid

Monomer energies	Low energy monomer	High energy monomer	Difference (LM-HM)
Excess energy (MM)	-1.24 kJ/mole	+9.82 kJ/mole	-11.06 kJ/mole
Heat of formation (MOPAC)	-284.03 kJ/mole	-260.60 kJ/mole	-23.43 kJ/mole
Monomer ⟷ Dimer	Low energy monomer	Dimer	Difference (D-LM)
Heat of formation (MOPAC)	2 × (-284.03 kJ/mole) = -568.06 kJ/mole	- 593.89 kJ/mole	-25.83 kJ/mole (dimer)

However, benzoic acid is known to exist as a dimer, stabilized by hydrogen bonds (4). This structure can be modelled by taking two instances of the low energy monomer into structure areas *A* and *B* of INTERCHEM, and *Merging* them, so that the two component parts are roughly aligned as in structure (4). Exiting the Merge facility to form a single structure and running MOPAC on this, gives us the heat of formation of the dimer. The formation of the dimer from 2 molecules of the low energy monomer is exothermic with heat of formation of 25.83 kJ/mole. (Note that MOPAC reports energies on kcal/mole, hence the figures provided by this program must be multiplied by 4.184 to convert to kJ/mole).

4.4 Getting structures from the published literature

Chemical research in its present form can be traced back to the middle of the nineteenth century, and has from that time been characterized by two, at first sight incompatible, principles; rivalry and collaboration. What this means is that when the research has reached a significant stage (and in an industrial setting, when the Patent Officials have bled the results dry), the findings are published in a (more or less) reputable scientific journal. The level of detail expected in the publication has hitherto been at a level such that the experimental work described can be repeated, by another competent researcher.

4.4.1 Getting structures from X-ray crystallographic data

If you compare publications of fifty years ago with those of today, there is a striking difference in the amount of experimental detail. Older publications will have the complete experimental detail but it will be less in total; recent publications frequently relegate the experimental part of the paper to 'supplementary publications'. This extra information is usually made available by the publishers as data downloaded *via* the internet. Whether such information will be available in a hundred years is yet to be seen.

Where such supplementary information includes X-ray crystallographic data, this has (since 1970) been collected and made generally available by the Cambridge Crystallographic Data Centre (CCDC). Most journal publishers throughout the world require X-Ray crystallographic data of this form to be deposited with the CCDC as a condition of publication. The workings of the CCDC are dealt with in more detail in Chapter 8 of this book. At this point we will outline the ways that structures may be abstracted from these data files using INTERCHEM.

INTERCHEM can handle three of the types of data files provided the CCDC:

(1) XR files; so called because their filenames have the two-letter extension 'XR'. These files need to be renamed so that they can be accepted by INTERCHEM. For example rename `abcdefg.XR` to `abcdefgX.DAT`.

(2) CIF files. These normally have the (single) extension .cif; but (because Windows uses files with a single 'cif' in other ways) to make them acceptable to INTERCHEM this extension needs to be duplicated! (For example. `abcdefg.cif.cif`).

(3) FDAT files. This type of file is not so easy to read by eye, because of its unusual format. This format comes from an era when data was supplied on punched cards. These have 80 columns of data and a limited number of rows. The X, Y, and Z coordinates are supplied as integers that have to be divided by 10,000 to convert them into Ångstrom Units. INTERCHEM can handle them, but they need to be renamed to `abcdefgx.DAT`. Because of all these complications, use of this type of file is not recommended

4.4.2 Information from X-ray crystallographic data files

It is important to recognize that the primary function of the CCDC is to store the results obtained from experiments conducted by crystallographers. However over the years chemists have realized that the database provides a wealth of information about chemical structure. In particular medicinal chemists have made great use of it in this way. While the use by medicinal chemists has primarily been at the early stages of drug design (looking for new lead compounds), problems, for example, associated with polymorphism of drugs leading in turn to problems in the production of tablets, can be addressed by studying crystal morphology. (This is dealt with in Chapter 7). A variety of type of information suitable for graphical display can be derived from CCDC files. Treatment of the three types of file described above is, from the user's point of view, substantially identical. Pressing the button *Load Structure* (Dual display R2/C7; Single/Quad Display R3/C1; and also other menus), followed by selection of one of the options of buttons 2, 3, or 4 in the subsidiary menu (to select the type of data file – *XR, FDAT* or *CIF*), and then an appropriate file, will give a display similar to that in Figure 4.5.

In the following description we use, as examples, the files
    `caffeine2.cif.cif` and `cholic_acid.cif.cif`

that are present in the folder (accessible from within INTERCHEM)
    `~\test_data\CIF_files`

The figure shows the preliminary reports for both of these cases, that give the basic information about the crystal(s) used in the experiments.

In the case of *caffeine* we see that the *asymmetric unit* has 15 atoms. The molecular formula of caffeine is $C_8H_{10}N_4O_2$, and this has a total of 24 atoms, of which 10 are hydrogen atoms. The explanation for this discrepancy is that hydrogen atoms are not included in the structure, but there is present an extra oxygen atom, due to water molecule, (caffeine occurs as a monohydrate). The absence of hydrogen atoms in the crystal structure is common in older data. It is only in recent years that hydrogen atoms are routinely reported, and in some cases their locations are calculated. The phrase 'riding on' in the text of a paper means just this.

The program finds the maximum number of unit cells that could be accommodated in the storage area, based on the number of atoms in the asymmetric unit and the maximum number of atoms (65536) allowed in the storage area.

In the case of *cholic acid* there are 137 atoms in the *asymmetric unit*. The molecular formula of cholic acid is $C_{24}H_{40}O_5$, and this has 69 atoms. It appears at first glance, that the *asymmetric unit* contains two molecules of the steroid with one atom (a hydrogen atom) missing; but the true explanation is only revealed when we display the structure.

Pressing the *Escape* key will clear this display and present you with a new menu offering seven options. These options range from providing the simplest *chemical* information, to those that model the *crystal habit*. In all the cases the data in the file, which are in crystal coordinate form, are suitably processed and transformed into Cartesian coordinates, and displayed on the screen as a simple wire frame display.

(1) *Simple structure (as supplied)*. This will process the data in the simplest way; no centering on the screen will occur, and, because of this, sometimes nothing will be visible. In this case press *Centre Structure* (R3/C5) and then *Wire Frame* (R3/C6).

(2) *Asymmetric Unit Only*. In this case, with an organic molecule, INTERCHEM will add hydrogen atoms if necessary to the structure as supplied, and centre the result on the screen window. This option provides the best starting point if you require a structure for further modelling.

(3) *Form One Unit Cell*. In each of the reports shown in Figure 4.5 there is a statement of the number of replications of the atoms in the asymmetric unit that will form the unit cell. Simple arithmetic will provide the number of atoms that will be formed. To avoid having a cluttered display, hydrogen atoms are not added at this stage, although any that are present in the original data will be preserved.

(4) *Unit Cell with Frame.* This is based on the preceding option. The addition of the lines showing the edges of the original crystal cell (coloured: a = *cyan*, b = *green*, c = *yellow*) gives an indication of the shape of the unit cell.

(5) *Nest of Unit cells.* In order to model the interactions of atoms in a crystal, it is necessary to take into account those atoms that are close to the unit cell boundaries. The minimum size of a cluster of cells that would provide this is 3×3×3. This is the default provided in the submenus that appears. Other sizes, (say 2 ×2× 10) could be more appropriate if, for example, a cylindrical cavity parallel to the c axis were present.

(6) *Nest of Unit Cells and Frames.* This adds frames to all the component unit cells. Bear in mind that adding frames (option 4 and this option) adds pseudo atoms and pseudo bonds to the structure. While the structures resulting from these additions can be stored as standard INTERCHEM 'D' files, the structures are not valid starting points for modelling with other techniques (*e.g.* molecular orbital calculations), because of the presence of these extra atoms and bonds.

(7) *Exit.* When you have exhausted all the other options!

We illustrate the uses of these various displays with reference to the two structures *caffeine* and *cholic acid*.

4.4.2.1 The crystal structure of caffeine

In the minimal display (option 1) no hydrogen atoms are present because none were present in the original data. A better picture is provided by option 2. Here the hydrogen atoms have been added correctly by the program, but INTERCHEM has made an incorrect assignment of bond types. To correct this, use the facilities of *Build Structure* (R9/C1).

Copy the structure into the *Base* area, using the appropriate first row button. First use *Delete Atoms* (R3/C3) to remove atom 15 (the oxygen atom belonging to the water of crystallization, to which no hydrogen atoms have been added!)

Now use *Alter Bonds* (R5/C2) so that bond 8 – 9 becomes *double*, 8 – 1 becomes *single*, and 5 – 1 becomes *double*. Finally save the structure (R2/C5) in a file.

In one sense the program was not at fault in classifying the three rogue bonds as delocalized (*i.e.* aromatic) bonds, but the ambiguity arises because there is no clear distinction based on *bond length alone* between aromatic and double bonds.

If the shape of the crystals and how the molecules of caffeine are disposed is what you want to see, the options 3, 4, 5, and 6 are suitable. Note that only in option 3 are hydrogen atoms added. In the two options that incorporate cell edges (4 and 6), the display of the cell edges is toggled on (or off) by the button *Extra Displays* (R5/C7).

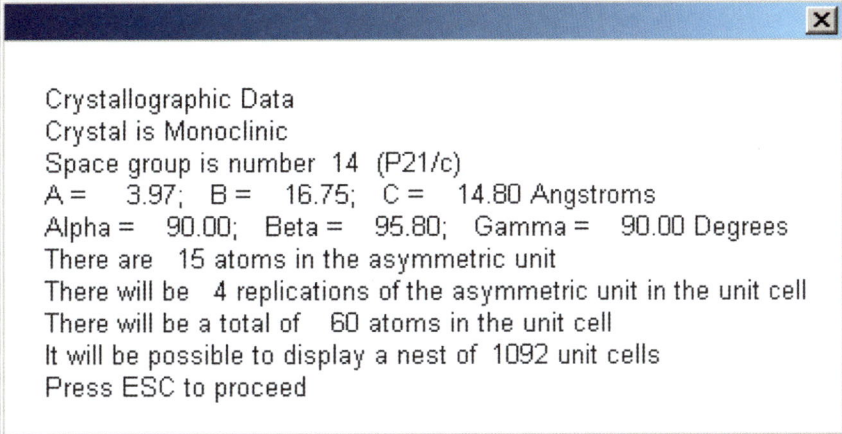

(a)

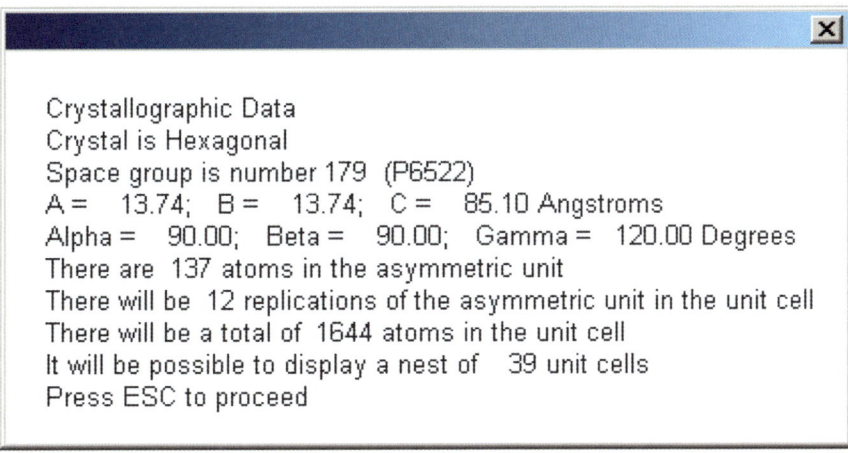

(b)

Figure 4.5
Preliminary Reports on Crystal Structures
(a) Caffeine hydrate from file caffeine2.cif.cif
(b) Cholic Acid Hemihydrate from file cholic_acid.cif.cif

4.4.2.2 The crystal structure of cholic acid.

This is a more complex structure than that of caffeine. It is instructive to look at the CIF itself using NOTEPAD. The compound is described there as a *monohydrate*. In fact there is one molecule of water for every two molecules of cholic acid. The better classification would be *hemihydrate*. This does go someway in explaining why there are two molecules of cholic acid in the asymmetric unit. Because hydrogen atoms are included in the original file, they will be included in all the six display options, unless suppressed by *Toggle Hydrogens* (R2/C7), and the only significant difference between option 1 and option 2 is that the latter will add extra hydrogen atoms if there are any missing in the original file! Assuming that the initial aim is to get a 'good' structure for (one molecule) of cholic acid, one way to proceed is as follows:

(a) Use option 1 (no extra hydrogen atoms are added).

(b) Use *Build Structure* and copy the structure to the *Base* window.

(c) Use *Delete Hydrogens* (R4/C3) to completely remove (not just hide) them.

(d) Use *Delete Atoms* (R5/C3) to delete atom number 11 (oxygen of water molecule).

At this stage there should be two separate molecules of cholic acid, complete except for hydrogen atoms. We can retain each of these structures separately and compare them, so:

(e) Use *Base -> Fragment* (R1/C9) to save the present structure (two molecules of cholic acid with the water removed).

(f) Press *Delete Segment* (R5/C6) and select the number for any atom in the *first* structure.

(g) Press *Delete Hidden Atoms*. The whole of the selected first structure is removed

(h) Use *Centre Structure* (R2/C10).

(i) Use *Copy Base to C* (and also store the remaining second structure in a file for safety!)

(j) Use *Base <-> Fragment* (R1/C10) to restore display of the two molecules.

(k) Press *Delete Segment* (R5/C6) and the number of any atom in the *second* structure

(l) Repeat steps (g) to (i) storing the remaining first structure in area *D*.

The result is that the first structure is in area *D*, and the second structure is in area *C*.

It is now possible to compare the two structures (in areas *C* and *D*) by using *Exit Building* (R8/C10) and then *Align D with C* (*Dual Mode*, R6/C8), followed by *Structure Merging* (R9/C4). Careful comparison of the two structures reveals that they are significantly different, especially in the way that the side chains bearing the carboxyl groups are arranged; the difference cannot be accounted for as a simple crystal symmetry transformation. The complete structure shows that the single water molecule is placed between the two cholic acid structures, and is probably bound to hydroxyl groups by hydrogen bonds.

The carboxyl group in one of the structures has two carbon-oxygen bond lengths that are equal (1.2265 Å) and typical of the lengths expected for a carboxylate anion. In the other structure the lengths are unequal (1.1988 and 1.3308 Å); values that are appropriate for a CO_2H group. If hydrogen atoms were added to it, this would be a better representation of the structure of cholic acid.

4.5 Analyzing structures

To completely understand and characterize a structure it is necessary to be able to make measurements to find, for example, the distances between pairs of atoms. It used to be the case that, when results of X-Ray crystallography were published, the authors of a paper would list extensive tables of bond lengths, bond angles etc., but relegate the basic results (*i.e.* the crystal coordinate data) to supplementary documents available only on request from the publishers or the authors. Nowadays, with a personal computer on every scientist's desk, it is far easier for readers of a paper to download a structure and make this sort of analysis for themselves. Recognizing this trend, software packages for molecular modelling now provide the basic tools for doing this. These tools are generally applicable; coordinate data can come from modelling experiments as well as from crystallography and other experimental methods.

In INTERCHEM the tools are presented as a series of orange coloured buttons on the seventh row of the menu in *Dual Mode,* on the ninth row in the *Build Menu,* and near the bottom of the menu in *Single Mode* or *Quad Mode*.

Several of the tools measure purely geometrical arrangements of atoms, and these are often the same measurements that are made during the operation of the molecular mechanics programs. We discuss elsewhere the details of the calculations and the algorithms. When invoked by pressing the appropriate button, these functions require the specifying of the appropriate atoms using popup menus. These menus are 'intelligent'; *i.e.* only the appropriate atoms are called for after the first menu. The tools are applicable to small molecule structures only; in particular they will not work with protein or nucleic acid structures (for which special facilities are provided), or with structures that have crystallographic axes *etc*.

4.5.1 Bond lengths

Use button *Bond Length* (R7/C1). If, from the first menu, you pick an atom that has only one connection (*e.g.* a hydrogen atom), the bond length is delivered forthwith. In a normal stable structure the length of the bond between two atoms should not differ appreciably from the standard bond length (see Appendix 2 for tables of standard bond lengths and bond angles).

4.5.2 Inter atomic distances

Use button *Inter Atom Distances* (R7/C2). In this tool you are prompted for specifying the serial numbers of two atoms in a single structure. While the same atom cannot be specified twice, the choice is otherwise unconstrained. This measurement will allow you to check the distance between two non-bonded atoms for unlikely close proximity, or to measure the distance between atoms on opposite sides of a molecule to estimate a diameter. Bond lengths and inter atomic distances are quoted in Ångstrom units (in spite of Ångstrom units not being in the S.I. canon).

4.5.3 Bond angles

Use *Bond Angle* (R7/C3). You will be prompted for selection of three atom serial numbers (N1, N2, N3) from popup menus. After selecting the first atom, the constraints applied to the menus are: N2≠N1, N3≠N2, N3≠N1, atom N2 must be bonded to atom N1, and atom N3 must be bonded to atom N2. The result will be the angle subtended by the bonds N1→N2, N3→N2 at atom N2. The angle will be unsigned, in the range 0 to 180.0°.

4.5.4 Torsion angles

Pressing the button *Torsion Angle* (R7/C4) will prompt you for selection of the serial numbers of four atoms (N1, N2, N3, N4). After selecting the first atom, the constraints applied to the menus are: N2≠N1, N3≠N2, N3≠N1, N4≠N3, N4≠N2, N4≠N1, atom N2 must be bonded to atom N1, atom N3 must be bonded to atom N2, and atom N4 must be bonded to atom N3. Putting this in another way, the atoms N1, N2, N3, and N4 are to be a series of non-identical, sequentially directly bonded atoms. The result will be a signed angle in the range −180.0 to +180.0°. Torsion angles have both magnitude and sign. The convention for the sign is stated thus; looking down the bond N2-N3, if the bond N1-N2 has to be turned clockwise so that it eclipses bond N3-N4, then the torsion angle is positive. If the necessary rotation is anti-clockwise, the angle is negative. Note that the sign of the torsion angle is not changed when the central bond is viewed from the direction N3-N2, and the motion test is applied to atom N4 moving to eclipse atom N1. Note that torsion angles are sometimes referred to as *dihedral* angles (some say incorrectly).

4.5.5 Pseudo torsion angles

Pressing button *Pseudo Torsion* (R7/C5) activates a calculation that is identical to the torsion angle calculation; the difference is in the constraints applied in the pop-up menus. With pseudo torsion angles these are relaxed so that there is no requirement that atoms N3 and N2 should be bonded. This tool allows you to examine, for example, how nearly a pair of bonds, remote in a structure, eclipse one another. The sign convention of torsion angles applies to pseudo torsions also.

4.5.6. Inter planar angle

Pressing the eponymous button (R7/C6) gives access to a tool that allows the measurement of the angle between any two planes in a structure. Each plane is defined by three atoms. The sequence of pop-up menus prompts for entry of six atom serial numbers; these are treated as two sets of three atoms, with constraints that within each set the atoms are all different, and that the two sets shall not be identical. However, the two sets may have up to two identical atoms. There are no constraints regarding how the atoms are bonded together. This tool is useful in a number of situations:

(a) Testing whether two distinct disjoint (typically aromatic) rings are coplanar or parallel (all six atoms different).

(b) Measuring the 'pucker' angle in cyclopentane rings (two atoms common to both sets).

(c) Measuring the angle between two 'spiro' bonded rings (one atom common to both rings).

(d) Measuring the non-planarity of rings that otherwise are expected to be planar (two atoms common to both sets). In this case the angle is often referred to as an 'improper torsion angle'.

4.5.7 Molecular mechanics calculations and geometric measurements

The measurements that are described in the preceding paragraphs are related to the measurements that occur in molecular mechanics optimization processes. When such optimizations fail to give the expected results, the failure can usefully be traced through such individual measurements. You should be aware that high energy structures *are* acceptable when they arise because of the presence of three-membered or four-membered rings. The apparent high energy arises because the molecular mechanics used in INTERCHEM does not have special atom types for atoms in small rings.

4.5.8 Molecular volume

This measurement is accessed by pressing button *Molecular Volume* (R7/C7). It uses an approximate method. In this the largest distance between two atoms in a structure is used to define a longest axis. Two other axes (next largest, and smallest) are defined at right angles to this. These three axes are then used to define a rectangular box that will enclose the structure with some space to spare. The box is then populated with a regular 3D array of points of known spacing. Then those points that are within a van-der-Waals radius of any atom in the structure are marked and counted. A simple calculation involving the number of marked points, the total number of points in the box, and the volume of the box gives the volume of the structure in units of cubic Ångstroms. The precision of this method depends on the grid spacing; this is fixed at 0.25 Å and means that the method is not suitable for either small molecules (because the grid spacing is too large to give an accurate result) or very large structures (because the number of grid points is so large that the calculation is very slow).

The accuracy of these measurements critically depends on having chosen suitable van-der-Waals radii for the atoms. For structures that are derived from crystallography, data in CIF form usually includes a measurement of the volume of the unit cell. Dividing this by the number of replications of the asymmetric unit in the unit cell (the Z value) will give the volume of the asymmetric unit. Note that the asymmetric unit may include solvent of crystallization, molecules of any counter-ions, and sometimes more than one molecule of the structure; further calculations may therefore be needed. If the figure for the 'crystallographic' volume is not reported, it can be calculated if the cell parameters are available. Table 4.4 shows comparisons of volumes of structures found by the two methods for typical natural products and drugs. The table also shows values for *Molar Refractivity* (see section 4.5.10)

4.5.9 Molecular formula and molecular weight

Pressing the button *Molecular Formula* (R7/C8) provides the formula of the structure in the active window area. In addition the molecular weight based on the normal atomic weight of the elements is shown, plus the masses for the expected ions in the parent ion region of a mass spectrum.

These ion masses are based on the IUPAC table of the masses of stable isotopes (Standard: $^{12}C = 12.000000$), and their normal abundances.[11] The ion masses do not include the contributions for the mass of the electron, so this value (5.485×10^{-4}) should be subtracted from masses for positive ion spectra, or added to masses for negative ion spectra, for each of the charges. A combinatorial process is used to calculate the molecular ion patterns. It is assumed that, although the masses are calculated to high accuracy, the *patterns* will be compared with experimental spectra obtained on instruments of moderate resolving power; so each peak in the calculated pattern is a combination of peaks that have masses closer than 0.1 mass units.

Table 4.4
Measurement of Molecular Volume

Name	Formula	F.W	S.G.	Z	V_{cr}	V_{ic}	MR
Caffeine hydrate	$C_8H_{10}N_4O_2$, H_2O	212.2	14	4	245.1	232.3	45.2
Diazepam	$C_{16}H_{13}ClN_2O$	248.8	14	4	344.3	319.5	81.1
2,6-Dimethoxyphenyl penicillin methyl ester	$C_{18}H_{22}SN_2O_6$	394.4	4	2	482.2	456.6	
Hexa-acetyl galactitol	$C_{18}H_{26}O_{12}$	434.4	14	4	529.0	507.5	
Hecogenin hydrate	$C_{27}H_{42}O_4$, H_2O	448.6	19	4	614.8	636.0	120.1
Reserpine	$C_{33}H_{40}N_2O_9$	608.7	4	2	784.9	782.3	167.6
Cholic Acid hemihydrate*	$2(C_{24}H_{40}O_5)$, H_2O	835.2	179	12	579.6	589.1	110.0
Cephalostatin-1†	$C_{54}H_{74}O_{10}N_2$, $2(C_5H_5N)$	1064.1	19	4	1468.0	1450.0	294.4
Triphenylmethyl Perchlorate‡	$C_{19}H_{15}^+$ ClO_4^-	342.6	210	16	6762.0	6966.3	

This table compares the volumes of structures measured using INTERCHEM with the volumes calculated from unit cell volumes reported in CIFs. All of the featured compounds are structures that are included as CIFs that are distributed with the INTERCHEM software.

Abbreviations:
  F.W. Formula Weight for the formula shown including solvent molecules.
  S.G. Space group.
  Z The number of replications of the Asymmetric Unit in the Unit Cell.
  V_{cr} The volume (except where noted) of the solvated structure found by dividing the unit cell volume (in Å^3) by Z.
  V_{ic} Volume of the unsolvated structure determined in INTERCHEM (in Å^3).
  MR Calculated Molar Refractivity (see reference 11). Units are ml/mole

* In cholic acid V_{cr} was found by dividing the cell volume by 24 (twice Z) to allow for the two molecules of the steroid in the unit cell.

† With Cephalostatin-1 the value of V_{ic} was corrected by including the volume of two pyridine molecules.

‡ Triphenylmethyl perchlorate is an ionic crystal. The figures to be compared refer to the unit cell volume.

4.5.10 Hydrophobicity

An important property that is used in Medicinal Chemistry to assess the ability of a drug to be transported around the human (or animal) body is its *hydrophobicity*. If a compound favours an aqueous environment it is said to be *hydrophilic*. Conversely if it favours a non-aqueous environment it is said to *hydrophobic*. Traditionally this property has been measured by the partition coefficient of the biphasic system *n*-octanol/water.[12] The compound to be tested is shaken with a mixture of the two solvents and after equilibration, the concentration (C) in the two phases is measured. The quantity logP is a measure of the hydrophobicity of the compound.

$$\log P = \log_{10}(C_{octanol}/C_{water}) \tag{4.1}$$

The measurements are quite difficult to make (and impossible if the compound has yet to be synthesized!), and so there are methods available to *calculate* logP by adding together parameters associated with structural features in organic compounds. (Note that the inherent inaccuracy and imprecision is hidden by quoting the logarithm of the partition coefficient!) INTERCHEM gives access to two of these methods. Press the button *Hydrophobicity* (R7/C9). The two methods are those due to Visvanadhan, Ghose, Revankar, and Robins,[13] and Wang, Fu, and Lai.[14] The first of these methods also calculates *Molar Refractivity*, a quantity that correlates well with (is proportional to) molecular volume (some comparisons are made in Table 4.3)

4.6 Stereochemistry

When structures are displayed by INTERCHEM as wire frame diagrams, there is the option (button *Chirality* R5/C4) that will cause the Cahn-Ingold-Prelog (CIP)[4] stereochemistry of carbon atoms to be displayed alongside the atom serial numbers. Associated with the facility is a button *Chirality Level* (R9/C9) that controls the *depth* to which searching of structures is extended to decide the priority of ligands attached to the (potentially) chiral atom. This is best explained with reference to the structure (5):

```
      H     (CH2)n-Cl
       \   /
        C
       / \
      X   (CH2)n-Br
```

(5)

The chirality of the central carbon atom (underscored) can only be determined if the chirality level is set to $\geq n+2$. For most purposes the default level (5) works well.

On the popup menu that controls the level there is a further parameter (*Backtracking*). This is normally set *ON*. When changed to *OFF* using the last (toggle) button in the submenu, the effect is to limit the loops in the depth search to avoid repeated searching on atoms already found in ring structures. This shortens the depth search, but has additional effect of showing pro-chiral atoms. Use this facility with caution!

CIP chirality of nitrogen and sulfur atoms can also be displayed if *Toggle Lone Pairs* (R1/C5) is pressed. Electron lone pairs have lowest priority when determining chirality, they are shown as cyan coloured 'ice-cream-cones' in ball and cylinder displays, as cyan coloured spheres in CPK displays, and as unlabelled lines in wire frame displays.

4.7 Geometric isomerism (cis/trans or E/Z isomerism)

The *cis/trans* form of *stereo-isomerism* associated with structures having double bonds, and now more systematically termed *E/Z*-isomerism, is shown in separate popup displays for each double bonds in turn. To access this press button *Show E/Z State* (R7/C10).

4.8 The use of random numbers in INTERCHEM

The methods that are used by Molecular Mechanics heavily rely on the use of *random numbers*. This is particularly true for the algorithms used in INTERCHEM. When INTERCHEM is started, the random number generator is also started, normally with the same fixed *seed*. This is done so that, if necessary, experiments are repeatable. To make use of this facility, it is well to remember that random numbers are used throughout INTERCHEM, for example to resolve difficulties involving singular matrices occasioned in rotating structures. To achieve reproducibility in an experiment it is necessary that the experiment be performed immediately after starting the program. However, the random number generator can be seeded with a number derived from the computer's clock. To instigate this press, the button *Seed Random Number* [Dual Display Mode, (R10/C7)]. The sequence of numbers generated thereafter, while not strictly random, is less easy reproduced, does go someway towards meeting the requirements of randomness. For discussion on the complexities involved in generating random numbers see Knuth.[15]

4.9 Problems for you to solve, and questions for you to answer

In many of these questions the first step is often the building of a structure. Sometimes you will be given a hint as to how to go about this, but in most cases there will be no hint, and part of the problem is choosing the most appropriate way to build the structure. However the final step in building a structure usually involves recourse to Molecular Mechanics methods.

[1] It was suggested in the text, that you should investigate the conformers of cyclododecane. If you have not done this already, do it now! Refer to section 4.2.5.6 for the suggested strategy. (*Hint.* Use SMILES to generate a series of structures and look for low energy conformers)

[2] Investigate the conformational preferences of di-substituted cyclohexanes. This is quite a big project, but the results you get will aid you in understanding a lot about the structures of small molecules. To give you help, we suggest that you use for the substituents: chlorine and bromine atoms, and then methyl and tert-butyl groups. Make sure that each structure that you analyse and compare has the cyclohexane ring in the chair conformation, (otherwise the results could be less meaningful). You should compare the results with 1,2-, 1,3-, and 1,4-disubstitution. (Hint. Use SMILES to generate series of structures and look for sets of low energy conformers having identical energies in the sorted list). Use MOPAC to subject

representative structures to further energy minimization. Check for cases where the relative energies of isomers or conformers seems to vary between the Molecular Mechanics and MOPAC results.

[3] This problem looks forward to our discussion of stereochemistry in Chapter 6. It would be prudent to look at that chapter before answering this problem. Certain derivatives of biphenyl having substituents at the four positions *ortho* to the bond joining the two rings as are capable of being resolved into enantiomeric forms. The formulae (16) and (17) in Chapter 6 are relevant. The question is this: what is the minimum diversity of substituents (R_1, R_2, R_3, and R_4) in those formulae that is necessary for chirality to occur? *Hint..* Construct a 3D picture of biphenyl in which the angle between the two benzene rings is 90°. Then use single atom substituents such as the halogens, F, Cl, Br, and I. Under what combinations of substituents is it impossible to superimpose a structure on its mirror image? Note that only a geometric solution is required; there is no need for computation.

[4] Hexahelicene is known to be resolvable into enantiomorphic forms, stable at room temperature.[16] Heating can cause racemization. The SMILES string for hexahelicene is shown in Table 4.2. Use this to generate a series of structures. From examination of these structures, infer a possible mechanism for the racemization process.

[5] In the text we used a succession of techniques for making the structure of buckminsterfullerene? Could it be done in its entirety with sketching? The answer to this is 'Yes', (a template structure is provided and can be used from within *Sketch)*. Could it be done using SMILES? The answer is 'Maybe?'. (If this were a book written by mathematicians, there would be something like: 'Construction of the SMILES for buckminsterfullerene is left as an exercise for the reader'. But we are not that cruel!). If you like puzzles, then this is for you. But we do give an answer.

References and Endnotes. Chapter 4.

[1] The designation '8.3' arises because the filename consists of eight characters before a (decimal) point, followed by three after it. Such filenames were acceptable to both DOS (the forerunner of Windows) and UNIX. INTERCHEM adopted this convention in the version for SGI IRIX systems. However it modified it to require that the eighth letter marked the type of file. Thus D was used for INTERCHEM's own format, X was used for the format generated from crystallographic files by Daresbury Laboratory, M for MOPAC input files. The three-character suffix was most often DAT. This convention still works with the version of INTERCHEM you are now using, although the requirements of the '8.3' standard have been relaxed, to allow filenames of any reasonable length. However, Windows still has problems with some suffixes; thus to make the crystallographic format CIF files work the names need to have duplicate suffixes (*e.g.* `biotin.cif.cif`. Remember also that Windows, in contrast to UNIX and LINUX is case-insensitive for filenames. A reasonable length filenames could go up to (say) '12.5'. Some of the files that are generated by INTERCHEM have long names that incorporate the dates and times when the file was created. You are recommended to rename such files if the information that they contain is useful to you.

[2] With steroids there is an established convention for drawing and numbering the 2D structures. This is shown in Appendix A3. 'Above' and 'below' the plane of the rings refers to this convention.

[3] D. Weininger, *J. Chem. Inf. Comput Sci.*, 1988, **28**, 31; D. Weininger, A. Weininger, and J. L. Weininger, *J. Chem. Inf. Comput. Sci.*, 1989, **29**, 97; D. Weininger, *J Chem. Inf. Comput.Sci.*, 1990, **30**, 237. SMILES is an acronym for **S**implified **M**olecular **I**nput **L**ine **E**ntry **S**ystem. As this implies, it was concerned with *entry* of (chemical) data; the simplification refers to other systems, notably Wiswesser notation, that preceded it, and had more elaborate rules. It was not initially designed for generating chemical *structures*, nor was the specification of *stereochemistry* defined in the published *papers*. When specifying a structure, the rules of SMILES are tolerant, in that any syntactically correct string is acceptable. The second of the three papers cited deals with generating **Canonical** SMILES. A canonical SMILES string is designed to be a unique specification for a connected structure. The third paper is concerned with an algorithm for generating a 2D formula from a SMILES string.

[4] R. S. Cahn, C. K. Ingold, and V. Prelog, *Experientia*, 1956, **12**, 81; V. Prelog and G. Helmchen, *Angew. Chem. Int. Ed.*, 1982, **21**, 567; E. L. Eliel and S. H. Wilen, *Stereochemistry of Organic Compounds*, Wiley, New York, 1994, p. 101ff

[5] U. Burkert and N. L. Allinger, *Molecular Mechanics*, American Chemical Society, Washington D.C., 1982, p.107.

[6] J. J. P. Stewart, *J. Comp.-Aided Mol. Design*, 1990, **4**, 1; J. J. P. Stewart, *Mopac93 Manual*, Fujitsu Ltd, 1993.

[7] T. S. Kaufman and E. A. Rúveda, *Angew. Chem. Int. Ed.* 2005, **44**, 854. This is an excellent review called: 'The Quest for Quinine: Those Who Won the Battles and Those Who Won the War'. It deals with the discovery of quinine, the history of its use as a prophylactic against malaria, Perkin's ill fated early attempts at synthesis, and modern successful syntheses.

[8] The discovery of ferrocene was reported in 1951 by T. J. Kealy and P. L. Pauson, (*Nature*, 1951, **168**, 1039). The now generally accepted structure was suggested by G. Wilkinson, M. Rosenblum, M. C. Whiting, and R. B. Woodward, (*J. Am. Chem. Soc.*, 1952, **74**, 2125). The aromatic character of this compound was recognized by R. B. Woodward, M. Rosenblum, and M. C. Whiting, (*J. Am. Chem. Soc.*, 1952, **74**, 3458). For a review of the early work see P. L. Pauson (*Quart. Rev.*, 1955, **9**, 391).

[9] The discovery of C_{60} was reported in 1985 (H. W. Kroto, J. R. Heath, S. C. O'Brien, R. F. Curl, and R. E. Smalley, *Nature*, 1985, **318**, 162. There is an enormous and ever expanding literature on buckminsterfullerenes (often abbreviated to 'fullerenes'). Without qualification, 'buckminsterfullerene' is taken to mean the molecule C_{60} (which is the structure that we have featured in this chapter). Other well characterized members of the fullerene family are C_{70}, C_{76} (D_2 symmetry, chiral), C_{78}, (two isomers having C_{2v} and D_3 symmetry), and C_{84}. The structures have been determined by ^{13}C nmr spectrometry. At present the only practical way of obtaining these *allotropes of carbon* is by 'resistive heating' of graphite in an inert atmosphere (W. Krätschmer, L. D. Lamb, K Fostiropoulos, and D. R. Huffman, *Nature*, 1990, **347**, 354). This involves use of high current arcs between carbon rods. A rational synthesis of C_{60} has been reported (L. T. Scott, M. M. Boorum, B. J. McMahon, S. Hagen, J. Mack, J. Blank, H. Wegner, and A. de Meijere, *Science*, 2002, **295**, 1500) although the yield is too small to compete with Krätschmer's method. One issue of *Accounts of Chemical Research* (March 1992, **25**, 97-175) is devoted to the early work on fullerenes and concentrates on the *chemical physics* aspects. The *organic chemistry* of fullerenes is dealt with in the book; R. Taylor, *Lecture Notes on Fullerene Chemistry, A Handbook for Chemists*, Imperial College Press, London, 1999. For a discussion of the nature of *Aromaticity* in Fullerenes (and the lack of it in C_{60} itself) refer to the paper: *Spherical Aromaticity in I_h Symmetrical Fullerenes: The $2(N+1)^2$ Rule*, (A. Hirsch, Z. Chen, and H. Jiao, *Angew. Chem. Int. Ed.*, 2000, **39**, 3915).

[10] B. Pniewska and A. S. Purzycka, *Acta Cryst. Sect. C, Cryst. Str. Comm.*, 1989, **45**, 638.

[11] I. Mills, T. Cvitaš, K. Homann, N. Kallay, and K. Kuchitsu. *Quantities, Units and Symbols in Physical Chemistry*, Blackwell Scientific Publications, Oxford, 1988, p. 85ff.

[12] C. Hansch and A. Leo. *Exploring QSAR, Fundamentals and Applications in Chemistry and Biology*, American Chemical Society, Washington D.C., 1995; C. Hansch, A. Leo, and D. Hoekman, *Exploring QSAR, Hydrophobic, Electronic, and Steric Constants*, American Chemical Society, Washington D.C., 1995.

[13] V. N. Visvanadhan, A. R. Ghose, G. R. Revankar, and R. K. Robins, *J. Chem. Inf. Comput. Sci.*, 1989, **29**, 163.

[14] R. Wang, Y. Fu, and L. Lai, *J. Chem. Inf. Comput. Sci.*, 1997, **37**, 615.

[15] D. S. Knuth, *The Art of Computer Programming,, Volume 2, Seminumerical Algorithms, Third Edition*, Addison-Wesley, Reading, Mass., 1997.

[16] M. S. Newman and D. Lednicer, *J. Am. Chem. Soc.*, 1956, **78**, 4765,

Chapter 5
Using INTERCHEM for Modelling of Proteins and Nucleic Acids

5.1 Introduction
In this chapter we show how the program INTERCHEM can be used to model the large macromolecular structures of *proteins* and *nucleic acids*, and how the program PRESTO can reveal facts about the *sequences* of *amino acid residues* in proteins.

INTERCHEM will display the structures of large molecules and allow you to examine their features. However, because of the size of these structures there are some restrictions on what the program can do. There are special features for displaying and analyzing these large molecules, but some of the facilities used with small molecules are not available.

5.2 The nature of proteins
The purpose of this section is to introduce some basic facts concerning these biomolecular structures, so that you may understand how molecular modelling programs tackle the jobs of displaying them. Proteins are often referred to as polymers. This term might seem to imply that they are similar to synthetic polymers, formed from a single or limited set of monomers, and having a range of, rather than a single definite, molecular mass.

This is not so. Proteins are molecules of definite molecular mass; they can be purified, crystallized, and analysed just as small molecules of low molecular mass can. They can also be made synthetically. True, the methods of purification, crystallization, analysis, and synthesis tend to be rather specialized. However, there is no clear dividing line between large organic molecules and small proteins based on size alone. There are important chemical entities like glucagon and insulin that, while they are most often (and correctly) termed proteins, are of intermediate size, and could be put into either category.

5.2.1 The structure of proteins
In simple molecules, structure is taken to mean how the atoms are joined together. With proteins, the complexity requires that structure be considered at several levels. The basic level does not involve individual atoms since the building blocks are in fact amino acid residues. The chains formed by joining these units together give what is called the *primary structure*. The next stage; *secondary structure* involves the coiling, bending, and twisting of these chains to form structures that are capable of exerting a specific biological activity. It is possible to define further levels of organization, but the exact division of these into *tertiary* and *quaternary* structures is questionable. Instead we concentrate of guiding you through the complex and sometimes confusing terminology.

5.2.1.1 The natural amino acids
What distinguishes proteins as a class is the fact that their main structures are constituted by joining together units (*residues*) of twenty naturally occurring alpha amino acids. These twenty units are shown in Table 5.1. They differ in their side chains (the R groups in the Table), which gives them distinctive properties (acidic, basic, polar, non-polar, etc).

5.2.1.2 The peptide bond

The junction between each pair of amino acids in a protein is known as a peptide bond. These bonds are formally made by a reaction between the carboxylic acid group of the first amino acid and the amino group of the second amino acid with the elimination of a molecule of water:

$$H_2N.CH(R^1).CO_2H + H_2N.CH(R^2).CO_2H \rightarrow H_2N.CH(R^1).CO.NH.CH(R^2).CO_2H + H_2O$$
    Residue 1   Residue 2   Dipeptide

The result of this elementary reaction is the formation of a dipeptide. The reaction, formulated in this way, would not work; there would be side reactions, a mixture of compounds would result. Special techniques are needed to ensure that a specific product is formed. The result is that there is a whole branch of synthetic organic chemistry devoted to making peptide synthesis work.[1]

5.2.1.3 Stereochemistry

The amino acids from which proteins are constituted are known as alpha amino acids because the amino group is bonded to the carbon atom adjacent to the carboxyl (CO_2H) group, known as the alpha carbon atom. In all these amino acids, this atom has bonds to four other atoms; the nitrogen atom, the carbonyl group (C=O), a hydrogen atom, and the first atom of the (variable) side chain. This is usually a carbon atom and then is known as the *beta* carbon atom. In the case of glycine, this is a second hydrogen atom. This makes glycine unique, since the presence of two hydrogen atoms attached to the alpha carbon atom makes this carbon *achiral*. In contrast, all the other nineteen natural amino acids have four different substituents attached to the alpha carbon atom, and this renders this carbon *chiral*. Proteins are (with a few exceptions) constituted from alpha amino acids of identical stereochemistry; and are referred to as *L*-amino acids.[2]

5.2.1.4 Protein chains

It is convenient to refer to an amino acid *residue* in a protein chain by number. The numbering of the residues in a protein starts at the end of the chain where there is (formally) the free amino (H_2N-) group and runs to the end where there is (formally) the free carboxyl (-CO_2H) group. We are being a little pedantic here; the words 'formally' are introduced because in fact the two end groups are ionized and exist as positively charged H_3N^+ and negatively charged CO_2^- groups respectively. This type of structure is known as a *zwitterion* structure. Figure 5.1 shows how the amino acid residues are numbered. The boundary between one residue and the next is considered to be the peptide bond.

5.2.1.5 Conformations of protein chains

Figure 5.2 shows a fragment of a protein chain concentrating on one residue, and shows how the important atoms are named, and how the torsion angles are defined.

There are several overriding factors that determine the shapes of protein chains. The first is the way that *hydrogen bonding* is deployed. The second is the nature of the *side chain groups*. A third is the positions of any *disulfide bonds* (see later).

Table 5.1
Amino Acid Residues in Proteins

Abbreviation		Name	Type	Side Chain	Mass	Colour
1-Letter	3-Letter			R		
A	Ala	Alanine	S,N	CH_3	71.1	Brown
B	Asx	Asp or Asn		See notes		
C	Cys	Cysteine	S,N	CH_2SH	103.2	Yellow
D	Asp	Aspartic acid	A	CH_2CO_2H	115.1	Red
E	Glu	Glutamic acid	A	$CH_2CH_2CO_2H$	129.1	Red
F	Phe	Phenylalanine		$CH_2C_6H_5$	147.2	Brown
G	Gly	Glycine	S,N	H	57.1	Brown
H	His	Histidine		$CH_2.C_3H_3N_2$ See below (1)	137.2	Blue
I	Ile	Isoleucine	L,N	*$CH(CH_3).CH_2CH_3$ *S conf.*	113.2	Green
J		Spare				
K	Lys	Lysine	L,B	$CH_2CH_2CH_2CH_2NH_2$	128.2	Blue
L	Leu	Leucine	L,N	$CH_2CH(CH_3)_2$	113.2	Green
M	Met	Methionine	L,N	$CH_2CH_2SCH_3$	131.2	Yellow
N	Asn	Asparagine	L,P	$CH_2CO.NH_2$	114.1	Magenta
O		Spare				
P	Pro	Proline	L,N	C_4H_7N See below (2)	97.1	Green
Q	Gln	Glutamine	L,P	$CH_2CH_2CO.NH_2$	128.2	Magenta
R	Arg	Arginine	L,B	$CH_2CH_2CH_2NH.C(=NH)NH_2$	156.2	Blue
S	Ser	Serine	S,P	CH_2OH	87.1	Magenta
T	Thr	Threonine	S,P	*$CH(OH).CH_3$ *R conf.*	101.1	Magenta
U		Used for unknown				
V	Val	Valine	L,N	$CH(CH_3)_2$	99.2	Green
W	Trp	Tryptophan	L,N	$CH2.C_8H_6N$ See below (3)	186.2	Brown
X		Used for unknown				
Y	Tyr	Tyrosine	L,P	$CH_2C_6H_4OH$ (*para*)	163.2	Brown
Z	Glx	Glu or Gln		See notes		

Intact Amino Acid: $H_2N-C^{\alpha}HR-CO_2H$; **Residue:** $-NH-C^{\alpha}HR-C(=O)-$ Mass refers to residue mass

1-Letter codes B and Z. These are used in notating sequences determined by (chemical) amino acid analysis where it is impossible to distinguish aspartic acid from asparagine, and glutamic acid from glutamine

Chirality. The *DL* nomenclature is used for alpha amino acids. The natural amino acids in proteins have (almost) always have the *L* configuration. For the reasons for the choice on nomenclature see reference.[2]

Chiral side chains. These occur (noted above in CIP nomenclature!) in isoleucine and threonine.

Amino acid types: S = small; L = large; A = acidic; B = basic; P = polar; N = non-polar; S = sulfur-containing. The colours listed in the seventh column are those used by INTERCHEM and PRESTO when displaying amino acid residues. (Note that other programs may use different colour coding since there is no recognised standard).

L-Histidine (1)

L-Proline (2)

L-Tryptophan (3)

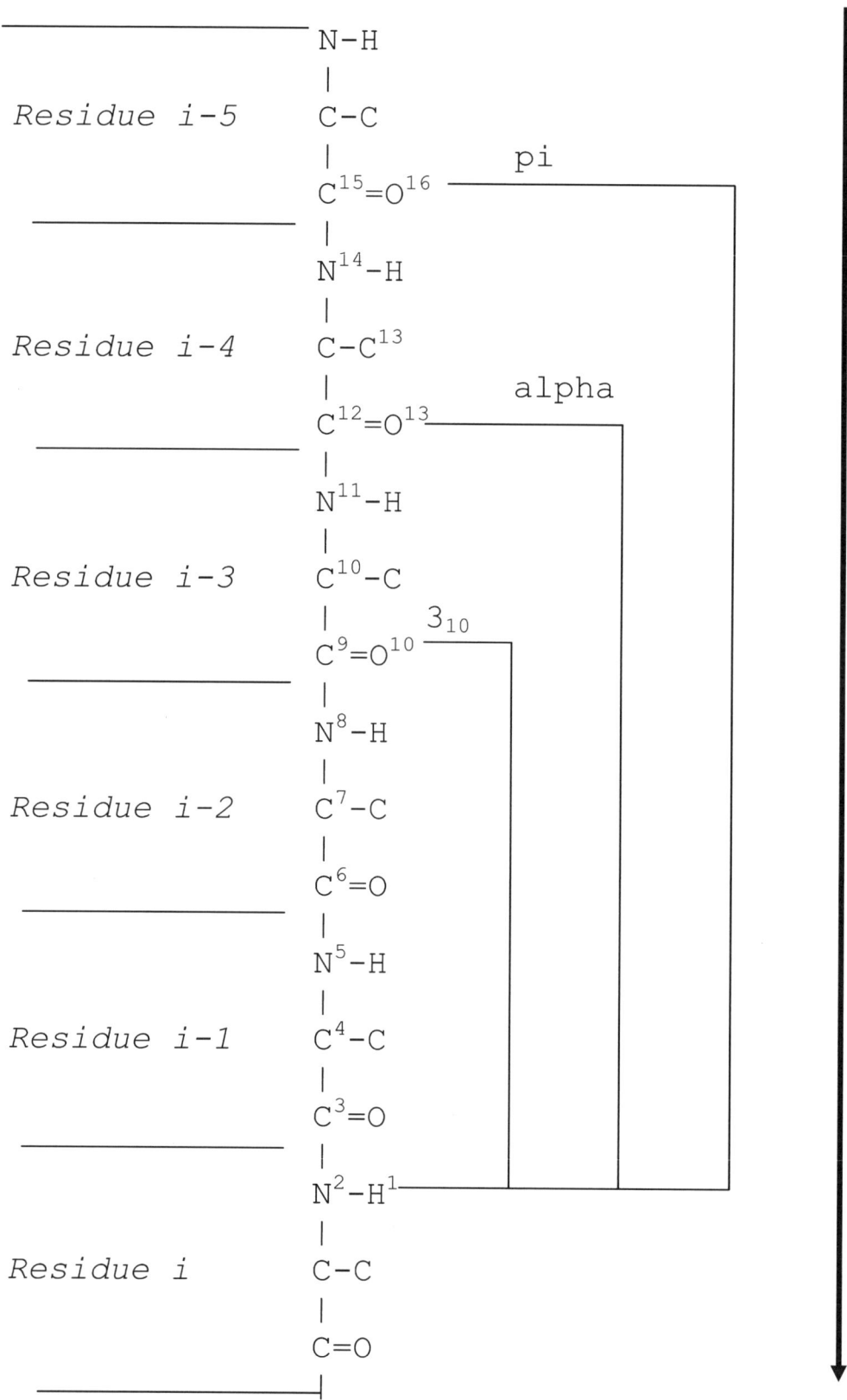

Figure 5.1
How the amino acid residues are numbered in a chain showing the hydrogen bond loops in helices
The looped lines show the hydrogen bond linkages for 3_{10}, alpha, and pi helices. The superscript numbers on the atoms show the numbering in the loops

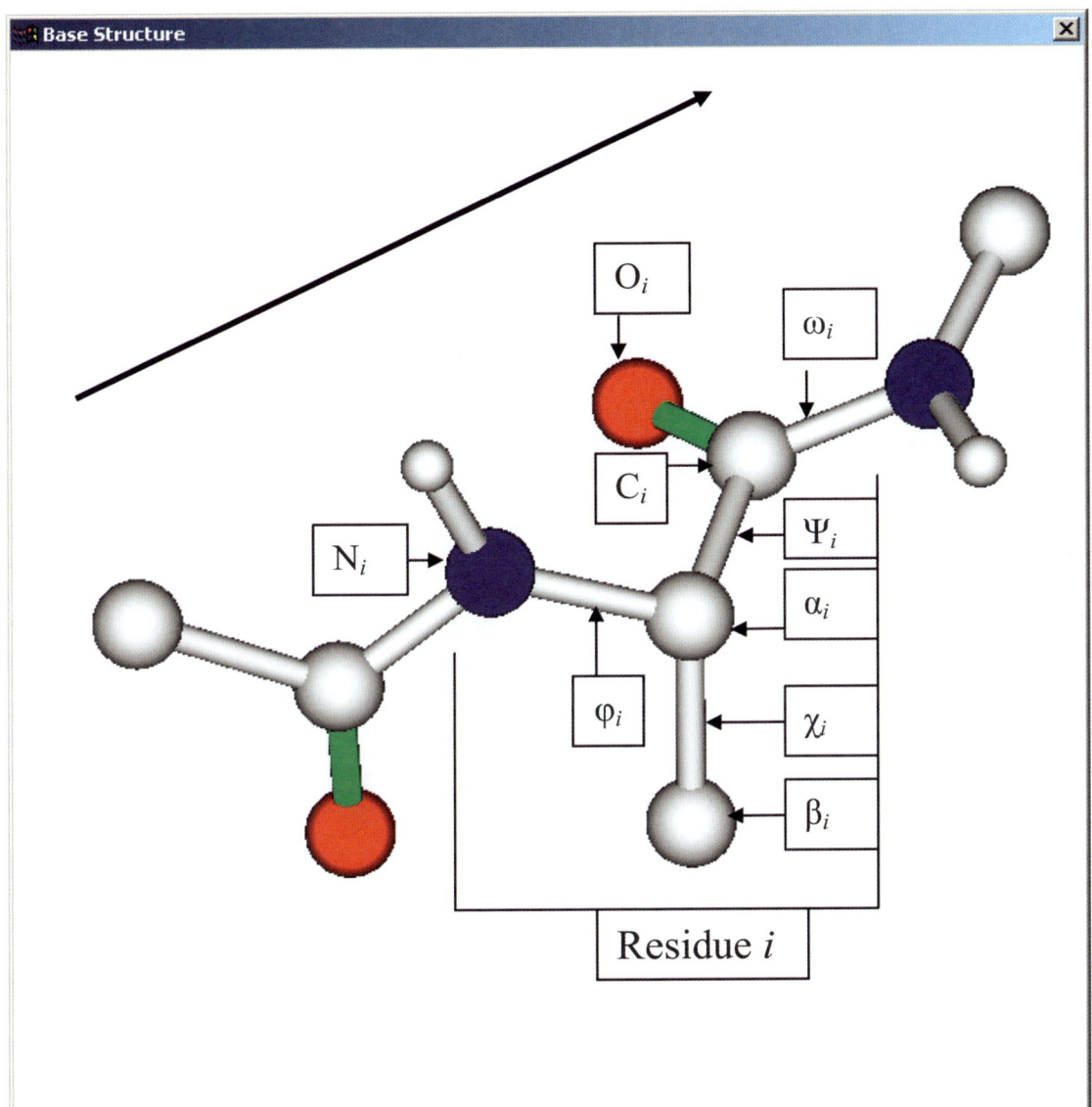

Figure 5.2
How the atoms in an amino acid residue are labelled and how the torsion angles are defined

The ith residue *(alanine)* is identified, in which only one hydrogen atom, that attached to the nitrogen atom, is shown (the small unlabelled sphere). The other atoms are, in order; N, the amino nitrogen atom; α, the alpha carbon; C, the carbonyl carbon; O, the carbonyl oxygen; and β, the beta carbon atom. All of these atoms are present in all of the 20 amino acid residues, with the sole exception of glycine. In glycine the beta carbon is replaced by a hydrogen atom.

The torsion angles associated with bonds in an individual residue are shown labelled with Greek letters:
Phi (φ) is defined as zero when the bonds H-N and α-C are *trans*.
Psi (ψ) is defined as zero when the bonds N-α and C=O are *trans*.
Omega (ω) is defined as zero when bonds α-C and N_{n+1}-H_{n+1} are *cis*.

There are two clearly distinguished elements of *Protein Secondary Structure*; *Helices* and *Sheets*. The most common helix structure is the right-handed *alpha-helix*; this is shown in Figure 5.3. Two other helix forms are possible. These are the 3_{10}–*helix* shown in Figure 5.4, and the π-*helix* shown in Figure 5.5.

The first of these (*3_{10}–helix*) occurs rarely while there is doubt that a complete π-*helix* has ever been seen.

The importance of 3_{10} and π helices rests in the geometry of their hydrogen bond structures, since these motifs do occur in *turns* in the chains.

The geometry of the helices is governed by the torsion angles φ and ψ (and to a lesser extent ω). They are characterized by other parameters: the length of a complete turn (measured in Ångstrom units), the number of residues in a complete turn, the number of atoms in the hydrogen bond loop (see Figure 5.1), and the distance between the residues involved. A hydrogen bond is described as the *donation* of a hydrogen atom *from* a *donor* atom (in this case nitrogen) *to* an *acceptor* atom (in this case the oxygen atom of a carbonyl group). Since in this case the donor atom is later in the chain than the acceptor atom, we characterise the alpha helix, (where the donor atom occurs in the nth residue), as an *n* to *n-4* hydrogen bond, (the negative sign meaning an earlier residue).

Sheet structure in proteins is characterized by the coming together of individual (more or less linear) *strands*. There are two varieties of sheets: *Anti-Parallel-Beta-Sheet*, and *Parallel-Beta-Sheet*. The features are shown in Figures 5.6, and 5.7 respectively. While these diagrams show only the interactions of pairs of strands, it should be realized that sheets are often composed of up to four or more strands in both parallel and anti-parallel arrangements. The strands (and consequently the sheets) are often twisted.

Two other features need to be taken into account. First, the presence of *turns* in a chain; it is clear that to accommodate anti-parallel sheets in a single chain, the chain must turn back on itself in some way. The conformations of these turns is less well characterized than the major features; helices and sheets. But this has not prevented their being named in various fanciful ways.[3] Turns often occur on the periphery of a protein where they are less subject to constraints. A consequence of this is that the crystal structures of these regions are often disordered.

Secondly, many proteins have covalent disulfide bonds (-S-S-) between sulfur atoms in cysteine residues. To go from the –SH group in cysteine to an -S-S- bond involves an oxidation step. Frequently the individual cysteine residues are a long way apart in the chain (or may be on separate chains). Consequently the presence of disulfide bonds can dominate the shape of a protein. In passing it is worth noting that the hair-dressing industry depends on the reduction of disulfide bonds and the oxidation of the resulting SH groups to form bonds in different places for (semi) permanent waving (or straightening) of the keratin of hair!

Protein tertiary structure is also controlled by hydrogen bonding. This term is used to describe how the three dimensional shape of a protein is developed; *i.e.* how the alpha helices and beta sheets, of possibly several discrete protein chains are packed together, so that the structure may fulfill its purpose The two terms *Secondary Structure* and *Tertiary Structure* are often referred to as *Protein Folding*.

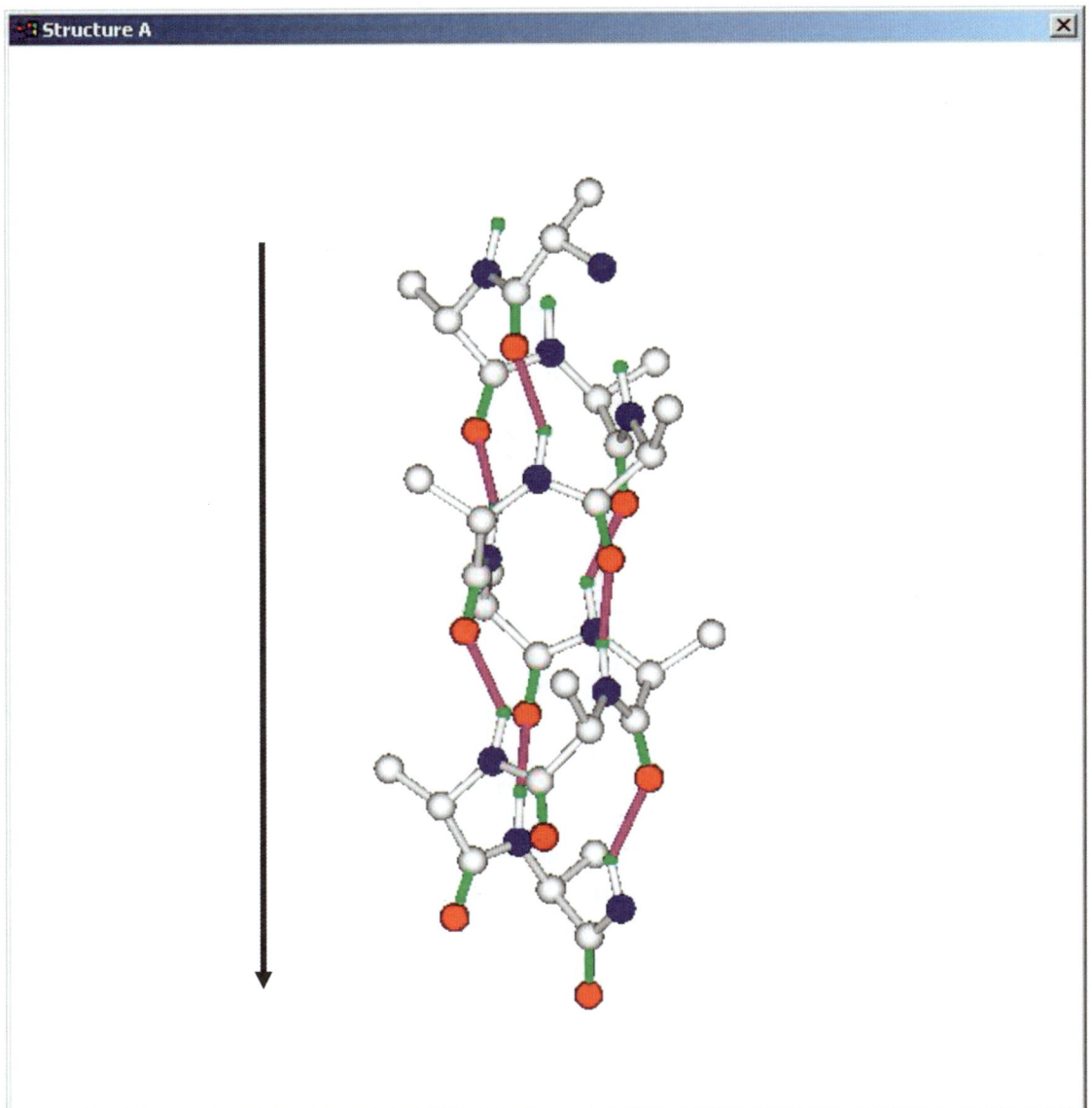

Figure 5.3
A portion of a poly-Alanine peptide chain that has the alpha-helix conformation.

The structure is stabilised by **hydrogen bonds** shown as **magenta** cylinders

These are formed by donation of the **hydrogen atoms (shown as small green spheres)** by the **amide nitrogen atoms (blue spheres)** of residues (*I*) to the **carbonyl oxygen atoms (red spheres) of residues (*I-4*).**

The conventional direction of the peptide chain, (low numbers at the amino group end to high numbers at the carboxyl group end) is shown by the arrow.

There are 3.6 residues per turn in an alpha helix. Another name for the alpha helix could be 3.6_{13} helix; the subscript number defines the number of atoms in the hydrogen bond loop (see Figure 5.1)

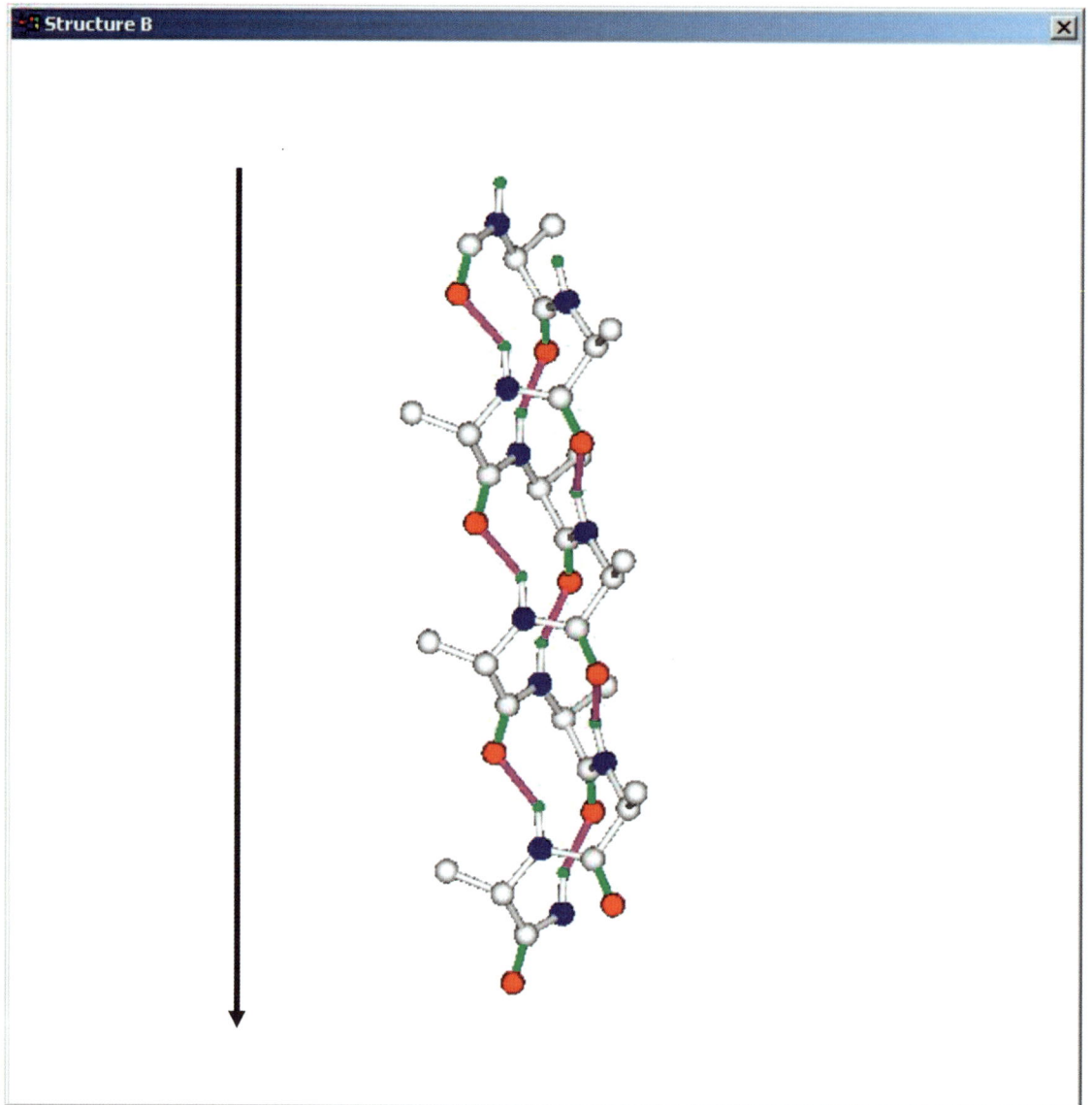

Figure 5.4
A portion of a poly-Alanine peptide chain that has the 3-$_{10}$ helix conformation.

The structure is stabilised by **hydrogen bonds** shown as **magenta** cylinders

These are formed by donation of the **hydrogen atoms (shown as small green spheres)** by the **amide nitrogen atoms (blue spheres)** of residues (*I*) to the **carbonyl oxygen atoms (red spheres) of residues (*I-3*)**.

The conventional direction of the peptide chain, (low numbers at the amino group end to high numbers at the carboxyl group end) is shown by the arrow.

The name (3_{10} helix) implies that there are 3 residues per turn, and that there are 10 atoms in the hydrogen bond loop (see Figure 5.1).

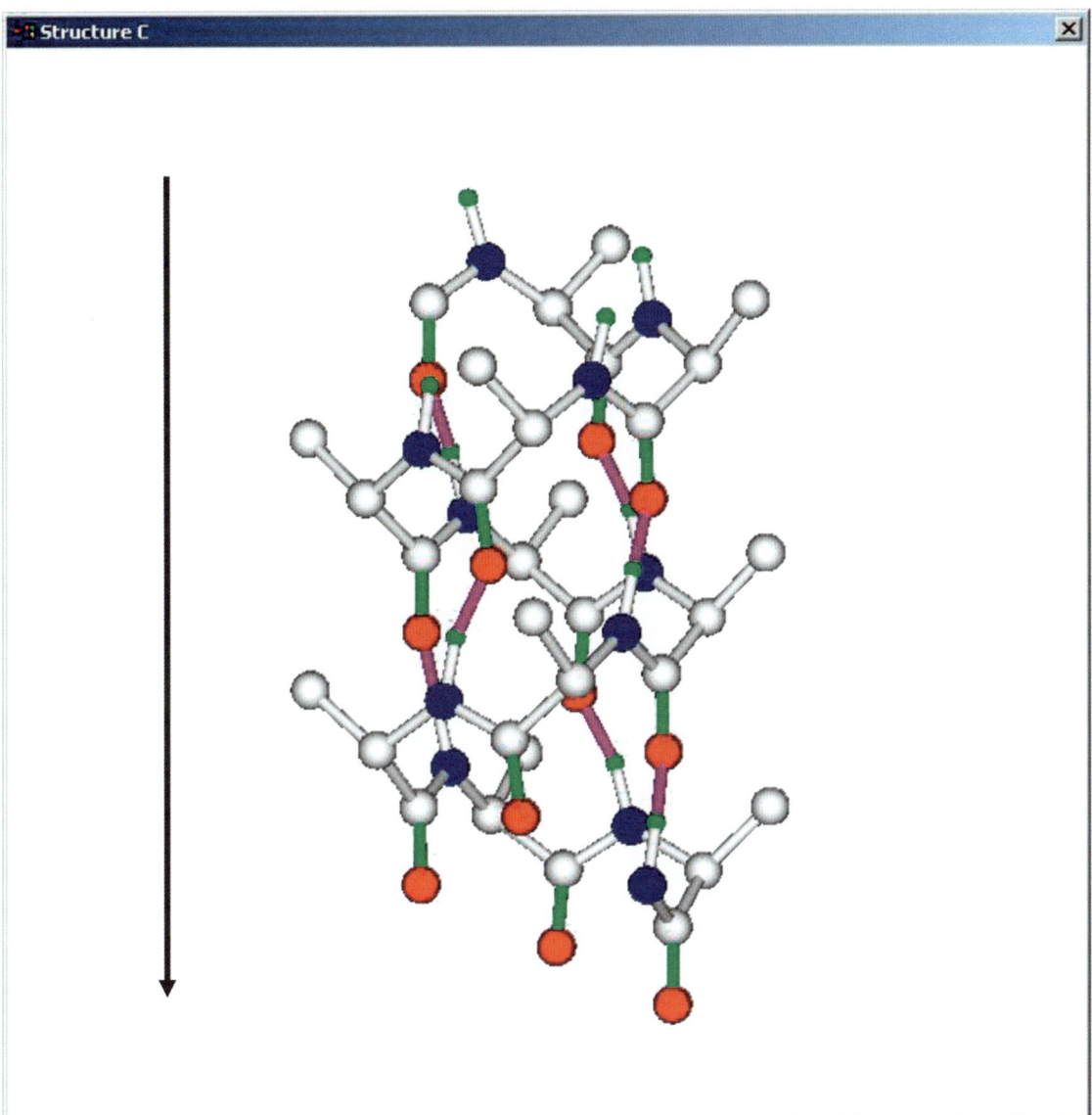

Figure 5.5
A portion of a poly-Alanine peptide chain that has the π-helix conformation.

The structure is stabilised by **hydrogen bonds** shown as **magenta** cylinders

These are formed by donation of the **hydrogen atoms (shown as small green spheres)** by the **amide nitrogen atoms (blue spheres)** of residues *(I)* to the **carbonyl oxygen atoms (red spheres) of residues *(I-5)*.**

The conventional direction of the peptide chain, (low numbers at the amino group end to high numbers at the carboxyl group end) is shown by the arrow.

Another name for the π-helix is 4.3_{16}–helix, implying that there are 4.3 residues per turn and 16 atoms in the hydrogen bond loop (See figure 5.1).

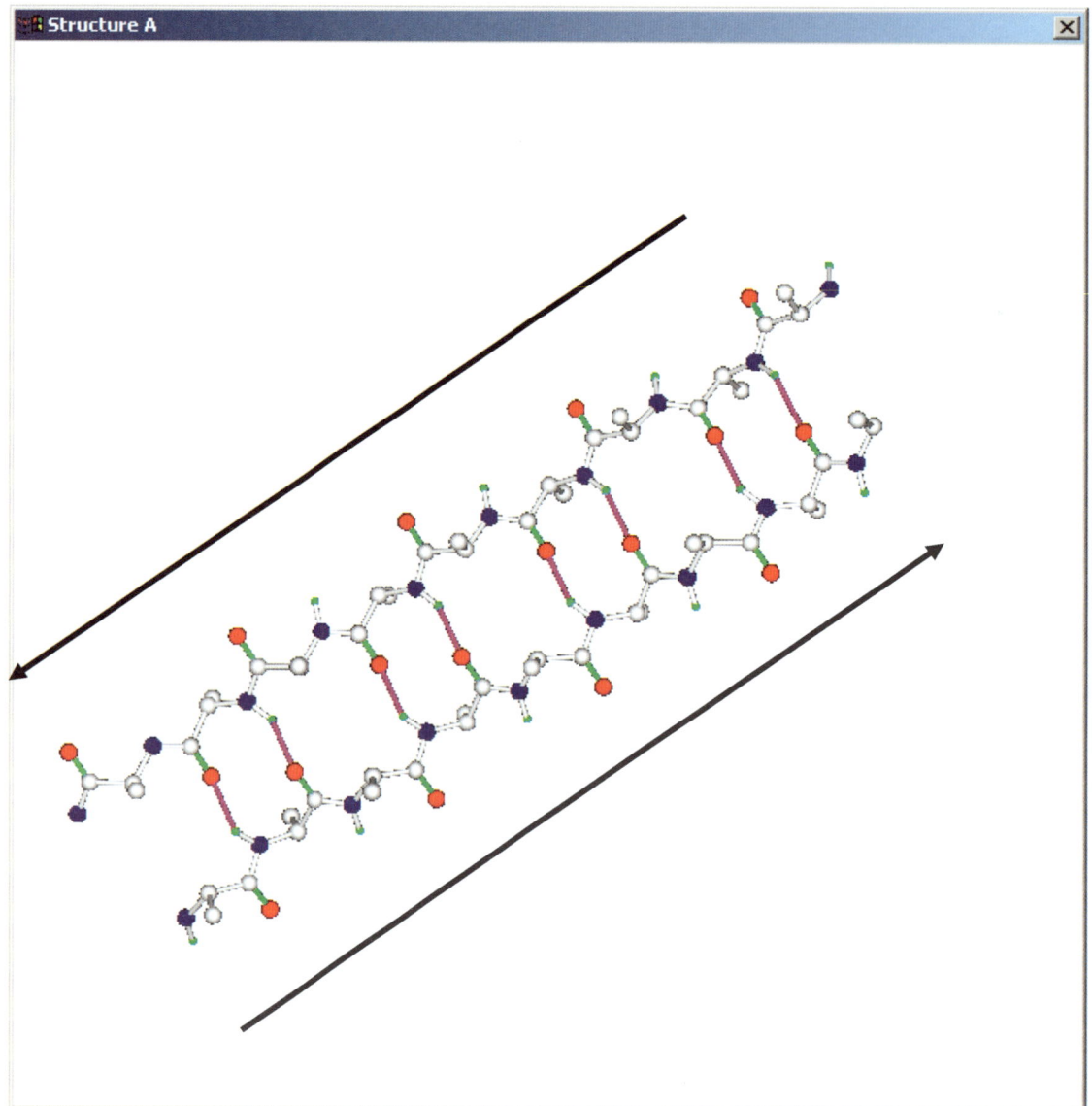

Figure 5.6

Hydrogen bonds in anti-parallel beta sheets, shown as *magenta* cylinders

These are formed by donation of the **hydrogen atoms (shown as small green spheres)** by the **amide nitrogen atoms (blue spheres)** to the **carbonyl oxygen atoms (red spheres).**

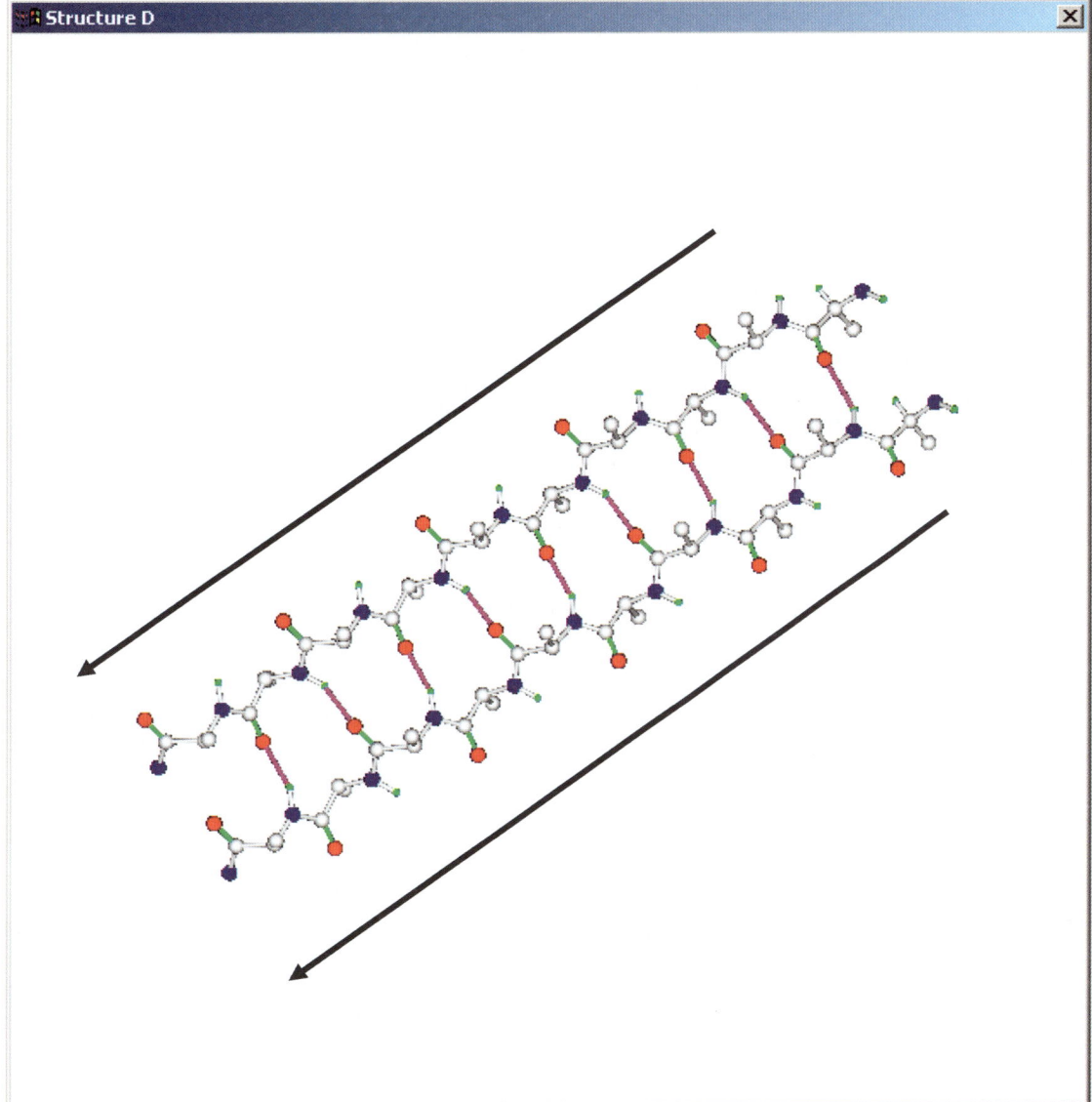

Figure 5.7

Hydrogen bonds in parallel beta sheets, shown as *magenta* cylinders

These are formed by donation of the **hydrogen atoms (shown as small green spheres)** by the **amide nitrogen atoms (blue spheres)** to the **carbonyl oxygen atoms (red spheres).**

To summarize so far: a protein consists of one or more chains. Each chain consists of a well defined *sequence* of amino acid residues, joined by peptide bonds. There may be extra *Disulfide* linkages between the sulfur atoms of cysteine residues. These help to make the structure more rigid. These main facts are well established. Yet one great mystery remains; what causes a protein to fold to the same conformation whether it has been isolated from a natural source or made synthetically.[4]

5.2.2 The structures of nucleic acids

In contrast to the complicated situation with proteins (the two types of structure; helices and sheets), the principal structure of the (polymeric) deoxynucleic acids is well known; the famous double helix shown in Figure 5.8. That this structure was arrived at by modelling, using essentially mechanical models, is one of the triumphs of scientific reasoning of the Twentieth century (if not for all time!). We have already touched on it (Chapter 1; reference[3]). It spawned a whole new scientific discipline in the form Molecular Biology, the repercussions of which are with us today.

The double helix structure (like the structures in proteins) owes its stability to the presence of hydrogen bonds, which are preferentially formed between specific purine and pyrimidine bases in the pairs of nucleotides on the antiparallel chains; (deoxyadenosine bonded to deoxythymidine and deoxyguanosine bonded to deoxycytidine). Figure 5.9 shows this hydrogen bonding in the bases. Figure 5.10 shows the same stretch of helix as Figure 5.8 in a form featuring the specific bonding of the bases.

The double helices of DNA duplexes have characteristic features; *Major Groove* and *Minor Groove*. These are labelled in Figure 5.10. The minor groove is important as the place where the anti-cancer drug cisplatin binds (preferentially to guanine). Disruption of this helical structure, by this sort of binding, is the way that drugs for treating cancer and viral diseases probably act.

5.2.3 Further reading

In this limited space we have given you sufficient information for you to progress to viewing the structures of these biomolecules. But to gain real insight we recommend your reading of the book by Schulz and Schirmer on Protein Stucture,[5] the book by Saenger on Nucleic Acid Structure,[6] and three books by Arthur M. Lesk on Protein Science,[7] Genomics,[8] and Bioinformatics.[9] To understand how the structures are determined the volume published by the International Union of Crystallography,[10] and the book by Würthrich[11] on the nuclear magnetic resonance methods should be consulted.

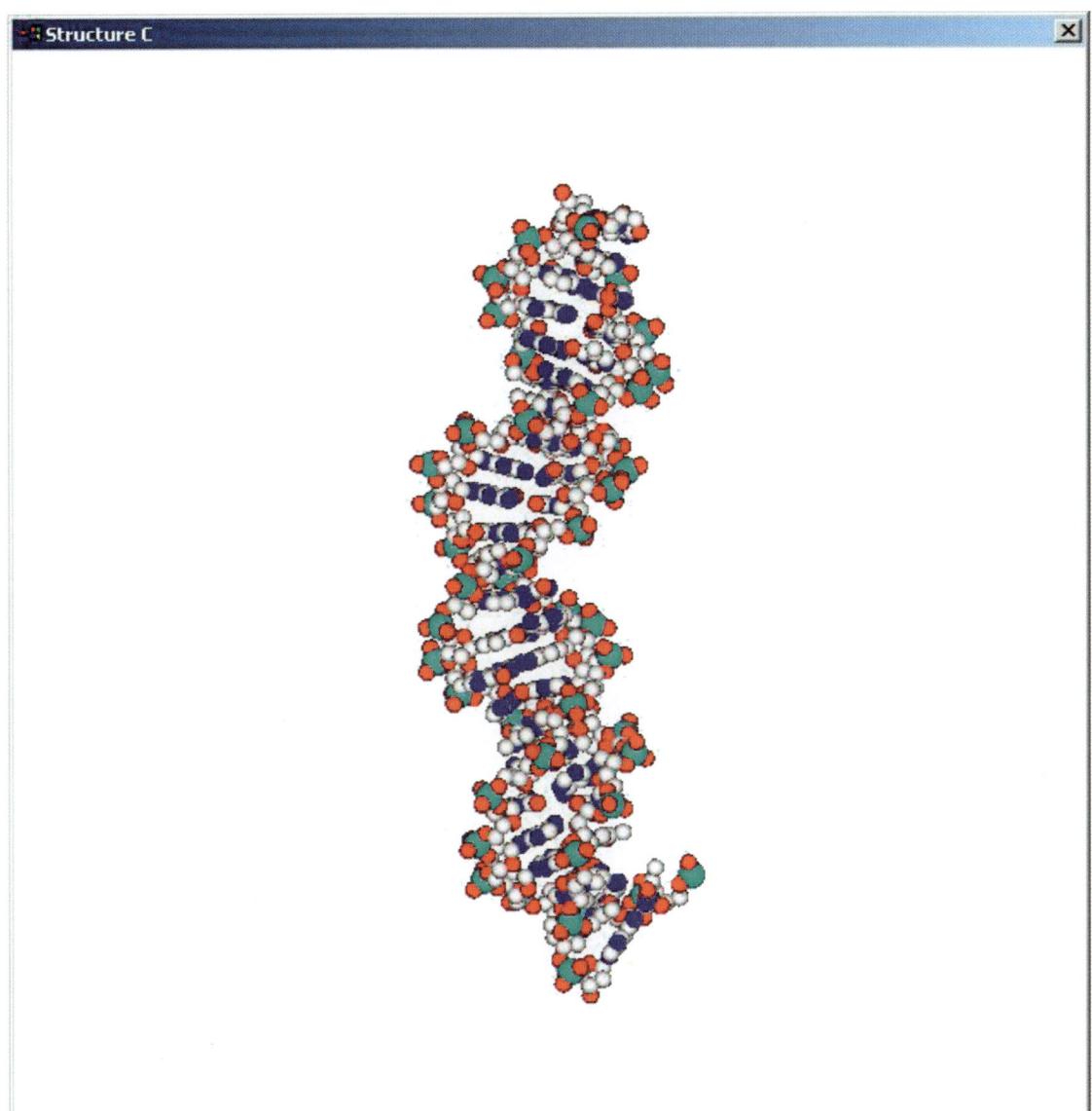

Figure 5.8
Twenty four nucleotide base-pairs cut from the DNA duplex of the protein-DNA complex in Protein Data Bank entry 1AOI.

Colour coding of atoms:
    Red: Oxygen
    Blue: Nitrogen
    Cyan: Phosphorus
    White: Carbon

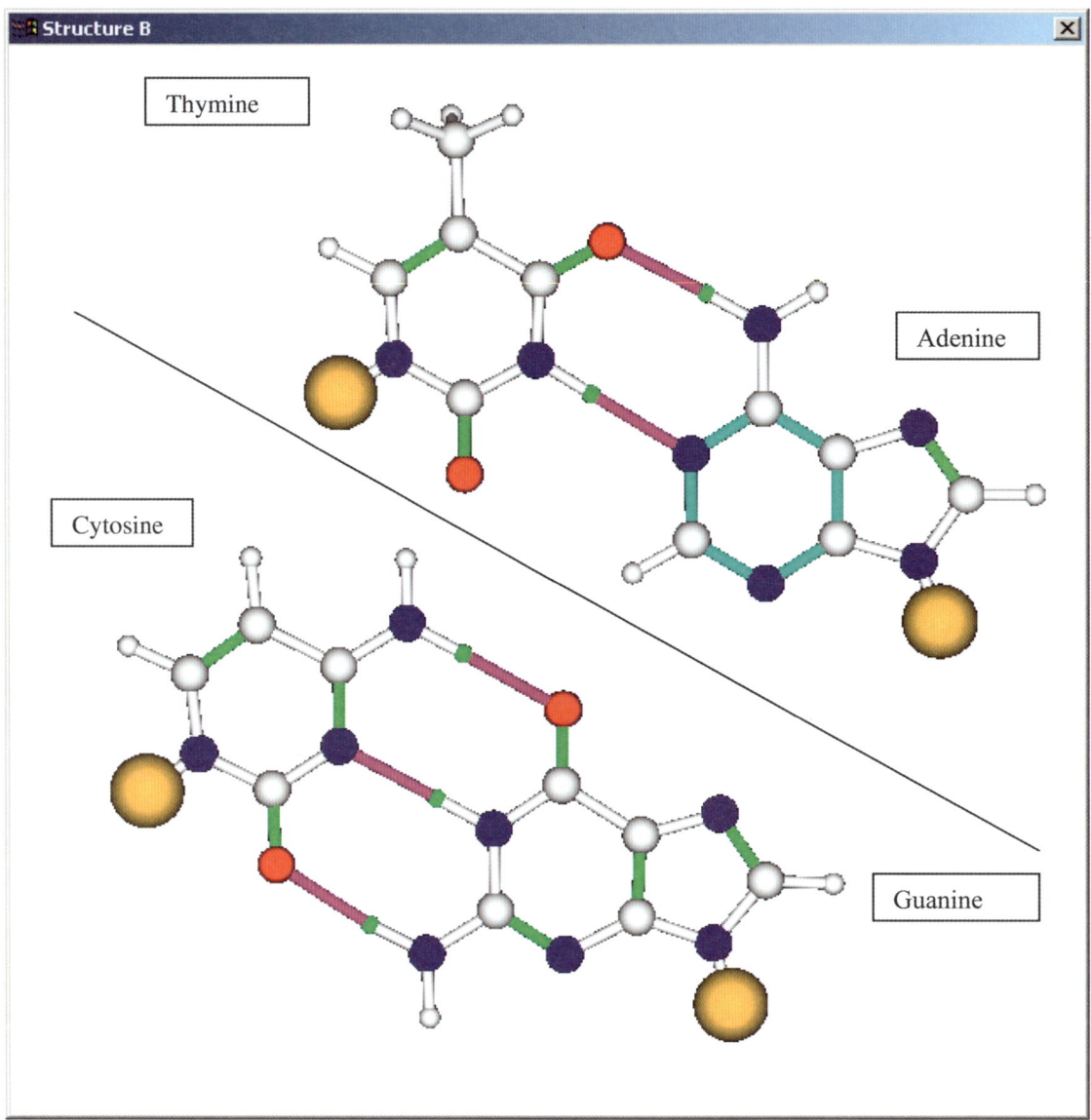

Figure 5.9
The hydrogen bonds that stabilize the Watson-Crick pairings of the nucleotide bases

Colour coding:
      Blue: Nitrogen atoms
      Red: Oxygen atoms
      Green: Hydrogen atoms involved in bonding/ Double bonds
      Cyan: Aromatic ring bonds
      Magenta: Hydrogen bonds
      Uncoloured: Carbon atoms and other hydrogen atoms/ Single bonds
      Yellow: Connecting Carbon atoms in the deoxyribose rings

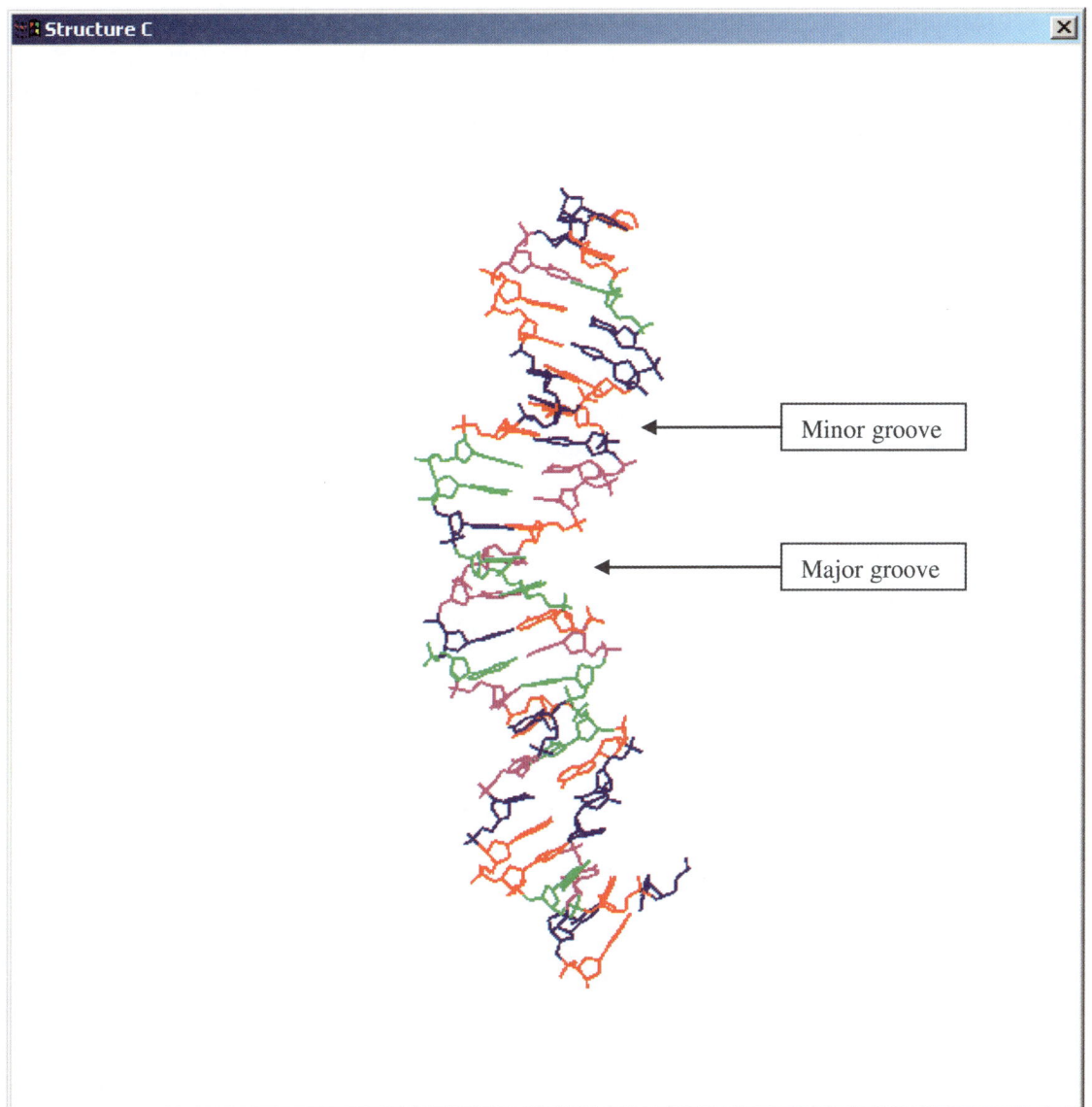

Figure 5.10
Twenty four nucleotide base-pairs cut from the DNA duplex of the protein-DNA complex in the Protein Data Bank entry 1AOI.

Colour coding of bases::
    Red: Adenine
    Blue: Thymidine
    Green   Guanosine
    Magenta   Cytidine

Note that the original DNA duplex is coiled; this results in the axis of the helix being curved.

5.3 Obtaining structures for proteins and nucleic acids.

Almost without exception the structures are obtained from the Protein Data Bank (PDB). This is an organization that holds structural data for these materials, obtained from X-Ray crystallographic and nuclear magnetic resonance (nmr) experiments. The way the organization works and the format of the data files is described in Chapter 8. Here we describe the way that INTERCHEM accesses and uses the data.

We assume that the program PROTEINS has been set up on your computer, and that clicking on its shortcut on your computer's desktop can start it. The data files distributed by the PDB are large and contain extra information that is not needed for visualization and analysis of the structure. Because of this, a decision was made to use the separate program PROTEINS to abstract the necessary information, and store it in files that are modified versions of standard INTERCHEM 'D' format files. These intermediary files are usually more compact than PDB files, but can if necessary hold extra information generated by PROTEINS (such as the coordinates of added hydrogen atoms). The program PROTEINS exists in two forms; one that is suitable if your local computer system holds a copy of the PDB files, and one that assumes that you will download, through the Internet, structures from the PDB (or a local mirror site). It is this second situation that we now describe.

5.3.1 Accessing the Protein Data Bank.

If INTERCHEM is running you can access the PDB directly by pressing the button *Protein Data Bank* (R9/C10), or you can directly access the web site:

http://www.rcsb.org/home/

The site is easy to navigate and there is extensive online help. It is also regularly updated, so the instructions we give here may need to be interpreted with care!

The essential information that you need is the four-character code for the particular structure in which your interest lies. Enter this in the search box (using either upper of lower case letters) and press *Search*. A snapshot picture of the structure will appear. Next press *Download Files*, and then choose *PDB File (text)*.

Warning.[12] *Do not be tempted to download a compressed version of the file. These versions of files have a suffix 'gz', and if your computer has a program called 'WinZip', it will take control of them and attempt to uncompress them. This can result in chaos!*

Your own computer will then respond by requesting the name of a folder in which to store the file. You can create a new directory, or if this is not the first file you have downloaded in this session, use an existing directory. With a good network connection, a file of 1 megabyte will be received in a few seconds. At this stage, if you do not intend to download further files, you can close the network connection, since the subsequent operations take place on your own computer.

We will assume that the structure that you have requested had the code `1AOI`, and that it has been placed in the folder:

`F:\PDB_Files.`

Using INTERCHEM for Modelling of Proteins and Nucleic Acids

INTERCHEM will still be running, but you need to suspend its operation for the time being so proceed as follows:

(Step 1) Click on the large button on the bottom corner of the *Dual Mode* menu. This will reveal four hidden buttons.

(Step 2) Click the red button *Exit the Program*, and then SUSPEND INTERCHEM.

(Step 3) Click on the shortcut for PROTEINS.

This will start the auxiliary program and show its main menu (Figure 5.11). PROTEINS requires that you enter a number to choose a menu option, so continue as follows:

(Step 4) Choose option (2) and enter the *full path* of the folder that holds the downloaded file. This will include a drive letter, for example: F:\PDB_Files.

(Step 5) Choose option (3) and enter the four letter code 1AOI.

This will allow you to read the contents of the file, a page at a time, to verify it. Reading the comments at the beginning of the file will tell you that the structure contains both protein and nucleic acid components. At any stage you can stop reading it, and, usefully, by entering sub option (4) check the number of chains and atoms in the structure. In this case, 12385 atoms and 10 chains.

(Step 6)   Next, because the structure is a protein *and* nucleic acid complex, the program PROTEINS requires that it be treated as a *nucleic acid*. So choose option (9) [rather than option (7)] from the menu. This will create a standard INTERCHEM file called NUC1xoiD.DAT. In order to allow for regions of a protein chain where there is disorder, the PDB files can offer alternative sets of coordinates. These alternatives are assigned alphabetical letter codes, and a question offering you a choice will appear; in order to answer this it is usual to enter the letter *A* (or *a*). The appearance of the question does not imply that there is disorder in the structure.

(Step 7) If this is the only file you wish to process, close down PROTEINS by choosing option (13). Then restart INTERCHEM by clicking on the cyan coloured button at the bottom of the screen. Alternatively, if you wish to convert other files, and go back to Step 5. You could also reconnect to the PDB and download further files.

(Step 8) Load this newly created file by pressing *Load Structure* (R2/C7) and then *Interchem 'D' Format* , and finally (*via* the succession of filtered submenus) the file itself NUC1xoi. The system generates the full filename NUC1xoiD.DAT.

The display that appears is a *Wire Frame* and is not very revealing. A better display option is *Ribbon Display* (R4/C10) followed by *Ribbons in Lighted Mode* from the sub menu. This will display only the protein part of the structure as a ribbon. To see the nucleic acid part click on *Extra Displays* (R5/C7). This will add two turns of a coiled up nucleic acid helix as CPK atoms, surrounding the protein.

5.3.2 The options provided by the program PROTEINS

The decision, made many years ago, to have a separate program to handle the conversion of PDB data into a form acceptable to INTERCHEM, has proved to be a wise one, since it has insulated the development of INTERCHEM from the many changes that have taken place at the PDB over the years.

The program PROTEINS allows for the PDB data to be processed into several sorts of files; this is shown in the main menu. This menu also allows access to a brief help display. Some knowledge of the type of data in a particular PDB file is needed before the program can be fully exploited   For example, data that comes from nuclear magnetic resonance (nmr) experiments are handled in a different manner by the PDB from data that comes from crystallography.

Sequence data for use with the program PRESTO can be extracted (Option 5). (PRESTO will be encountered in section 5.5). Simple files for use by INTERCHEM that contain only coordinates of the alpha carbon atoms of the amino acid residues (Option 6) can be useful in showing the shape of a protein.

It is important to recognize that the PDB is primarily a storage site for the results of scientific experiments; these results are not necessarily in a form that is easy to use for modelling. For example, many structures derived from X-Ray crystallography are incomplete; whole sections of a structure may be missing.

How this is handled is to some extent left to the people depositing the structure, but there may be discontinuity in the numbering of the amino acid residues. With disordered structures, the authors may have included alternative data for sections of a protein chain. INTERCHEM can handle both these types of situation, *but not necessarily in the same way that other programs do!* Consult the relevant parts of Chapter 8 to see how the data is stored in PDB files.

5.3.3 How the extra data is stored in INTERCHEM 'D' files

Except in one important detail, the format that INTERCHEM uses for holding the compact versions of protein and nucleic acid structures is identical to that used for storing small molecule structures. The difference is that the 32 bit integer, that stores the atomic masses of atoms for small molecules, is made use of to store the extra information provided in PDB data files. If it is accepted that chemical elements with atomic mass greater than 127 (the maximal integer that can be stored in 7 bits) are not likely to be encountered, then the remaining 25 bits can be used to store this extra information.

The formula is:
$$\text{IATN2} = (((((\text{IRES} \times 64) + \text{IAMA}) \times 64) + \text{IC}) \times 128) + \text{IATNO}) \quad (5.1)$$

Thus in this 32 bit Integer (`IATN2`) the bits have the following significance:

Bits 0 to 6 (7 bits)	The true atomic number of the atom:.	IATNO	Maximum	127
Bits 7 to 12 (6 bits)	Chain number:	IC	Maximum	63
Bits 13 to 18 (6 bits)	Amino acid or nucleic acid residue type:	IAMA	Maximum	63
Bits 19 to 30 (12 bits)	Amino acid or nucleic acid residue number:	IRES	Maximum	4095
Bit 31	Sign bit (not used).			

(Because of total storage requirements, the number of amino acid or nucleic acid residues is limited to 1023 in any one chain).

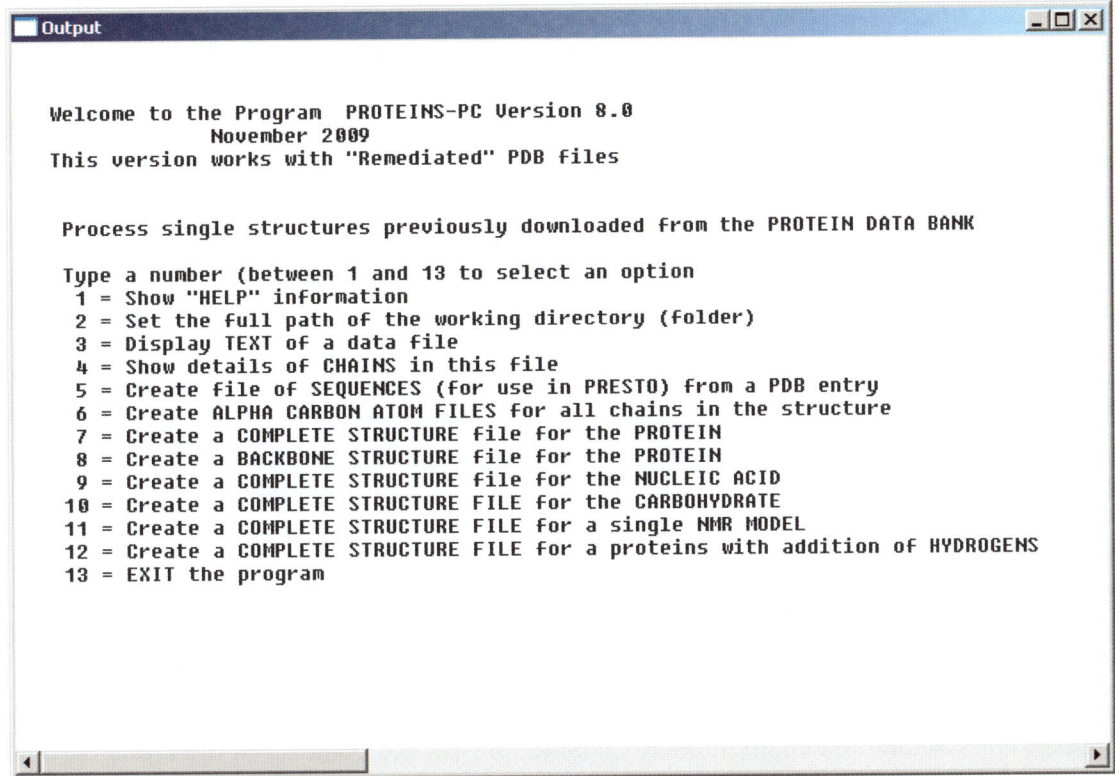

**Figure 5.11
The main menu of the program PROTEINS**

If, in a PDB file, there is more than one chain present, then the chains are labeled with letters that can be arbitrarily chosen by the person who submits the data. In the program PROTEINS, if more than one chain is found the chains are assigned sequential numbers. (If only one chain is present it is assigned the number zero). Consequently there is no way of deducing what the PDB chain *letter* is from the INTERCHEM chain *number*.

Amino acid types are assigned integers that are derived from the ASCII codes of their single letter codes. Nucleic acid types are treated in a similar way

When a protein file is loaded by INTERCHEM the extra information is extracted and stored by the program ready for use. The overhead associated with the method of packing the data is only encountered at this time. If you need to look at the extra data items stored in an INTERCHEM protein file, the utility program PROTOCHECK is provided. This will derive an expanded printable version of the file, where the values of IC, IAMA, and IRES can be examined.

5.3.4 Displaying protein and nucleic acid structures in INTERCHEM
Since protein structures are large there can be problems in displaying them. All of the display formats are useable but rotation of structures can be slow if either *Cylindrical Bonds*, *Ball and Cylinder*, or *CPK* modes are used. You will notice that protein structures are very 3-dimensional! It is virtually impossible to visualize the whole structure without one part of the structure obscuring another. *Wire Frame* and in particular *Anaglyph Stereo* modes are suitable however. There are also four special display modes that are only applicable to protein (and to some extent nucleic acid) structures.

(1) With *Coloured Chains* (R4/C7) the wire frame display is modified so that if there is more than one chain, a series of four pastel shades is applied cyclically.

(2) With *Residue Coded* (R4/C8) display, the bonds of amino acid and nucleic acid residues are colour coded according to the scheme shown in Table 5.1 (amino acids) and Figure 5.10 (nucleic acids)

(3) If *Highlighting* (R4/C9) is used, then the initial display is a wire frame, but pressing *Set Highlights* (R4/C5) opens a separate menu that allows the highlighting colours to be applied selectively.

The choices in highlighting are *Toggles* and are applied using *Boolean Logic*. Thus if *Arginines* is selected followed by *Basic Residues* then when *Accept/EXIT* is pressed, the residues highlighted are *Lysines* and *Histidines* but not *Arginines*.

Whole chains (*Highlight Chains*) or sequences of residues identified by number (*Highlight Range*) may be highlighted. Normally these last two options may be applied additively, but the buttons *Residue exclusive* and *Chain exclusive* have the effect of limiting the options to single sequences of residues and single chains. In addition, when any residue is highlighted already, *Toggle Labels* will add the appropriate residue and chain number to the display at the position of the alpha carbon of the residue.

(4) We have already encountered *Ribbon Display* (R4/C10 in *Dual* Mode). This is only applicable to protein structures, but there are various options that may be selected from a subsidiary menu, that are useful in some circumstances. The most pleasing however is *Ribbons in Lighted Mode*. In this the chains are differentiated by pastel colourings.

To generate the ribbons it is first necessary for the program to analyze the structure to find where the two principal motifs that characterize proteins occur; alpha helices and beta sheets. While this information is nearly always provided as comments in the original PDB files, the algorithms that are used by INTERCHEM do not make use of it. Instead the decisions are made based on measurements of the torsion angles Φ and Ψ. These characterize the conformations of the protein chain at the bonds that join the amino nitrogen atom to the alpha carbon atom (Φ), and the bond that joins the alpha carbon atom to the carboxyl carbon atom (Ψ) (see Figure 5.2, and the discussion in section 5.2.1, and also Figure 5.17).

5.3.5 Editing protein and nucleic acid structures

The power of the highlighting tools is exploited further to selectively delete or retain parts of a protein of nucleic acid structure. This is best illustrated with an example using the following steps:

(1) Load the file `NUC1aoiD.DAT` into area A, and switch to *Single Mode*.

(2) Then display the composite structure (protein plus nucleic acid components) with *Highlighting* (R20/C1 in the *Single Mode* menu).

(3) Press *Set Highlights* (*Highlight Menu*, R15/C2). (The extra menu will overlay the display but this does not matter).

(4) Choose *All Deoxy Bases* followed by *ACCEPT/EXIT*. The nucleic acid portion

of the structure that surrounds the protein part will be highlighted.

(5) Press *Set Highlights* (R15/C2)

(6) Click *Retain* → *C* (Retain the highlighted part in *C*).

(7) Press *Set Highlights* (R15/C2)

(8) Click *Eliminate* → *D* (Eliminate the highlighted part and retain what is left in *D*).

(9) Select *Quad Mode*

The display will now show: the original structure in area *A*, the nucleic acid component in area *C*, and the protein component in area *D*. The two component structures can be stored in separate files for further work.

5.3.6 Analyzing protein structures

The simple fact that the protein (and nucleic acid) structures that are obtained from the PDB are so large precludes the use of the analysis tools that we described in section 4.6. Any attempts to use them will produce an error message. You *can* use the *Atom Numbers* button (R5/C2) with a wire frame display of a protein; the result is that the residue numbers and residue types are displayed, instead of the individual atom numbers, reflecting the way that, in these structures, the constituent *Amino Acid Residues* are treated as *'Super Atoms'* (of *'Super Elements'* classified in a *'Super Periodic Table'*).[13]

However, an appropriate set of analysis tools is made available from the button *Protein Analysis* (R8/C10). The initial response is to report the number of separate chains in the protein, and the number of residues in each chain. After this, the main menu uses a set of 'radio' buttons to access the individual tools. These are chosen by clicking first on one of the radio buttons followed by use of the *Escape Key* (sometimes a further choice is needed). (Figure 5.12).

It is possible to analyse a protein structure, either in terms of the properties of the amino acid residues in the chains, or to detect the close spatial proximity of residues in a protein chain that are not close together in the sequence. The properties of such clusters of residues can then be investigated.

5.3.6.1 Simple Intra-Sequence Distance Analysis

The first choice 'Simple Int(er)(ra)-sequence Distance Analysis' prompts for selection of further parameters in the next menu. (Figure 5.13)

In this facility the program looks for close interactions between amino acid residues. You are next asked to select the numbers for two chains. These are called the `THIS` and the `THAT` chains. The radius of the 'Sphere of Interaction' must also be defined (you can either use the slider bar, or type a value in the box, or choose to accept the default value of 7 Ångstroms).

In the general case where the two chains are different, the program takes each of the residues in the `THIS` chain and finds all of the residues in the `THAT` chain that are within the Sphere of Interaction. These interacting residues of the `THAT` chain are a *cluster*. The program then looks at the residue numbers in the cluster, and finds the highest (`HI`) and lowest (`LO`) of these numbers. It also looks for the greatest gap (`GG`) in serial numbers in the cluster. The program then displays three plots: `HILO` (= `HI`-`LO`), `GG`, and `DIFF` (=`HILO` - `GG`) against residue number (of the `THIS` chain).

The case where the `THIS` and `THAT` chains are the same corresponds to SID (Simple *Intra*-sequence Distance) defined in the paper by L. Pritchard, L. Cardle, S. Quinn, and M. Dufton.[14] This paper should be consulted on interpretation of the results. This situation arises also in cases where there is only one chain in the structure; the program recognizes this, and there is no prompting for a second chain number.

The general case where the `THIS` and `THAT` chains are different (Simple *Inter*-sequence Distance) is an extension to the ideas defined in that paper. It can yield plots devoid of any information, but when an interaction is seen, it is possibly highly significant.

In this analysis section, distances between residues are defined as distance between their alpha carbon atoms. Note that the cluster includes all residues falling within the Sphere of Interaction; there is no exclusion of those residues within ± 2 residue positions (this is in contrast to the next analysis method).

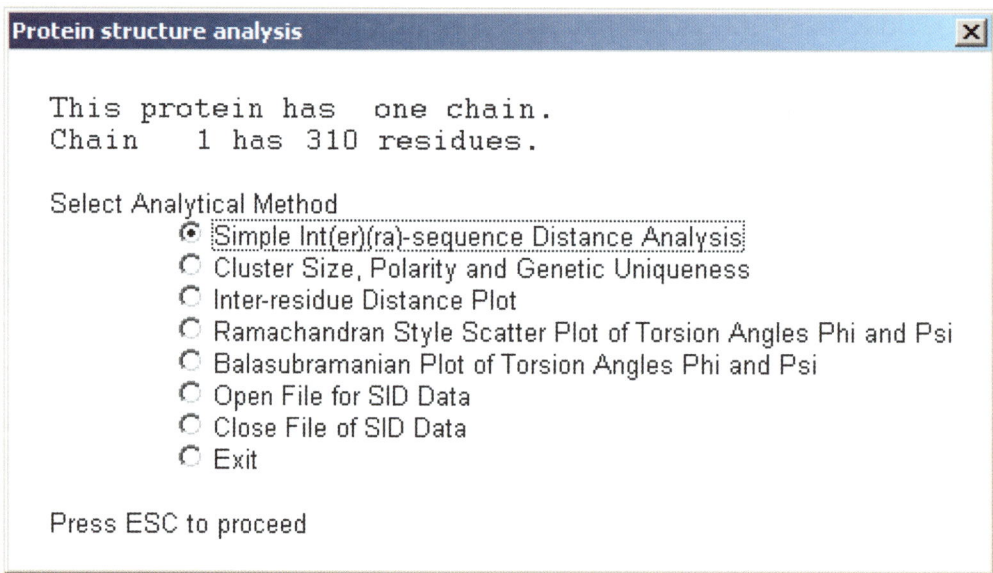

Figure 5.12
Menu for the Protein Analysis Module

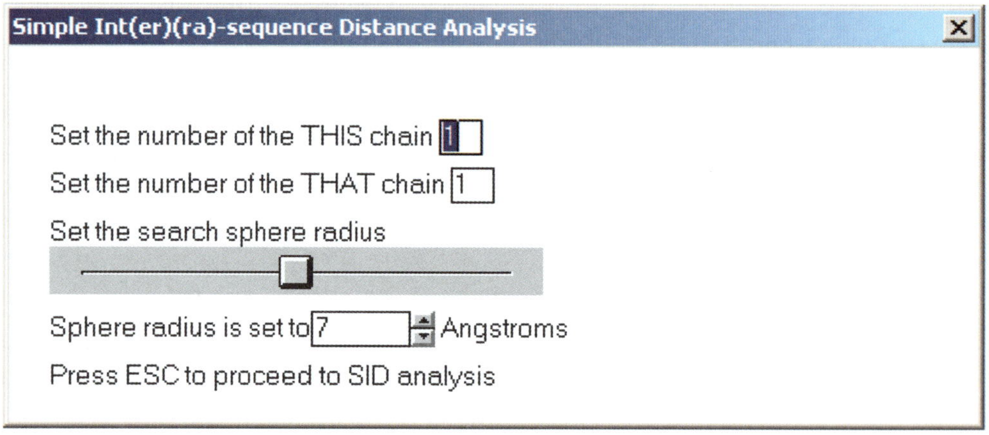

Figure 5.13
Setting Parameters for SID Analysis

5.3.6.2 Cluster analysis

The second choice in the first menu also produces a sub menu that requires the selection of a two chain numbers and a 'Sphere of Interaction'. The output consists of three plots.

The first of these is a plot of total *Cluster Size* against the residue number of the THIS chain. In calculating the cluster size, the current residue and those residues within ±2 positions on either side of it are not counted. This contrasts with the way clusters are defined in the SID plots (section 5.3.6.1).

The second plot is of *Total Cluster Polarity* against the residue number in the THIS chain. This is calculated by adding together the polarity values of the individual residues in the cluster, and dividing by the number of residues in the cluster. The result can be either, a *positive number* for a lipophobic cluster or, a *negative number* for a lipophilic cluster. The formula (for a cluster of *n* residues) is:

$$POL = (\Sigma P_n)/n \qquad (5.2)$$

The third plot is the *Genetic Uniqueness* of the cluster against the residue number of the THIS chain. The genetic uniqueness is calculated by the following formula (for a cluster of *n* residues):

$$GU = (3.5^n)/(\Pi G_n) \qquad (5.3)$$

The parameter *G* is the *effective* numbers of DNA triplets coding for the particular amino acid. The quantity '3.5' in the equation is the *average* number of DNA triplets over the whole range of 20 amino acids. The polarity parameters *P* are obtained from the hydrophobicity parameters as defined by J. Kyte and R. F. Doolittle[15] by negating and multiplying by ten. The values of the two parameters *P* and *G* are shown in Table 5.2.

Table 5.2
Polarity and Genetic Uniqueness Parameters for 20 Amino Acid Types

Amino Acid	P	G	Amino Acid	P	G
A	-18	4.71	M	-19	1.45
C	-25	1.13	N	+35	2.60
D	+35	3.13	P	+16	3.27
E	+35	3.73	Q	+35	2.53
F	-28	2.34	R	+45	3.38
G	+4	4.27	S	+8	4.40
H	+32	1.40	T	+7	3.71
I	-45	3.11	V	-42	4.01
K	+39	3.42	W	+9	0.89
L	-38	5.65	Y	+13	1.91

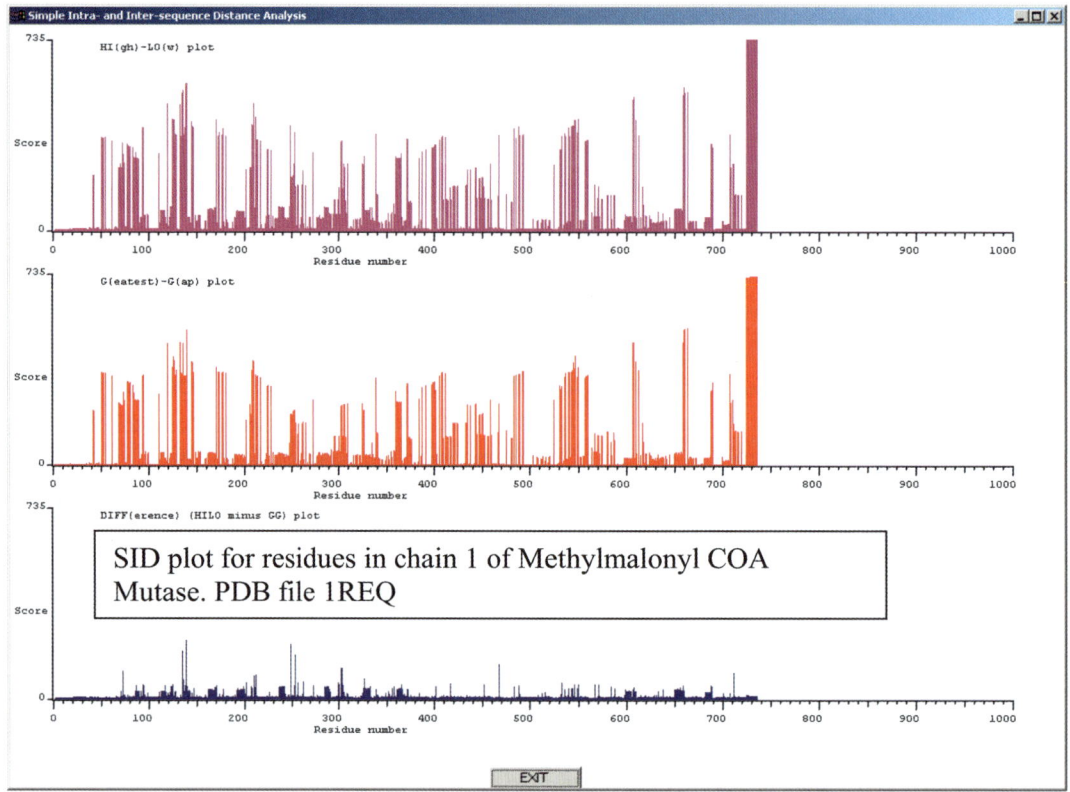

Figure 5.14

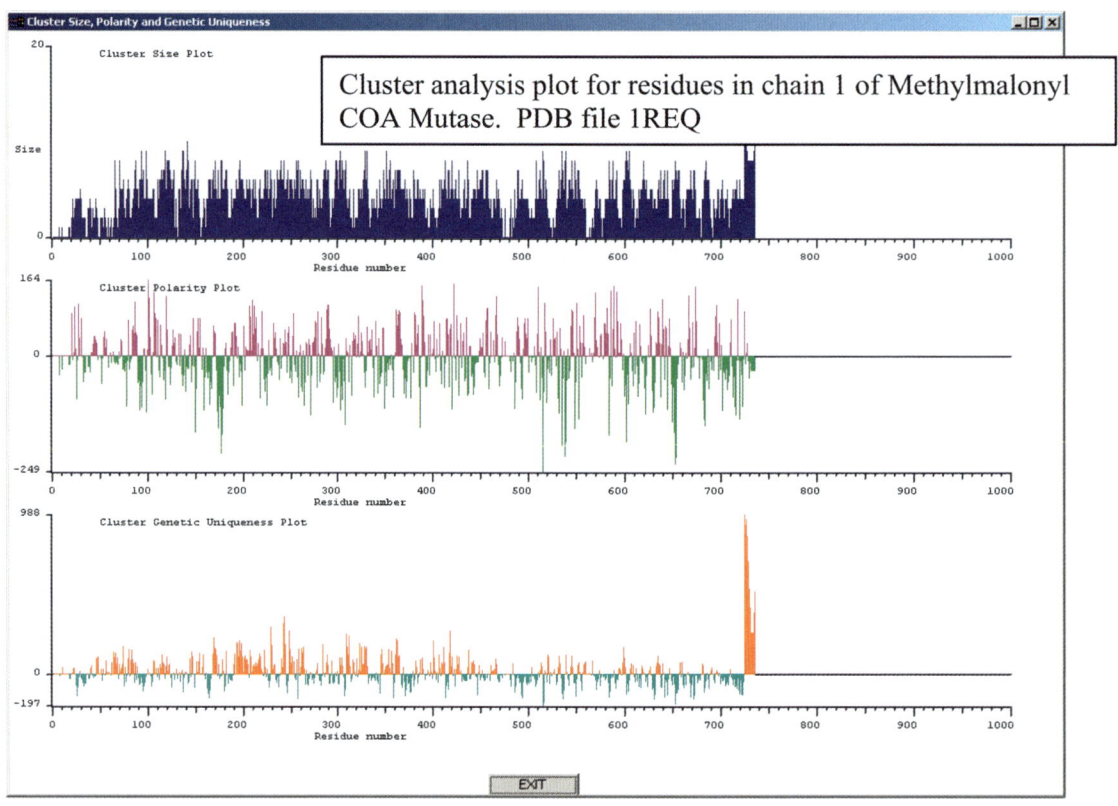

Figure 5.15

5.3.6.3 Inter residue distance plots.

A different type of plot is catered for in the third choice in the first sub menu. With this, a square diagram is obtained that shows, by means of colour coding, the distances between all pairs of amino acid residues in a single chain.[16] This represents the most useful form of this plot, and because it is symmetrical about the diagonal, only the upper triangular portion needs to be drawn. A rectangular plot is produced when the distance between residues in different chains is required. An example of this display is shown in Figure 5.16.

5.3.6.4 Ramachandran plots of φ and ψ angles

The fourth choice from the submenu will give the scatter plot of the torsion angles φ and ψ. This well known type of plot due to Ramachandran[17] allows a check on the quality of the protein structure to be made. The combinations of the values of the angles are very characteristic of α helices and β sheet structures; only rarely do residues other than glycine occur outside the characteristic regions. There are various optional colouring schemes available for this type of display; an example in shown in Figure 5.17.

5.3.6.5 Balasubramanian plots of φ and ψ angles

The fifth choice affords a less well known plot.[18] In this α helices are characterized by short vertical bars (coloured *magenta*) connecting torsion angles values; β sheets are characterized by longer (*green*) bars. An example is shown in Figure 5.18.

5.3.6.6 Retaining the results from the SID and cluster analyses.

The raw numerical results obtained in the plots described in sections 5.3.6.1 and 5.3.6.2 can be stored in a file. To do this select the radio button labeled *Open File for SID Data* before drawing the plots. Select the radio button labeled *Close File for SID Data* at the end of plotting. The file will be named:

```
SID-data-dd-mm-yyyy-hh-mm.DAT
```

where 'dd-mm-yyyy-hh-mm' is the date and time when the file was created. The file may be used by other programs including spreadsheet programs.

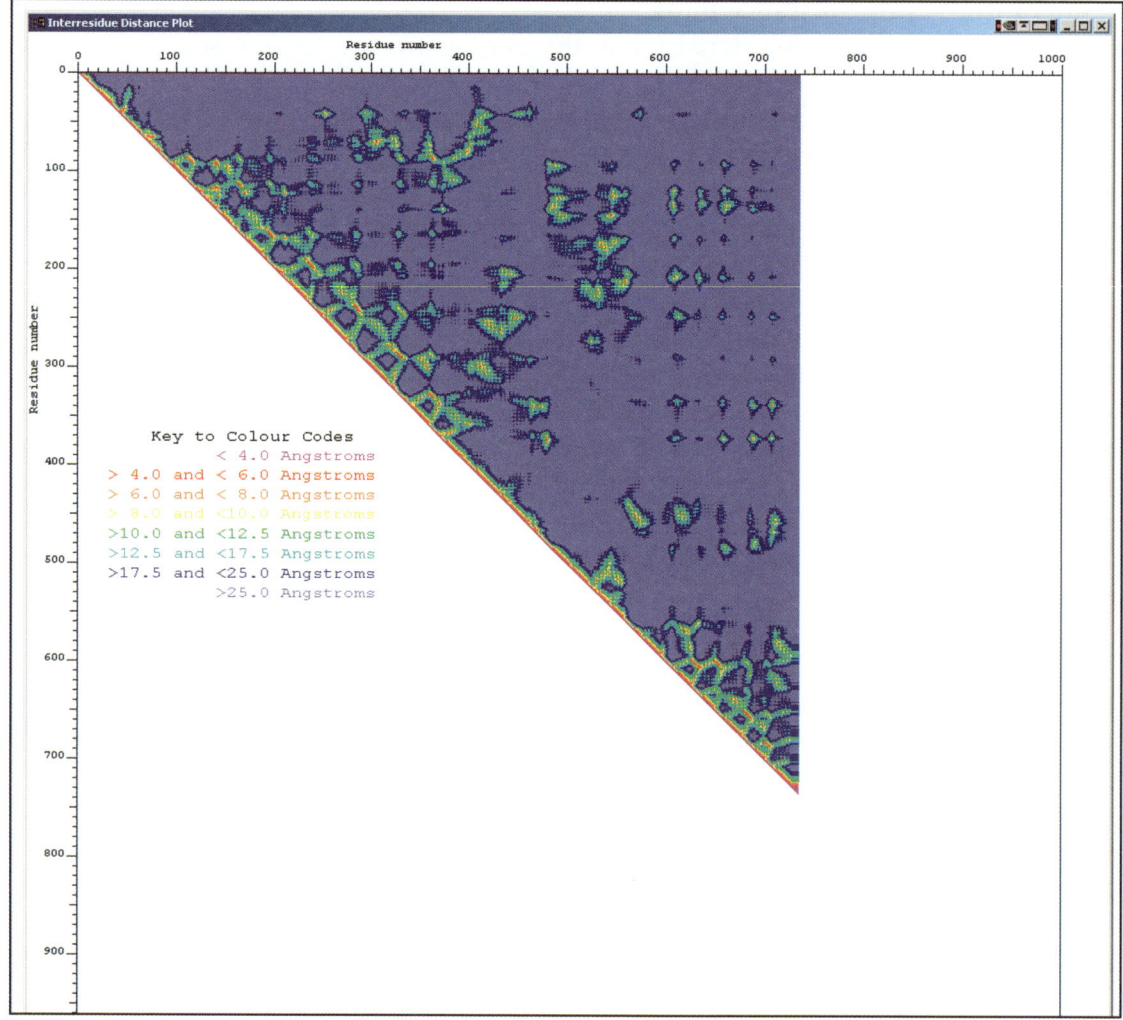

Figure 5.16
Distance plot for residues in chain 1 of Methyl-malonyl Coenzyme-A Mutase. PDB file 1REQ

The distances between the alpha carbons of all the residues in a single chain are plotted as colour-coded contours. The main structural motifs are identifiable. Helices show as thickened red bars along the diagonal. Parallel beta sheets show as green coloured regions parallel to the diagonal. Anti-parallel beta sheets show as green coloured bars at right angles to the main diagonal. The turns involved are revealed as the points where these bars join the diagonal.

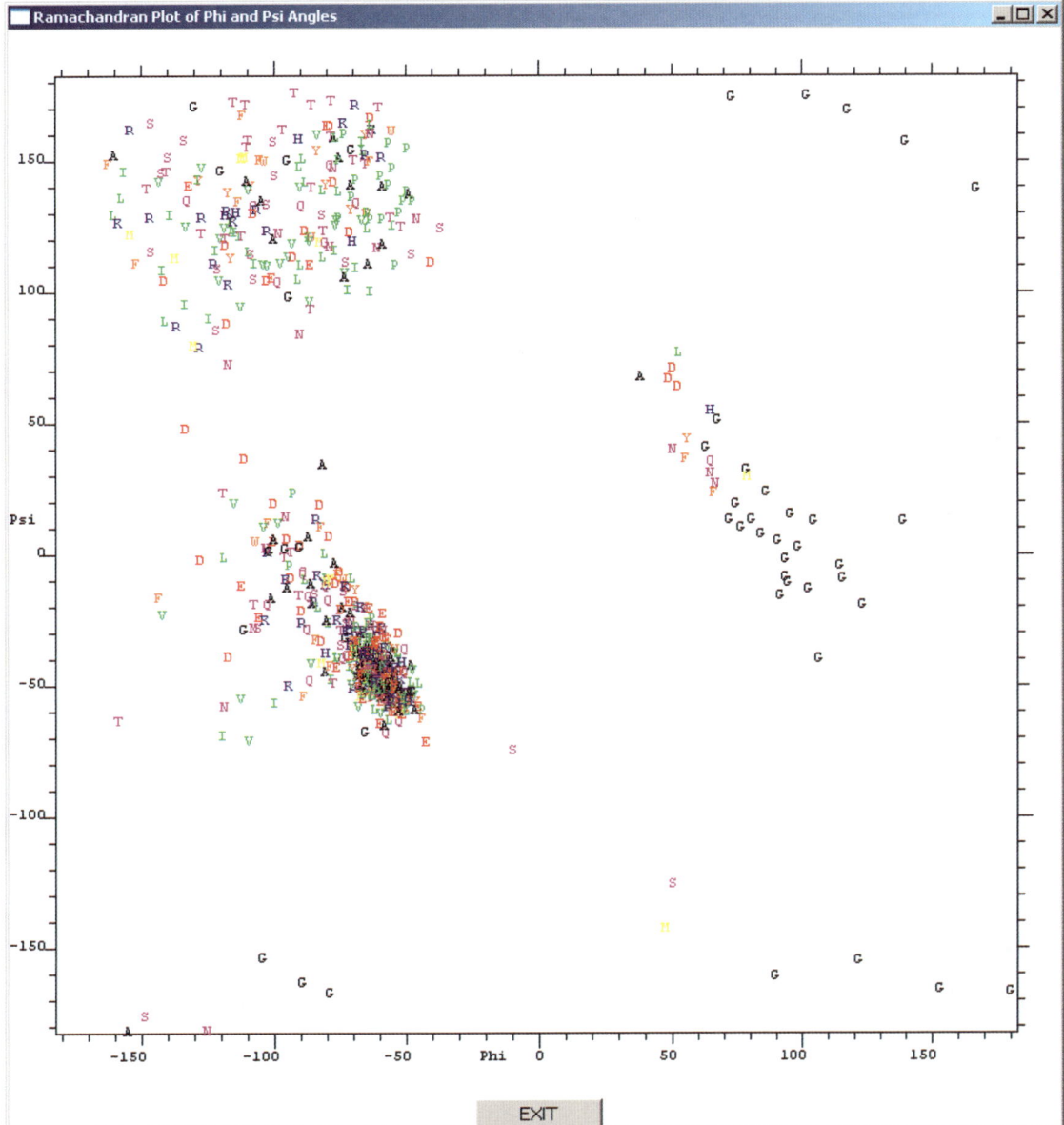

Figure 5.17
Ramachandran Plot for Methyl-malonyl Coenzyme-A mutase chain 1.
PDB File 1REQ

This figure shows how the two torsion angles Φ (Phi) and Ψ (Psi) determine the geometry of a protein chain.
Alpha helices are characterized by a densely packed area centered at Φ = -55°, Ψ = -45 °, with a more diffuse extension..

Beta sheet strands (both parallel and anti-parallel) are defined by diffuse area centered at Φ = -100°, Ψ = +125°. This region wraps round onto the bottom of the plot (Ψ = -170°).

Glycine residues, because they lack bulky side chains, can be found throughout the plot. Other residues that occur outside the helix and sheet regions are probably located in turns

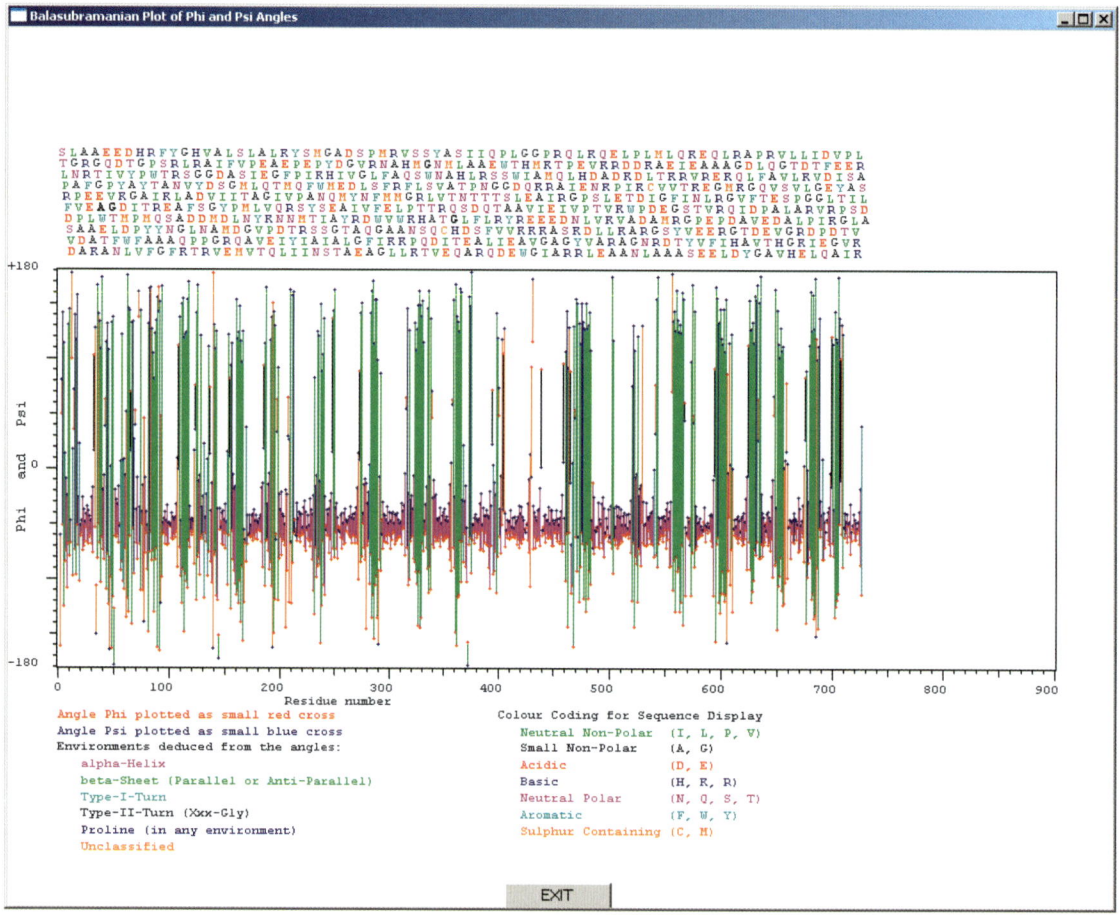

Figure 5.18
**Balasubramanian Plot for Methyl-malonyl Co-enzyme-A mutase, Chain 1.
PDB File 1REQ**

5.4 Protein sequences

The number of proteins for which the 3D structure is known is steadily increasing; the number in the PDB (on 11th November 2010) was 69,162. This databank is updated weekly each Wednesday. In contrast, the number of proteins for which a sequence is known is much larger. For example the Protein Information Resource (PIR) held data on 283,152 sequences in 1st March 2002.[19] The reason for the relative abundance of sequence data is not hard to understand; it is much easier to find a sequence than to determine a structure by crystallography. With a sequence to hand it is possible to see whether a protein is already known (in which case the structure is possibly known), or whether it is similar to a known one (in which case it may be possible to infer a structure). Handling sequences requires software tools that have become part of the modeller's toolkit. Invariably the sequences that are used are those that are constructed using the single-letter abbreviations for the amino acid residue names (See Table 5.1).

5.4.1 Some definitions
Before introducing software for handling sequences, it is important to define some of the important terminology used in this area:

(1) Two sequences are said to be *homologous* if they have arisen, either (a) by one having being derived from the other, or (b) both having been derived from a common ancestor, by *mutation* in the course of (*Darwinian*) *evolution*.

(2) *Homology* is not quantifiable; sequences are either *homologous*, or they are *not homologous*, or the question has not been satisfactorily answered.

(3) When quantification is needed, it is permissible to measure the similarity of two (or more) sequences by the percentage of identities of amino acid residues at corresponding places in the sequences. In practice a more elaborate *scoring* of the similarity is likely to be used.

5.4.2 Sequence matching
There are two related algorithms for aligning *pairs* of sequences; the first one due to Needleman and Wunsch,[20] is capable of aligning all the matching parts of the sequences; to yield a so-called *global alignment*. The second one that is usually attributed to Smith and Waterman,[21] is a modification that will only retain those parts of the sequences that are related; a so-called *local alignment*. Both algorithms work by inserting gaps into one or other of the sequences so that there is maximal alignment of similar (or identical) residues. When it comes to scoring an alignment, the introduction of a gap in one sequence where the other has a residue results in a penalty.

A crucial set of parameters is needed in protein sequence alignments. This is the *alignment matrix*. This is the measure of how the twenty standard amino acid residues are seen to be related. If an *exact* match is required, this is a square matrix with entries of 1.0 on the leading diagonal, zeros elsewhere. If account of the similarities of residues is to be taken into account, then, for example the diagonal would still be set to 1.0, but the off-diagonal elements would be set to fractional values (less than 1.0) representing the similarity of pairs of residues.

The two types of alignments are used in different situations. Global alignments are chosen when it is necessary to look at the homology of sets of sequences; local alignments are suitable when (say) parts of a new sequence need to be matched against the contents of a database.

Alignment of whole sets of sequences presents problems. While pair-wise alignment is recognised to have a unique solution,[20] (but see below), this is probably not so with multiple alignment. What is not in doubt is the fact that multiple sequence alignment can be very costly in computer time. Using the Big-oh notation; matching a pair of sequences of lengths m and n requires time and memory $O(nm)$. For multiple alignment, if we make the assumption that there are k sequences, all of approximately the same length (m), the time requirements could be $O(m^k)$, with memory requirements $O(km)$. This assumes that a naïve extension to the basic algorithm would be used, and it would seem that alignment of many sequences might be prohibitively expensive in computer time. We shall see later that if some manual intervention to the automated process is allowed, a much less costly method can result.

5.4.3 Background of aligning protein sequences

Both the Needleman-Wunsch and Smith-Waterman algorithms use techniques that are known as *dynamic programming*. This term is not very appropriate, since it implies that the computer is still in the process of being programmed! In this context it means that the final solution is arrived at by a step-by-step process, (and this could be said to be the way many computer programs work). The important point is this; when matching a **pair** of protein sequences, using a particular similarity matrix, **and *starting from the amino group end of the sequences,*** there is one unique alignment. However, (and this point is rarely remarked on), there is often a ***second different alignment if the process is started from the carboxyl ends.***

At this point it is appropriate to mention books that treat sequence handling in depth. For an easily understood explanation of the Needleman-Wunsch algorithm the little book by Russell Doolittle[22] is recommended. A more detailed explanation is given in the book by T. K. Attwood and D. J. Parry-Smith.[23] The mathematical background is covered in three books, [24, 25, 26] while two volumes of *Methods in Enzymology* have review articles on various aspects of analysing protein and nucleic acid sequences. [27, 28] Alignment of character strings is a computer application that is not confined to chemical biology; it has applications in the fields of linguistics, textual comparison, and forensic science.

5.5 The program PRESTO

PRESTO is a program for handling sets of protein sequences. It allows them to be edited, compared, and aligned. It also allows for a single sequence to be compared with sequences in databases.[15] A copy of PRESTO will be found on the DVD that accompanies this book. PRESTO has one main screen display (Figure 5.19). This comprises a large scrolling window that displays the sequences. At any one time there are visible 80 sequences, with up to 90 residues. (The sequence / residue array is dimensioned for 200 sequences of length 2100 residues; the numbers of sequences and amino acid residues displayed at a time depends on the display dimensions of the screen being used). The functionality of PRESTO is catered for by one main menu to the right of the sequence display window. We follow the usual scheme for giving the menu button coordinates. Note that the menu has heading lines; these are not counted as rows. Instead of listing all of the functions of PRESTO here, we have designed the exercises to introduce those that are needed progressively. There is a complete manual in PDF format on the DVD.

5.5.1 Introductory exercise

To show you how to use PRESTO, we start with an exercise that verifies the assertion made in section 5.4.3, concerning the difference between alignments made on a pair of sequences starting from the amino and the carboxyl ends of the chains. There are several steps.

(1) Locate and click on the shortcut to the program PRESTO. This should be on the desktop of your computer. You will be presented with a *Welcome Page*, that you should dismiss by pressing the *Escape* button. The general arrangements for starting the program are similar to those in INTERCHEM (Restarting, Suspending *etc.*).

(2) You will first be invited to define a working folder (directory), and in this case a good place would be `C:\reversal`. If your disc system is partitioned, or you have more than one disc, replace `C:\` with (say) `F:\`. This directory will hold any files that are created in subsequent operations.

(3) Packaged with the program there is a set of test files. These are located in the folder `~\presto\test_data`. In the main menu at the right of the screen, click on the button *Load Sequence Set* (C1/R9), and then from the popup menu choose *PIR Unspaced Format*, find your way to the test data folder, and then click on `steroidI.DAT`. A set of twenty-nine sequences will be loaded into the main window area starting at the top. The main area has, at the left hand side, truncated titles of the sources of the sequences. If you wish to see the full titles click on *Toggle Titles* (C1/R7). Repeat clicking on this button will alternately reveal the full title or reveal the sequence display. For the purposes of the first part of this exercise we need copies of two of the sequences.

(4) You will notice that immediately after the rows of sequences that there is a blank numbered line (30) that is highlighted in green. This is the line at which subsequent loads of the sequence array will be made, and its position can be adjusted by clicking on *Set Entry Position* (C1/R8). You will then be prompted to click on the appropriate row, (say row 35).

(5) Click on *Copy Sequence Set* (C2/R12). Then, when prompted, click on the two adjacent rows 16 and 17. The two lines from the original array will be copied to lines 35 and 36.

(6) Now set the entry position to line 40, and make a further copy of the two line 35 and 36 at this position. You should now have two copies of a pair of sequences of *Zebra fish steroid receptor protein* starting at rows 35 and 40 respectively.

Now at last comes the experiment!

(7) Click on *Align Two Globally* (C1/R20). Then when prompted click on lines 35 and 36; then (from the popup menu) *Blosum62 Matrix*. First there will appear a statistical summary of the alignment (a copy of this will be made in your working directory), and then when you click on *Escape*, the pair of aligned sequences will appear.

(8) Click on *Reverse Sequences* (C2/R15), and then when prompted on lines 40 and 41.

(9) Repeat step (7) on lines 40 and 41, to align the reversed sequences

(10) Repeat step (9) on lines 40 and 41 to reverse the aligned-reversed sequences.

You should now have:
(a) In lines 35 and 36 a pair of sequences aligned globally starting (in the normal way) at the amino ends of the sequences.
(b) In lines 40 and 41 the same pairs of sequences aligned globally starting at the carboxyl group end.

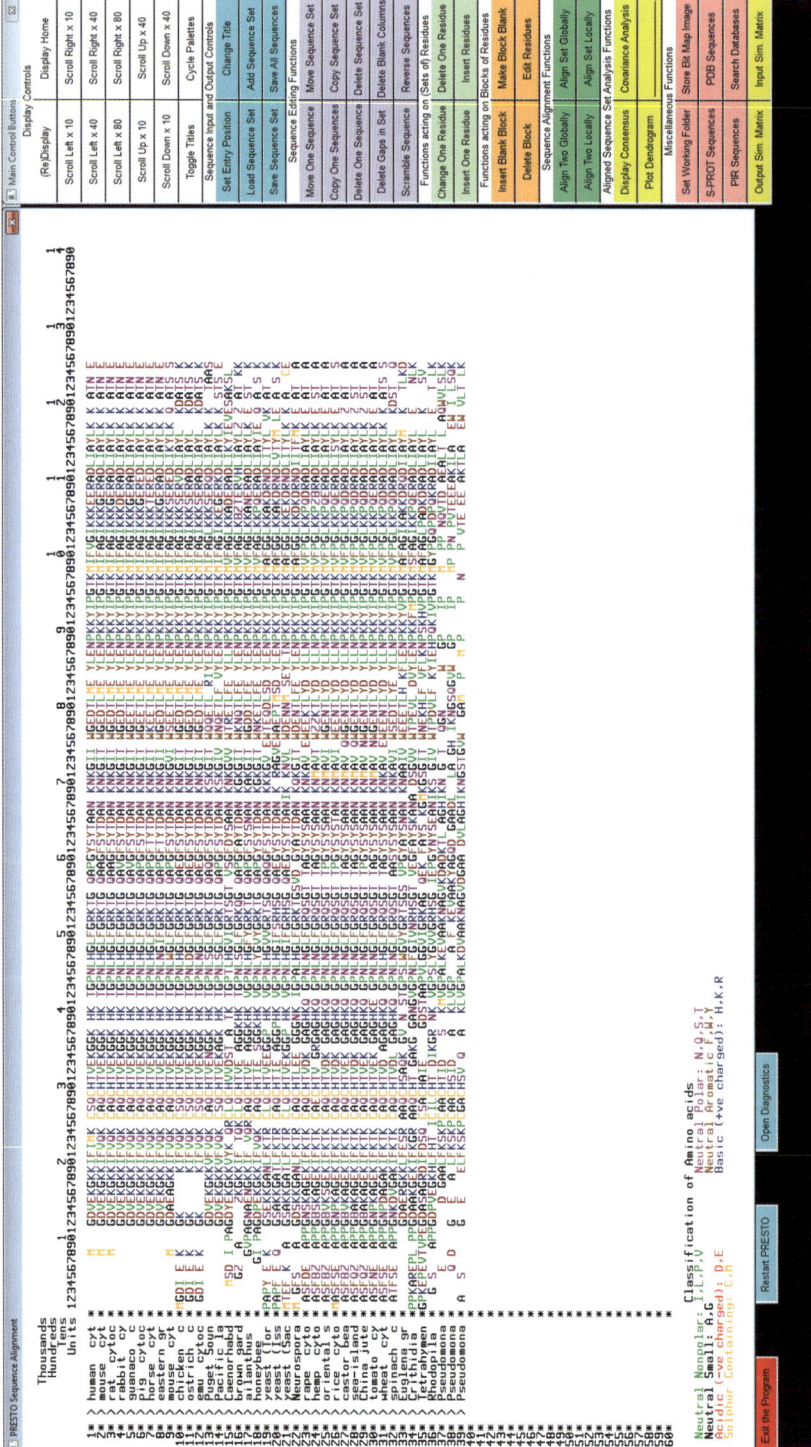

**Figure 5.19
PRESTO**

The display shows the *global* alignment of a series of Cytochrome-C sequences using the *Needleman-Wunsch* algorithm

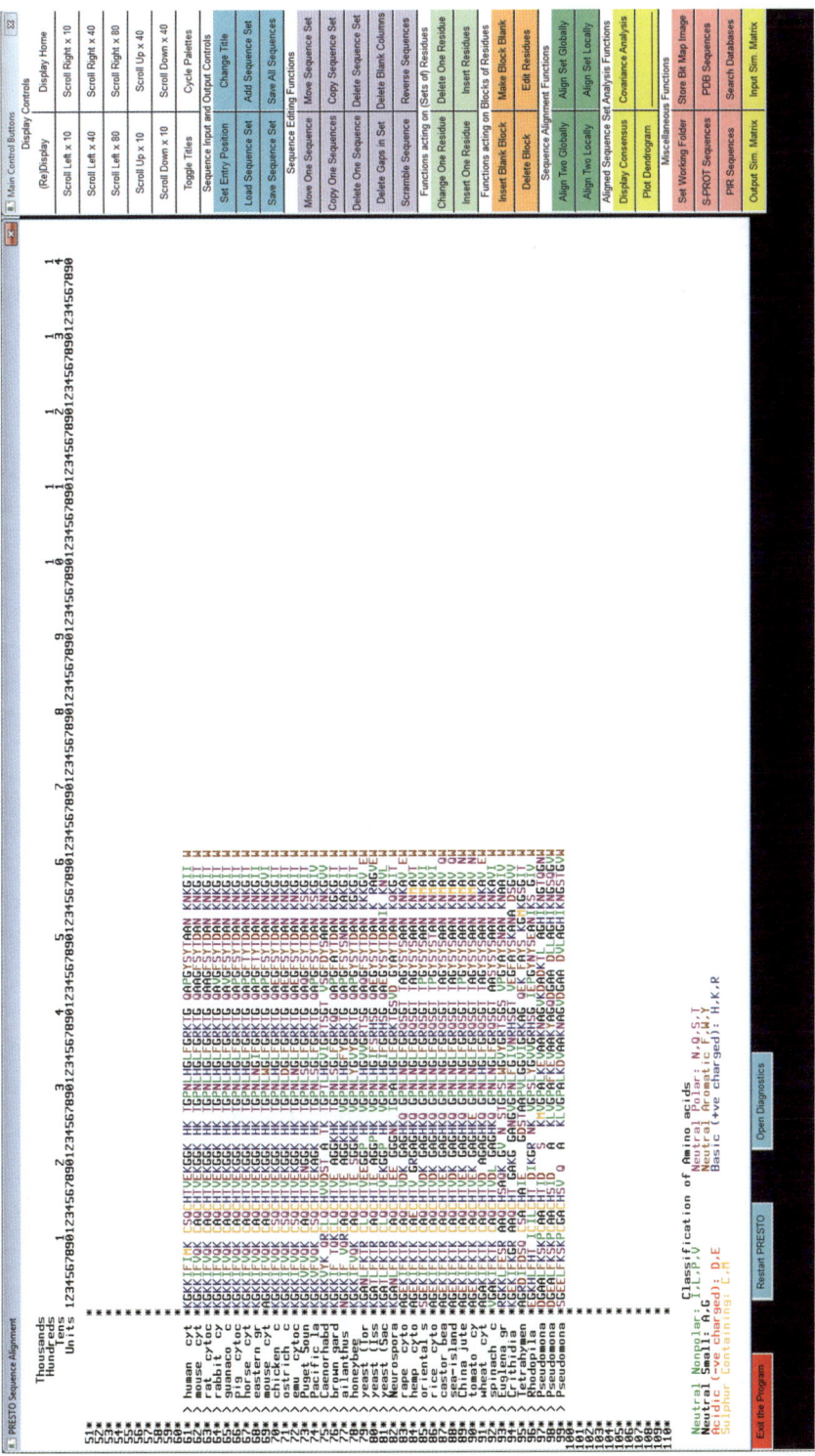

Figure 5.20 PRESTO

The display shows the *local* alignment of a series of Cytochrome-C sequences, using the *Smith-Waterman algorithm.*. Note than, in contrast to the results shown in Figure 5.19, the parts, at the starts and ends of the sequences, that are not well aligned, are deleted.

Carefully compare the two alignments, by scrolling the whole array sequence, right or left, using the buttons at the top part of the menu (Rows 2, 3, and 4). You should see that, while the alignments are both good, and very similar, that there are minor differences in the regions of residues 20 to 40, 58 to 60. It is not normal practice to align sequences starting from the carboxyl end, but it can be useful.

5.5.2 Aligning sequence sets globally and locally

For this test we use a smaller set of sequences that has been made by pruning the set used in the previous experiment. This is contained in the file steroid2I.DAT and contains eight sequences.

(1) When loaded the sequences will occupy rows 50 to 57 of the sequences display. Use *Set Entry Position* (C1/R8) and *Copy Sequence Set* (C2/R12) to make **two** well separated copies starting at rows 2 and 14, then delete the blank row number 1, using *Delete One Sequence* (C1/R13).

(2) Subject the first of these copies of the set to *Align Set Globally* (C2/R20). There will now be a succession of prompting messages and submenus. You will first be asked to select *Master Sequence*. For this choose the last sequence (number 8). Then click on the first and last sequences of the set (numbers 1 and 8), and finally select a similarity matrix; choose *Blosum Matrix62*. After a few seconds the sequences will be aligned *in place*. Use the scrolling buttons to find where the similarities are (residue 118 to 198). This is a good result showing that the sequences are indeed homologous. This is an example of the use of the Needleman-Wunsch (NW) algorithm.

(3) Repeat this process but using *Align Set Locally* (C2/R21), on the second copy of the set (rows 14 to 21), the same *Master Sequence*, and similarity matrix. Again there results a well aligned set, *but all the poorly aligned residues at the beginning and end of the set have been deleted.* This is an example of the use of the Smith-Waterman (SW) algorithm.

Looking at the results of the first alignment, it is possible to see that there may be other regions of the sequences that might show similarity. This might be investigated as follows. We first need to get rid of the well aligned region in that set that was aligned using the NW algorithm.

(4) Copy the whole of the array from the first (Needleman-Wunsch) result to row 27 onwards.

(5) Click on *Delete Block* (C1/R19) and when prompted define a block of residues by clicking on two pairs of sequence/residue coordinates at diametrically opposed corners of a rectangle. If you have followed the original instructions exactly these will sequence row 27, residue 118 and row 34, residue 199. Then, when prompted, click on any point in the white defining rectangle. All the previous well aligned residues within the rectangle will be deleted

(6) Working on the new edited copy, click on *Delete Gaps in Set* (C1/R14), and then select when prompted sequences 27 and 34. All the gaps in the previous alignments will be removed.

(7) Make a copy of the sequences 27 to 34 beginning at row 37.

(8) You can now proceed to subject the two copies (rows 27 to 34, and 37 to 44) of the edited set to further alignment tests.

5.5.3 Questions arising from the alignment experiments
From the pair wise alignments:

(1) How can you account for the differences in the results from aligning in the two directions?
From alignments of sequence sets

(2) What are the two or three most highly conserved types of residue?

(3) What is the significance of this?

(4) When the most highly aligned regions are removed, what do you infer from the subsequent alignment experiments?

5.5.4 Making inferences from alignments
This second experiment seeks out information about a set of Cytochrome-C sequences.

From the test data folder, load the file `cytoced3I.DAT` and globally align the protein sequences in it, using the longest sequence as the *master sequence,* and the `Blosum32` alignment matrix.

The cytochrome-C group of proteins occurs widely in micro-organisms, plants, and animals, and is highly conserved. The alignment is almost embarrassingly perfect. If you click on *Display Consensus* (R22/C1), the program will deliver two extra lines below the sequence set display. The first of these lines show integers 1 to 9 showing the percentage identity of the amino acid residues for each position, rounded up to the next ten percent. If there is 100% identity, the single letter code for the amino acid is shown. The second extra line shows a *consensus sequence* calculated by the program.

With a set of sequences like this, clicking on *Plot Dendrogram* (R23/C1) and following through the submenus (include the whole of the sequence set, *but not the consensus lines,* and choose *Use Averages of Distances, and Penalize Residue-Gap Alignments*), will yield a tree showing the taxonomy of the organisms. It is perhaps reassuring that we (*Homo Sapiens*) are more closely related to horses and pigs, than to a roundworm (*Caenorhabditis elegans*)! To go back to the sequence display press the *escape* button on the keyboard.

The normal coloured display that you see can be modified by repeatedly clicking on *Cycle Palettes* (R7/C2). There are alternative displays that feature white backgrounds suitable for printing. Perhaps more interesting is a display that colours the residue symbols green, magenta, or white, according to whether the residues are favoured in beta sheet or helix environments, or are unselective.

See if you can relate these preferences to the 3D structures of Cytochrome-C protein structures obtained from the Protein Data Bank. You will need to make allowances for possible differences in the numbering of the residues (introducing gaps will have altered it). One way of finding those sequences in the PDB that are most related to the aligned sequences, is provided for in PRESTO. To do this, first look at the well aligned parts of the PRESTO display and note down a string of eight or more residues that are conserved well.

Now click on *PDB Sequences* (R25/C2), and use this sequence to search this database (this consists of all the sequences that are in the separate chains of proteins in the PDB). The fourth option in the submenu will get you a list of all the sequences that have the string, just enter the short string you have noted down. The display will show you how this (relatively short) string occurs relative to other parts of a sequence. (Note that the search tool allows you to enter two separate strings (that is strings that are disjoint; do not use this facility). You could be more selective by extending the string. Then if you use the fifth option with the same search string, a new set of sequences will be created in your working folder. You can now add that set to your PRESTO display by clicking on *Add Sequence Set* (R9/C2).

Finally, select one (or more) of the PDB *Structures* that you have just found, download it from the PDB, and use PROTEINS to get the INTERCHEM D structure. Then use INTERCHEM to display it. How well does the secondary structure compare with the predictions in the PRESTO display?

5.5.5 Storing images from the screen in PRESTO
To make a file that can be printed, click on the window that holds the image, then hold down the *Left Alt Key* while pressing *Print Screen*. Then on the main PRESTO menu click on *Store Bit Map Image* (R25/.C2). A file called:

```
Presto_pc_NNN.bmp
```

will be created in the current working folder. The three-digit code (NNN) will start at 001 and will be incremented by 1 for each new image. It is best to have an image that has a white background [use *Cycle Palettes* (R7/C2)], otherwise if you have a black background you might use a lot of toner or ink when you use a printer!

5.6 Racemic protein crystals as sources of protein structures.
In what is written in this chapter, you might think that only proteins derived from the *L* forms of amino acids are of interest. It happens that some natural proteins do not crystallize easily, and so their structures cannot be determined by crystallography. A solution to this problem can be to prepare racemic proteins. This can be done by synthesizing the enantiomeric versions of the protein from the natural *L* amino acids on the one hand, and from the unnatural *D* amino acids on the other. The native scorpion toxin BmBKTx1, a thirty one residue peptide with three disulphide bridges (1), provides such an example.[29]

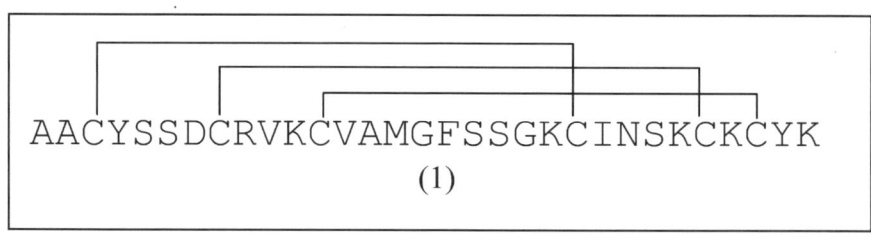

While neither the *L* nor the *D* proteins separately crystallized, an equimolar mixture of them both did form crystals. Unusually, the crystals were in a high symmetry tetragonal space group (Space Group 88, Hermann-Mauguin symbol $I4_1/a$). The crystallographic results from the experiments were deposited in the Protein Data Bank as structure 3E8Y. Because it was designed only to cope with natural proteins, the PDB format does not cope well in presenting the results, neither in the traditional format, nor in the CIF format. The file contains only coordinates for the chain formed from the natural *L*-amino acids, and the assumption is made (correctly) that the display will use the symmetry properties of the space group to generate the coordinates of the *D*-chain.

5.7 Problems for you to solve, and questions for you to answer

All of the problems for this chapter will involve your access of the Protein Data Bank, and retrieving data from that source. You can do that while you are using INTERCHEM, or you can logon to the websites:.

http://www.wwpdb.org
or
http://www.rcsb.org/pdb/home/home.do

using a second machine.

5.7.1 General instructions applicable to most of these problems.

These instructions are framed to be directly applicable to problem [1], but also are relevant in connection with other problems

Go to the PDB website and use its search tool to display data for the class of protein, by specifying (for problem [1]) *Cytochrome P450* in the search box.

Associated with this tool the PDB provides indexing software that will create a downloadable spreadsheet for entries meeting the specified criteria. The spreadsheet file can be very detailed but if you limit the columns to **PDBID** (obligatory), **Chain ID**, **Macromolecular Name**, **Source**, and **Chain Length**, it will be sufficiently small to handle yet detailed enough for the problem.. Download the file and print it out (best in landscape format).

You can now examine the spreadsheet, and choose the PDB identity codes that give you a range of sources and macromolecules. Note that all the chains for an entry are listed in the spreadsheet, and that frequently the chains for a given entry have the same length. Do not worry about this at this stage. The aim is to download a representative selection of the chosen PDB structure files. You should create a suitably named working directory to hold these files.

[1] This problem concerns an important family of proteins going under the name: *cytochrome P450*. These are enzymes, widespread among all forms of living matter, - *bacteria*, *plants* and *animals*, that have the function of the elimination of materials toxic to their parent organisms, and some other oxidation processes. In fact we can investigate two problems.

[1a] The objective is to look for similarities and differences between the same enzyme associated with different species. The easy way is to manually mark-up the PDB codes on the printout of the spreadsheet, using coloured highlighters.. Then go to the PDB site and downloaded the chosen structures into an appropriate directory.

[1b]. The objective is to search for similarities and differences between different versions of (cytochrome p450 enzymes) all found in one species (*Homo sapiens*). Again download the appropriate files into a different directory.

For both parts of this exercise, use the program PROTEINS to derive two sorts of file: alpha-carbon structure files (option 6 in the menu), and sequence files (option 5 in the menu). To make things more manageable put the various sorts of files into different subdirectories.

Some editing needs to be done on these files. The sequence files have an entry for each chain in a structure, and these are frequently identical. Use the program NOTEPAD for this, and remove all chain sequences after the first one. Each sequence file has a (not very informative) title. Change this to the enzyme name (for problem [5.1a], or the species name for problem [5.1b]. The title must be preceded by the single character > followed by a space.

The alpha carbon files also need to be edited to hold just one chain. This is done by using the *highlighting* tool in INTERCHEM.

The comparisons of the sequences are done using PRESTO. Load the first sequence with the command *Load Sequence Set* (R9/C1), and the subsequent sequences using *Add Sequence Set* (R9/C2). (Despite the names of these commands they also work on single sequences). When all the sequences are loaded, you can usefully save the work done so far with *Save Sequence Set* (R10/C1). This command allows you to save sets of sequences at any subsequent stages of your work. Next try to get the best alignment of sequences using *Align Set Globally* (R20/C2). Then use *Plot Dendrogram* (R23/C1). This will show you those sequences that are most closely related.

INTERCHEM can then be used to do comparisons of pairs of structures deemed to be closely related in the sequence comparisons. For this use *Align B with A* (Dual Mode, R6/C7), and then *Structure Merging* (R9/C4). With a certain amount of ingenuity, you could overlay up to six structures.

Using alpha carbon structures for comparative displays and overlays is preferred since helices are clearly identifiable, and the smaller number of atoms makes the pictures clear. (Of course the 'bonds' between the successive alpha carbon atoms are not real bonds!). In suitable cases where structures are very similar, the PROTEINS option *Backbone Structures* (number 8 in the menu) is useful. With this option all amino acid residues except *glycine* are replaced by *alanines*. The advantage of this last device is that such files can be used to generate *Ribbon* pictures, for purposes where a clutter-free display is needed, or where main chain torsion angle information is needed.

[2] A review[30] dealing with the chemistry of the enzymes classified as 2-oxo-glutarate dependent oxygenases has been published recently, listing the four-chararacter PDB access codes of the proteins. Using the techniques that you have learned from the preceding problem investigate this series of compounds, in order to see the variations in the structures and amino acid sequences, that cover the range of their activities

[3] Search the Protein Data Bank to find other examples of proteins for which crystallographic structure determination has only been achieved with racemic proteins (analogous to the case discussed in reference.[29]).

References and Endnotes. Chapter 5

[1] W. C. Shan and P. D. White (Eds.), *Fmoc Solid Phase Peptide Synthesis (A Practical Approach)*. Oxford University Press, Oxford, 2000,

[2] Although the Cahn-Ingold-Prelog (CIP) system of designating stereochemistry (R. S. Cahn, C. K. Ingold, and V. Prelog, *Experientia*, 1956, **12**, 81; V. Prelog and G. Helmchen, *Angew. Chem. Int. Ed.*, 1982, **21**, 567) is preferred for systematic nomenclature, it has some disadvantages in connection with amino acids. This arises because the (CIP) precedence rules involving sulfur, cause the replacement of the side chain oxygen atom in serine by a sulfur atom (to give cysteine), to demand an apparent change in configuration. So cysteine is unique among the natural *L* amino acids in having an *R* configuration in the CIP system. For all the **other common amino acids** this rule holds: ***L* ≡ S, *D* ≡ R**.

[3] We have omitted any detailed discussion of *turns* in protein chains, because the nomenclature is confusing. It is clear that turns are a necessary part of protein structure, but frequently their structures are disordered, particularly when they are near the surface of globular proteins. But look out for terms such as *hairpin*, *loop*. Other more fanciful terms such as *Greek-Key* and *Jellyroll* are used in describing tertiary structure.

[4] There are really two problems associated with the folding of proteins. Firstly, there is what *causes* proteins to fold in the way that they do. It has so far not proved possible to predict the conformation of proteins from their sequences. Whole books have been written on the subject; G. D. Fasman (ed.), *Prediction of Protein Structure and the Principles of Protein Conformation*. Plenum Press, New York, 1989. There are periodical competitions, in which sequences of proteins for which crystal structures have been determined (but not yet published) are released. Competitors are invited to predict the structures using any method to hand. The success rate, measured as correctly predicted environment, is usually about 70%.

The second problem is the *mechanism* by which proteins fold to their *native* state. To understand this problem we need the experimental evidence. Most proteins have a disordered so-called *denatured state*. If, for example, a protein containing several disulfide linkages is subject to mild reducing conditions, these bonds will be broken, and the protein will enter a (modified) *denatured* state. Placed back in oxidizing conditions, the protein will rapidly go back to the native form. However, C. Levinthal (J. T. P. Debrunner, J. C. M. Tsibris, and E Münck (eds.), *Mossbauer Spectroscopy in Biological Systems,* University of Illinois Press, Urbana, 1969. pp. 22-24) pointed out what has become known as the *Levinthal Paradox*. Assuming that a protein has two flexible single bonds in each peptide linkage (the φ and ψ bonds), and that each of these has three low energy states. Then for a protein chain of 101 amino acid residues there could exist $(2 \times 3)^{100}$ configurations. This number is approximately 7×10^{77}. Knowing the rate at which molecular conformations can be explored (10^{13} per second), it follows that it would take 2×10^{44} years for the whole conformational space to be explored. *The universe has not existed for that long!* This is the paradox; it is discussed (but not resolved completely) by R. Zwanzig, A Szabo, and B. Bagchi, *Proc, Natl, Acad. Sci.*, 1992, **89**, 20). See also J. T. Ngo, J. Marks, and M. Karplus, *Computational Complexity, Protein Structure Prediction, and the Levinthal Paradox* in K. Merz Jr., and S. LeGrand (eds.) *The Protein Folding Problem and Tertiary Structure Prediction.* Birkhauser, Boston, 1994.

[5] G. E. Schulz and R. H. Schirmer, *Principles of Protein Structure*. Springer-Verlag, New York, 1979.

[6] W. Saenger, *Principles of Nucleic Acid Structure*. Springer-Verlag, New York, 1984.

[7] A. M. Lesk, *Introduction to Protein Science, Architecture, Function, and Genomics*. Oxford University Press, Oxford, 2004.

[8] A. M. Lesk, *Introduction to Genomics*, Oxford University Press, Oxford, 2007.

[9] A. M. Lesk, *Introduction to Bioinformatics*, Oxford University Press, Oxford, 2002.

[10] M. G. Rossmann and E. Arnold (Eds.) *International Tables for Crystallography, Volume F, Crystallography of Biological Macromolecules*. Kluwer Academic Publishers, Dordrecht, 2001.

[11] K. Würthrich, *NMR of Proteins and Nucleic Acids*. Wiley-Interscience, New York, 1986.

[12] The warning only applies if you are working on a machine running under a Windows operating system. If you are using a Linux machine, or a Windows machine that has a Cygwin emulation of Unix (Linux) loaded (see Chapter 2), and are downloading a large number of protein files, then downloading files that have been compressed using *gzip* will save time. If you wish to download a complete distribution of the Protein Data Bank to your system, then this should be done using the files that are compressed with *gzip*.

[13] This pragmatic approach to science (looking at the bigger picture) is sometimes necessary for progress to be made; it is really at the heart of molecular modelling. However, there are occasions when the biochemist, looking at a protein, should consider the details of the chemistry, just as a chemist, when analyzing a nuclear magnetic resonance spectrum, has sometimes to consider the nuclear physics!

[14] L. Pritchard, L. Cardle, S. Quinn, and M. Dufton, *Prot. Eng.*, 2003, **16**, 87.

[15] J. Kyte and R. F. Doolittle, *J. Mol. Biol.*, 1982, **157**, 107.

[16] J. B. R. Dunn and I. M. Klotz, *Arch., Biochem. Biophys.*, 1975, **167**, 615

[17] G. N. Ramachandran, C. Ramakrishnan, and V. Sasisekharan, *J. Mol. Biol.*, 1963, **7**, 95; G. N. Ramachandran and V. Sasisekharan, *Adv. Prot., Chem.*, 1968, **23**, 283.

[18] R. Balasubramanian, *Nature*, 1977, **266**, 856.

[19] The UniProt database has been formed by amalgamation of the databases from separate organisations; SwissProt (Swiss Institute of Bioinformatics – SIB) and Protein Information Resource based (PIR) in the United States. Now there is a major contribution from the European Bioinformatics Institute based in the United Kingdom.

[20] S. B. Needleman and C. D. Wunsch, *J. Mol. Biol.*, 1970, **48**, 443.

[21] T. F. Smith and M. S. Waterman, *J. Mol. Biol.*, 1981, **147**, 195.

[22] R. F. Doolittle, *Of URFS and ORFS, A Primer on How to Analyse Derived Amino Acid Sequences*. University Science Books, Mill Valley, CA.1986.

[23] T. K. Attwood and D. J. Parry-Smith, *Introduction to Bioinformatics*. Pearson Education Ltd., Harlow, 1999.

[24] D. Gusfield, *Algorithms on Strings, Trees, and Sequences; Computer Science and Computational Biology*. Cambridge University Press, Cambridge, 1997.

[25] G. Valiente, *Algorithms on Trees and Graphs*. Springer, Berlin, 2002.

[26] R. Durbin, S. Eddy, A. Krogh, and G. Mitchison, *Biological Sequence Analysis, Probabilistic Models of Proteins and Nucleic Acids*. Cambridge University Press, Cambridge, 1998.

[27] R. F. Doolittle (Ed.), *Methods in Enzymology, Volume 188, Molecular Evolution: Computer Analysis of Protein and Nucleic Acid Sequences*. Academic Press Inc, San Diego, 1990.

[28] R. F. Doolittle (Ed.), *Methods in Enzymology, Volume 266, Computer Methods for Macromolecular Sequence Analysis*. Academic Press Inc, San Diego, 1996.

[29] K. Mandal, B. L. Pentelute, V. Tereshko, A. A. Kossiakoff, and S. B. H. Kent, *J. Am. Chem. Soc.*, 2009, **131**, 162.

[30] N. R. Rose, M. A. McDonough, O. N. F. King, A. Kawamura, and C. J. Schofield, *Chem. Soc. Rev.*, 2011, **40**, 4364.

Chapter 6
Essentials of Stereochemistry and Conformational Analysis

Confusion can sometimes arise in the understanding of the *words* used in describing stereochemistry. What follows here is largely a description of the way we have used the *words* in this book. The description is necessarily brief and for a comprehensive description the reader is referred to the book by Eliel and Wilen.[1]

6.1 Chirality.
Chirality is the term that is used to describe the *handedness* of a structure or a compound that can exist in two mirror image forms. As far as organic chemical compounds are concerned, this usually arises when a carbon atom (or sometimes a nitrogen, silicon, or sulfur atom) is attached to four different atoms or groups. The naming of compounds in this situation is handled by the rules of the Cahn-Ingold-Prelog (CIP) system.[2] This requires the four attached groups to be assigned *priorities* (*precedence*) according to a set of rules. The rules are complex in detail, but the principles are relatively simple. Chirality is assigned for each carbon (nitrogen, sulfur) atom in the following way:

(1) If there are less than four atoms attached then the atom in question is achiral (not chiral). For this test applied to nitrogen or sulfur atoms, lone pairs of electrons are considered as pseudo atoms of atomic number equal to zero

(2) If there are four directly attached atoms, these, considered as atoms of the first sphere, are assigned priorities based on their atomic numbers; higher atomic number implies precedence over lower atomic number.

(3) If two (or more) of these atoms (of the first sphere) have the same atomic number, then atomic mass is used; higher atomic mass implies precedence over lower atomic mass.

(4) If no distinction can be made using rules (2) and (3), then the more remote atoms attached to the atoms of the first sphere are examined. If the attachment uses multiple bonds (double, triple, or aromatic type bonds), then extra (phantom) atoms are created. Then the real and phantom atoms are examined and assigned precedence using rules analogous to rules (2) and (3). Priorities for these atoms of the second sphere are applied to the atoms of the first sphere. Phantom atoms introduced to cope with multiple bonds are considered to have atomic number equal to zero, and are not considered further in calculating priorities.

(5) If a decision cannot be made with atoms of the second sphere, then further more remotely connected atoms are considered in the same manner (a third sphere), and so on, until a decision can be made.

(6) When a decision on priority has been made, the atom under consideration (AUC) is viewed with the attached atom or group of *lowest* priority most remote from the viewer. The three other atoms or groups then appear at the vertices of a triangle. If the descending order of priorities of these atoms is *clockwise*, the chirality of the

AUC is assigned the symbol *R*, (*Rectus*, Latin. right-handed); if the arrangement is *anticlockwise* the AUC is assigned the symbol *S* (*Sinister*, Latin. left-handed). See Figure 6.1.

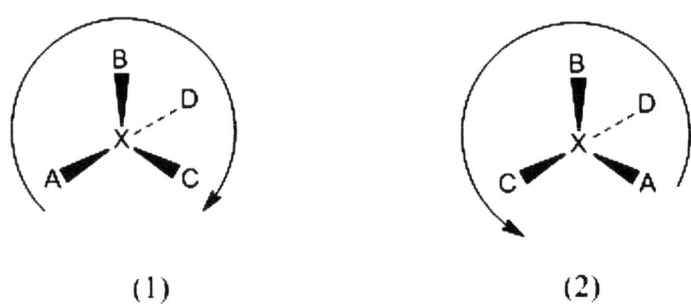

Figure 6.1
Definition of the Chirality of the Atom X (C, N, or S)
The attached atoms have the precedence A>B>C>D.
For structure (1) the chirality is ***R***. For Structure (2) the chirality is ***S***.

For convenience a list of functional groups in descending order of priority is shown. In this list, the group R is to be read as an unspecified alkyl group:
I > Br > Cl > PR_3 > SO_3H > $SO_2.R$ > SO.R > SH > F > $O.SO_2R$ > $O.CO.CH_2R$ > $O.CO.CH_3$ > OC_6H_5 > $OCHR_2$ > OCH_2R > OCH_3 > OH > $N^+(CH_3)_3$ > $N(C_2H_5)_2$ > $N(CH_3)_2$ > $NH.CO.C_6H_5$ > NH_2 > CO_2R > CO_2H > $CO.C_6H_5$ > $CO.CH_3$ > CHO > CH_2OR > CH_2OH > C≡N > CH=NR > CH_2NR_2 > CH_2NH_2 > C_6H_5 > C≡CR > C≡CH > $C(CH_3)_3$ > cyclohexyl > $CH(CH_3)CH_2CH_3$ > $CH=CH_2$ > $CH(CH_3)_2$ > $CH_2C_6H_5$ > $CH_2.CH=CH_2$ > $CH_2.CH(CH_3)_2$ > C_2H_5 > CH_3 > T > D > H > lone electron pair

When there is only one chiral atom in a structure this gives rise to two stereoisomeric forms known as ***enantiomers.*** They can be separately described as having the ***R* or *S* configuration.** When fully naming the compounds then the *R* or *S* label may be applied to the name as a whole, or to the individual chiral atom

When there is more than one chiral atom in a structure further isomers are possible, and in general for a structure with *n* chiral atoms there is a ***maximal possibility*** of 2^n isomers, that can be grouped into *n* pairs of mirror image forms. The relationship between members of different mirror image pairs is described as ***diastereoisomeric***.

Figure 6.2 shows formulae for examples where there are two and three chiral carbon atoms. The structure (1), (2), (3), and (4) show the stereoisomeric tartaric acids. The structures are shown as what is known as Fischer projection formulae. By convention the vertical chain of carbon atoms is presented to the viewer with the top and bottom carbons more remote. (*I.e.* as a convex vertical curve). This is shown explictly in the first row of formula, by using dotted bonds. (The same convention is used in the last eight structures, but is not shown explicitly). There are two other rules or conventions; (a) all atoms of groups attached by single bonds, are capable of free rotation with out changing the identity of the compound represented; (b) rotation of any structure around an axis perpendicular to the plane of the paper, does not cause a change in the structure. Fischer projection formulae have been favoured by chemists as the easiest way of *correlating* the stereochemistry of carbohydrates.

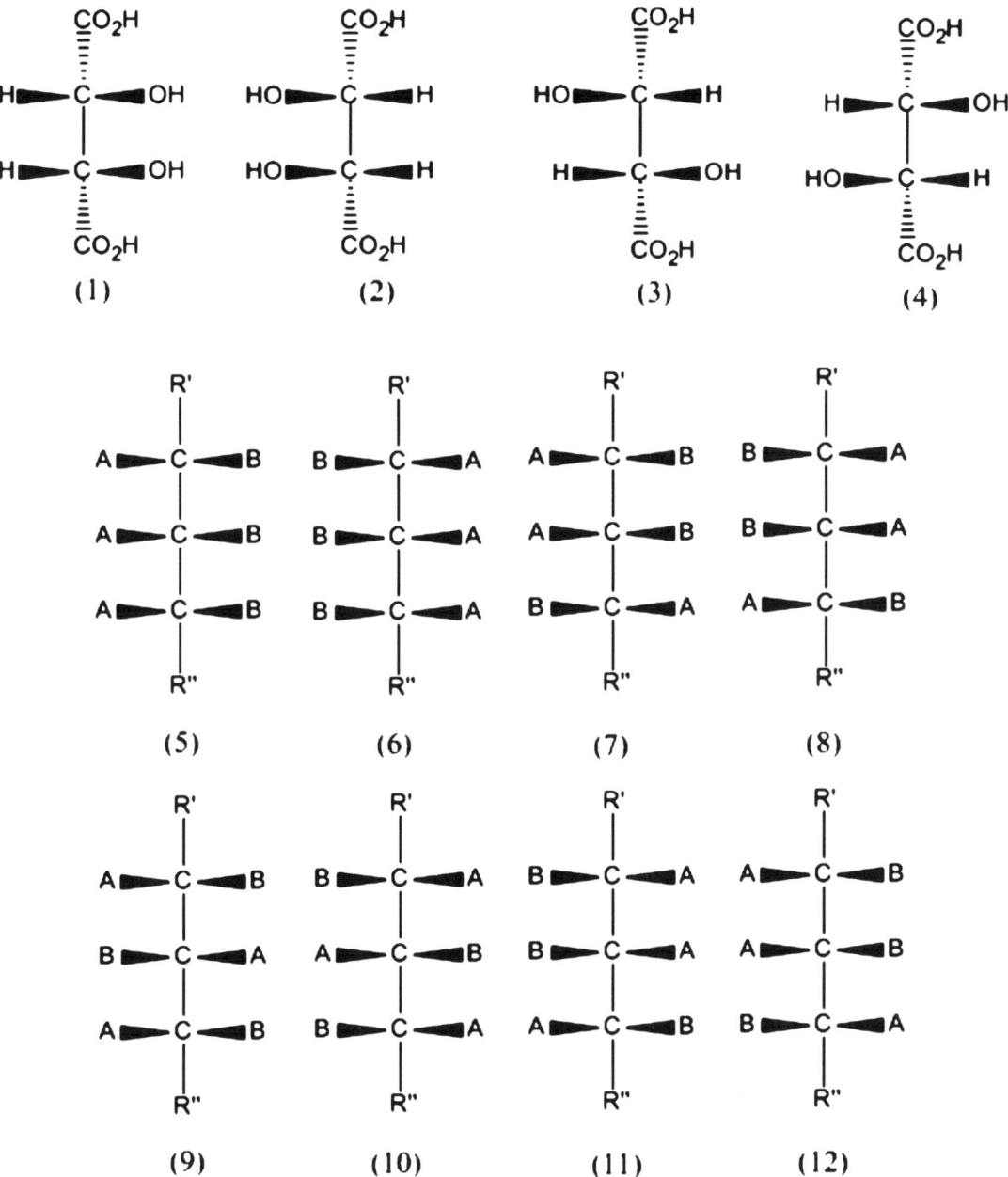

Figure 6.2
Stereochemistry of Compounds with Two and Three Chiral Centres

With the tartaric acid structures, applying rule (b) reveals that structures (1) and (2) are identical, the two chiral atoms have opposite configurations [the top chiral atom in (1) is *S* the lower chiral carbon atom is *R*]; the structure is said to be *internally compensated* and the actual compound is known as *meso-tartaric acid*. Structures (3) and (4) represent **D-tartaric acid** and **L-tartaric acid** respectively.

When there are three chiral centres as in structures (5) to (12) there are at maximal total of 2^3 (= 8) structures. This number is reduced if the groups shown as R' and R" are identical. In that case structures (5) and (6) are identical and so are (9) and (10). These are two distinct *meso* forms. Furthermore, (7) and (12) are identical and are the mirror image (*enantiomer*) of (8) or (11) (also identical). The cases where the groups denoted as A and B in these structures are a hydrogen atom and a hydroxyl group respectively, and R' and R" are either both carboxyl groups, both -CH$_2$OH groups, or are -CHO and -CH$_2$OH respectively are biologically significant compounds (See Table 6.1).

Table 6.1
Significant Chiral C$_5$ Compounds in Figure 6.2

Compound	R' = R" = CO$_2$H Trihydroxyglutaric acids	R' = CHO R"= CH$_2$OH	R' = R" = CH$_2$OH
(5)	Meso form A	*aldo*-D Ribose	Ribitol = Adonitol
(6)		*aldo*-L-Ribose	
(7)	Identical to (12)	*aldo*-L-Lyxose	L-Arabitol
(8)	Identical to (11)	*aldo*-D Lyxose	D-Arabitol
(9)	Meso form B	*aldo*-D-Xylose	Xylitol
(10)		*aldo*-L-Xylose	
(11)	Identical to (8)	*aldo*-D-Arabinose	D-Arabitol
(12)	Identical to (7)	*aldo*-L-Arabinose	L-Arabitol

It sets of compounds such as those shown in Figure 6.2 and Table 6.1, pairs of compounds that are described as ***meso*** are chemically different from the others; they are likely to have different solubilities, melting points, *etc.* and are capable of being distinguished by methods based on spectroscopy. The compounds described as being members of a pair of mirror image structures have identical properties except for those properties that depend on interactions with polarised light **in solution or in the liquid phase.** Such compounds are described as being ***optically active.*** Meso compounds are devoid of optical activity.

Being able to draw a pair of structures that are mirror images of each other does not, in itself, mean that the corresponding compounds can be separately made. Many such compounds are unstable. The process by which one member of such a pair can be converted to the other is known as ***racemization***. The mixture of the two stereoisomers that results from this process is known as a ***racemate***. It will, in general, be easily distinguished from the separate enantiomeric forms, by melting point, solubility, and lack of optical activity. Compounds that are easily racemized frequently have structures with carbonyl groups adjacent to the chiral centre, allowing loss of chirality through an enolization process:

$$R'(R")C^*H\text{-}C(=O)\text{-}R''' \rightleftarrows R'(R")C=C(OH)\text{-}R'''$$

(The labile chiral centre is marked *). Any synthetic chemical reaction starting with achiral compounds will (in the absence of chiral auxiliaries or catalysts) invariably end with a racemic mixture as a product

This brief discussion has not mentioned other classes of compounds where chirality can occur. For example, allenes, biphenyl derivatives, and metallic coordination compounds. Some of these are mentioned later. The way that chirality affects the crystal habit of compounds is dealt with in chapter seven.

Essentials of Stereochemistry and Conformational Analysis

It is important to make a distinction between *structures* obtained through modelling, and the existence of the *real compounds* that they represent. Indeed, just because a model can be made of a compound, it does not imply that the compound can be synthesised.

6.2 Conformation and conformational analysis

While the recognition of optical activity, and chirality as its underlying cause are shrouded in the history of Chemistry of the Nineteenth Century, **conformational analysis** is of the Twentieth Century and will be forever associated with name of D. R. H. Barton. In his paper, called "The Conformation of the Steroid Nucleus,"[4] there is an important footnote: "The word conformation is used to denote differing strainless arrangements in space of a set of bonded atoms. In accordance with the tenets of classical stereochemistry, these arrangements represent only one molecular species." It is in that sense that the word *conformation* is used in this book.

Barton's paper deals not only with steroids, but builds up the ideas on conformation, starting with isolated cyclohexane rings, gradually building up to various systems involving fused rings. To understand the purely mechanical properties of these systems, it is best to revert to the use of 'hardware' models, preferably those that allow freely rotating joints. (While this book concentrates on the use of computers for modelling, it has to be admitted that the old-fashioned models do have their uses!). With a model of a single cyclohexane ring it is easy to see that conversion between the two conformations shown in Figure 6.3 is quite easy, merely by twisting slightly. It is also easy to see that while the **chair** conformation is quite rigid, the **boat** conformation is not so, and that is can flow between numerous more or less equivalent states, *via* so-called **twist-boat** conformations.

Barton recognised the need for terms to describe the bonds associated with a chair form cyclohexane ring. The six bonds that were roughly parallel to the axis through the centre of the ring, and perpendicular to the average plane of the ring are termed **axial** bonds, the other six bonds that lie in the average plane, are termed **equatorial** bonds. In general, when a hydrogen atom in cyclohexane is substituted by another atom or group, there is a preference for the substituent (usually larger than a hydrogen atom) to be joined by an equatorial bond. When there is more than one substituent group the disposition of these groups depends on energies associated with the spacial interactions of the groups. This aspect of stereochemistry is explored in the exercises included in Chapter 4.

When two cyclohexane rings are fused together the structure is that of decalin. Decalin exists as two stereoisomers: *trans*-decalin and *cis*-decalin. (The prefixes *trans* and *cis* refer to the hydrogen atoms on the bond common to both rings, being on opposite sides or the same sides of the bond respectively). The compounds are quite distinct, and can be separated by fractional distillation or gas chromatography. For the *cis* form two mirror image structures can be made; the compound is potentially chiral although there is no identifiable chiral centre. However, because the ability of the rings to flip over into boat forms is still present in the fused cyclohexane rings, converting between the chiral structures is very easy. (This is easily shown with 'hardware' models). Consequently it is not possible to separate *cis*-decalin into stable enantiomers.

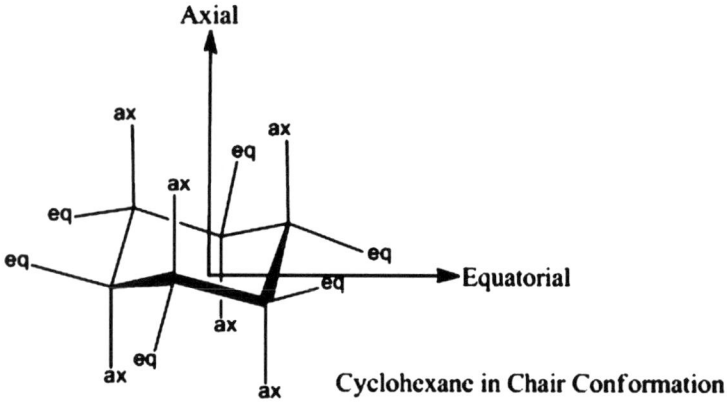

Cyclohexane in Chair Conformation

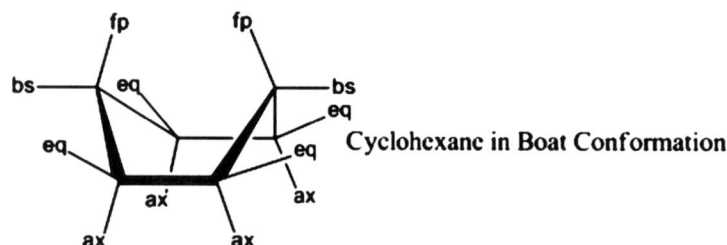

Cyclohexane in Boat Conformation

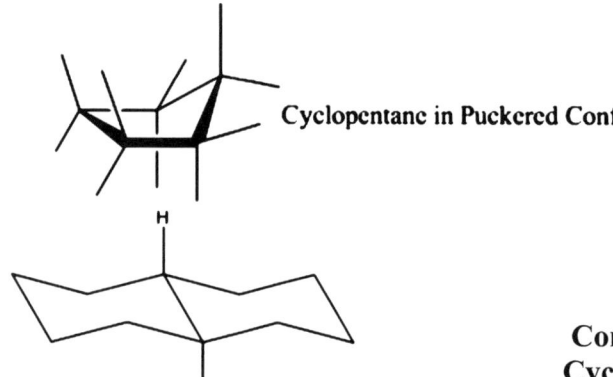

Cyclopentane in Puckered Conformation

trans-Decalin

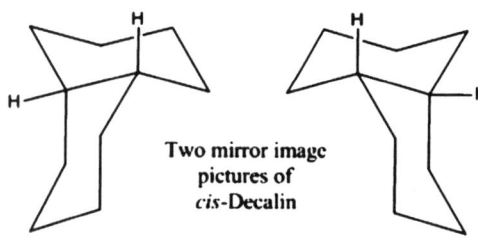

Two mirror image pictures of *cis*-Decalin

Figure 6.3
Conformations of Cyclohexane, Cyclopentane Rings, and Decalins

In the *chair* form of cyclohexane, the two sets of bonds, *axial* and *equatorial* are shown. In the (idealized) boat form of cyclohexane, names for two other sorts of bonds, *flagpole* (fp), and *bowsprit* (bp) are sometimes used.

The structure of cyclopentane is shown with the puckering somewhat exaggerated. In real compounds the actual conformation depends on the nature of the substituent groups.

Essentials of Stereochemistry and Conformational Analysis 137

We return to the stereochemistry of the tartaric acids, and show, in Figure 6.4, three low energy conformers that resulted from using the SMILES facility in INTERCHEM. The SMILES string used was:

$$O=C(O)C(O)C(O)C(O)=O$$

and a total of 512 conformers were generated. The molecular formula of tartaric acid is $C_4H_8O_6$; (a total of 18 atoms). To model this system of conformers would need a coordinate space of $(3 \times 18) - 6 = 48$ space coordinates (plus one extra to show the energy!).

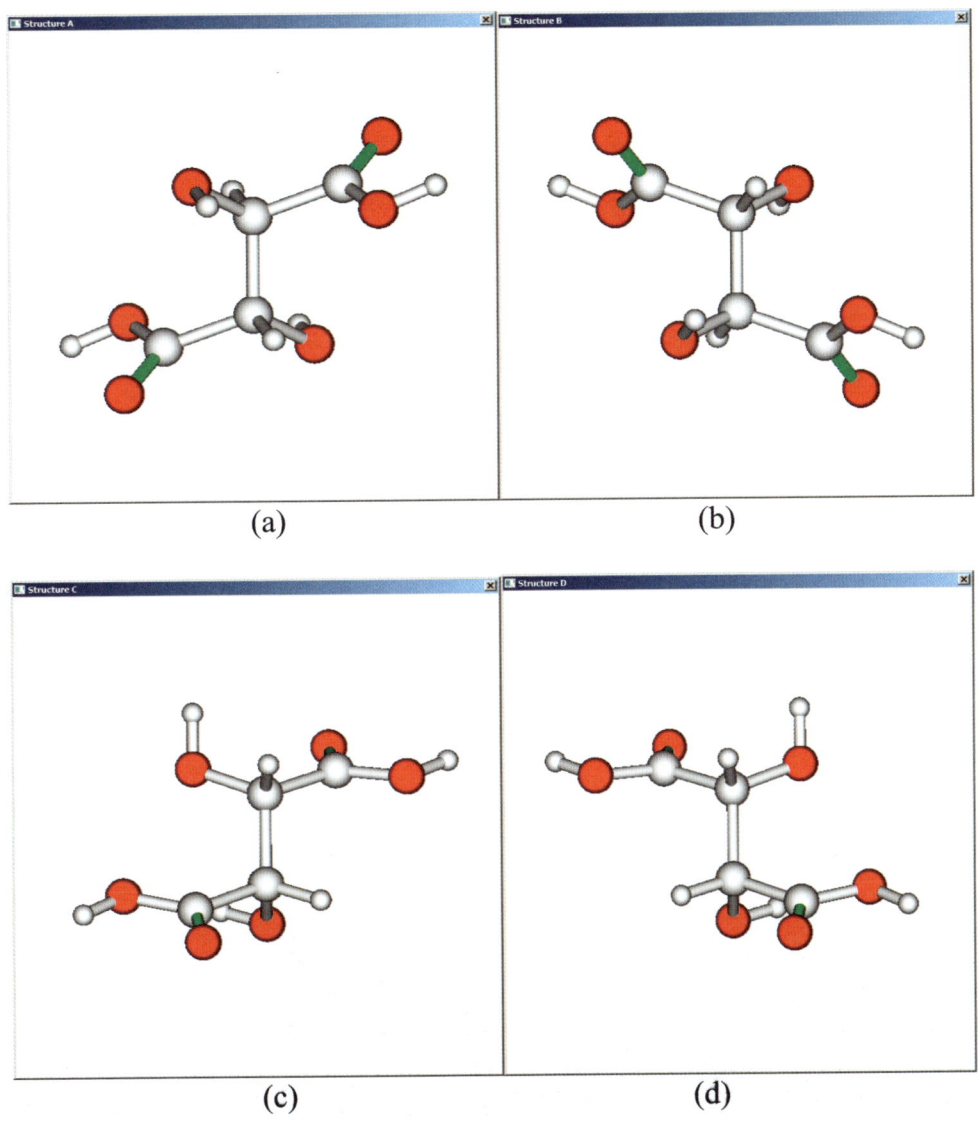

(a) (b)

(c) (d)

Figure 6.4
3D Pictures of Tartaric Acid Stereoisomers

Pictures (a) and (b) are mirror images of the internally compensated *meso* isomer of Tartaric Acid. Appropriate rotation of (a) will allow it to be superimposed on (b).

Picture (c) is of *S,S*-Tartaric acid. Picture (d) is *R,R*-Tartaric acid. The two pictures are mirror images of each other, and so are the compounds that they represent. Picture (c) cannot be superimposed on structure (d)

6.3 Isomerism involving double bonds and rings

There is another form of isomerism that arises when there are different substituents attached to double bonds. In the substituted ethylene (13), if the groups R_1 and R_2 are different, *and* the groups R_3 and R_4 are also different, then two isomers can exist.

Note that it is not a condition that R_1 should be different from R_3 or that R_2 should be different from R_4. In fact there are often identities; thus with $R_2 = R_4 = CO_2H$, $R_1 = R_3 =$ hydrogen atoms, we have maleic acid (14), while with $R_2 = R_3 = CO_2H$ and $R_1 = R_4 =$ hydrogen atoms, the isomer is called fumaric acid. These two compounds differ in their physical and chemical properties. For example, maleic acid readily forms an anhydride on heating, while fumaric acid is converted to *maleic* anhydride on much stronger heating. The older nomenclature for compounds such as these used the prefixes *cis* for the case where the identical groups are on the same side of the double bond, and *trans* when they are on opposite side of the double bond. Problems in nomenclature arise, however, when all four groups are different. It was realised that the priority rules used in the CIP system, could be applied to solve the problem.

The rules are applied separately to the pairs of substituents (groups) at both ends of the double bond. The outcome when the rules are applied to structure (13) are best viewed as a truth table (Table 6.2)

Table 6.2
Truth Table for Deciding *E/Z/* Configuration in Structure (13)

Priority R_1 versus R_2	*Priority R_3 versus R_4*	*Outcome*
$R_1 > R_2$	$R_3 > R_4$	Z
$R_1 > R_2$	$R_3 < R_4$	E
$R_1 < R_2$	$R_3 > R_4$	E
$R_1 < R_2$	$R_3 < R_4$	Z

The letters ***E*** (from the German word **Entgegen** = opposite) and ***Z*** (German **Zusammen** = together) are used to denote the cases where the priority 'vectors' are either 'antiparallel' or 'parallel'.

The use of the *E/Z* method of naming stereoisomers consequent upon the presence of double bonds is also applicable to compounds containing C=N bonds such as oximes.

The use of the terms *cis* and *trans* is now reserved for distinguishing isomers in ring compounds where there are two substituents on different ring atoms, for example in *cis*- and *trans*-1,2-difluoro cyclohexane.

6.3.1 Chirality in biphenyl derivatives and in allenes
The biphenyl structures (16) and (17) in which there are four substituents in the 2, 6, 2', and 6' positions are capable of existing in two mirror image forms (as shown). The necessary condition is that the substituents (R_1, R_2, R_3, R_4) are bulky enough to prevent free rotation about the single bond joining the two benzene rings. There is no real chiral centre. There must also be a lack of symmetry in the structure, and this can be from the non identity of both pairs of substituents; $R_1 \neq R_2$, $R_3 \neq R_4$, but can arise elsewhere in the rings (*e.g.* differences in substituents in the 3,5 and 3', 5' pairs).

A similar situation arises in the case of allenes (18) and (19). In both the biphenyl and allene structures the lack of symmetry arises because of lack of identity in a set of four groups, that have an elongated tetrahedral arrangement. However, the tetrahedrons concerned lack the regularity of those found 'traditional' chiral compounds.

When the two adjacent double bonds in allenes are further extended to a third, we have compounds called cumulenes, and these can exhibit *cis/trans* isomerism, that for nomenclature puposes can use the *E/Z* system.

140 *Molecular Modelling: Computational Chemistry Demystified*

References and Endnotes. Chapter 6

[1] E. L. Eliel and S. H. Wilen. *Stereochemistry of Organic Compounds*, Wiley, New York, 1994.

[2] R. S. Cahn, C. K. Ingold, and V. Prelog, *Experientia*, 1956, **12**, 81; V. Prelog and G. Helmchen, *Angew. Chem. Int. Ed. Engl.*, 1982, **21**, 567.

[3] Optical activity in solution is usually measured by an instrument known as a **polarimeter**. Traditionally, this involved visually measuring the angular direction and extent, by which a beam of plane polarised light was turned, by a solution of the compound being investigated. The measurement was confined to monochromatic light at one wavelength, usually the sodium D line. The formula applied is

$$[\alpha]_D^t = angle\ /(length \times concentration) \qquad (6.1)$$

where:
- *angle* is the measured angular deflection (clockwise is positive; anticlockwise is negative).
- *length* is the length of the cell (polarimeter tube)
- *concentration* is measured in grams/litre (mg/ml)
- t is the temperature of measurement (°Celsius)
- D refers to the wavelength of the sodium D line and can be replaced by the wavelength in nm for measurements at other wavelengths.

For purposes of comparison between compounds of different molecular mass (M) the molar rotation $[\varphi]$ is used:

$$[\varphi] = [\alpha] \times M/100 \qquad (6.2)$$

Optical **R**otatory **D**ispersion (ORD) refers to measurements of the optical activity as either $[\alpha]$ or $[\varphi]$ through a wide range of wavelengths. See the book: C. Djerassi, *Optical Rotatory Dispersion*, McGraw-Hill, New York, 1960.

Circular **D**ichroism (CD) refers to measurements of the differences between the visible/ultra-violet light absorption of optically active compounds, for right and left handed circularly polarised light. See the book: P. Crabbé, *Optical Rotatory Dispersion and Circular Dichroism in Organic Chemistry*, Holden-Day, San Francisco, 1965

[4] D. H. R. Barton, *Experientia*, 1950, **6**, 316. In this seminal paper, the bonds and axes, that are now labelled *axial*, were called *polar*. The terminology was very shortly afterwards changed to avoid confusion, since the word *polar* is used frequently in connection with the *electronic* structures of molecules.

Chapter 7
Molecular Modelling and the Solid State of Materials

7.1 Introduction

Here we address the application of molecular modelling techniques to the solid state of matter. To tackle the subject exhaustively is a massive undertaking and beyond the scope of this modest chapter. However, a number of topics will be discussed that are, from a historical perspective, seminal to the subject. Initially let us consider the sorts of questions about solids one may like to address through molecular modelling. High value-added chemical products, often referred to as speciality materials, are predominantly produced as solids[1]. Hence, one perspective to consider is that of process engineers. Tasked with synthesising and formulating solids, they want to know how such materials will behave under processing conditions and how to safely and efficiently scale-up processes to produce the desired quantities of a given product. Another perspective is that of the physicist, chemical physicist or materials scientist who wants to understand the relationship between the underlying, atomic-scale structure of solids and their properties and characteristics as functional materials, *e.g.* as electronic devices, sensors, pharmaceuticals, foodstuffs, and pigments to name just a few types.

Molecular modelling is being employed to address perhaps the most fundamental question about solids: how to predict the atomic-scale structure of crystalline solids from first principles. Over twenty years ago, John Maddox[2] described our lack of ability to predict, routinely, the internal structure or packing arrangement of molecular solids solely from knowledge of the molecular structure as "one of the continuing scandals in the physical sciences". Academic and commercial entities now have access to levels of computing power far greater than Maddox could have envisaged when he made his statement. Nevertheless, although a lot more is now understood about the reasons for the difficulty and many of the obstacles have been overcome,[3, 4] for the general case the problem remains unsolved.

A more modest ambition, and still a challenging problem, is to use molecular modelling to answer this question: what change in energy accompanies the formation of a crystalline state of a material, assuming its structure is known from experiment, from the liquid or gaseous state? The relative magnitude of this energy change reflects the physical stability of the solid state. Related to this problem is the ability to predict the heat capacity of a solid, as a function of temperature, and the temperature at which the solid will melt at a specified pressure. On the one hand you can argue why not just measure melting points experimentally but, at this point in the 21st Century, should we not be able to predict, accurately, and from first principles, a melting point on the click of a computer mouse?[5] Other questions we may wish to answer relate to the mechanical properties of solids for example: what is the bulk modulus or can we predict the elements of the elastic tensor describing a material? The properties mentioned here are examples of ones which process engineers are interested in knowing and molecular modelling has a useful role to play in their estimation.

When it comes to understanding the electronic properties of solids for designing novel functionality and new devices we enter into extensive areas of current research activity. For example, modelling the response of crystalline solids to incident electromagnetic waves, by predicting an absorption or emission spectrum, can reveal how solid particles interact with and adjust to their environment. Hence molecular modelling can help to interpret experimental data and determine the surface structure of solids. Usually surface structures are not the same as structure in the bulk. Since many processes in which we are interested occur at interfaces (arguably the whole of biology), the surface structure of solids (or more generally self-assemblies of molecules) is of the upmost significance. However, we begin by considering how to construct models of the interior structure of crystals.

7.2 Classification of solids

When we start to think about states of matter almost invariably we begin by defining two hypothetical states; the perfect gas and the perfect crystalline solid. The reason we do this is that the definitions of these states lead to the simplest possible descriptions which are, nevertheless, still useful as points of comparison with real states of matter. Of course there is the liquid state as well, but this provides less opportunity to make elegant, simplifying assumptions although, in terms of thermodynamics, we may talk of a perfect solution. So far the molecular modelling concepts introduced in this book relate either to isolated molecules or molecular adducts and in both cases a gaseous state is implied because the positions of neighbouring molecules are not being taken into account. Hence we have seen that we can calculate the enthalpy of formation of a molecule from its elements in their standard states using MOPAC (Section 4.3.1) but not, so far, how we could calculate the enthalpy changes associated with melting a crystalline solid or with boiling a liquid.

The crystalline solid is a state of matter in which the relative positions of all the constituent ions, atoms or molecules remain the same throughout time and in which a characteristic structural-unit is repeated exactly in every direction from the centre to the surfaces of the crystal. With an architecture based on exact repetition of a characteristic structural-unit, crystals possess long-range order in all directions through space, they are fully periodic. The perfect crystal is an idealised state of matter in the same way as the perfect gas. However, the perfect crystal is almost always the starting point when modelling crystalline materials. The next section of this chapter will provide an overview of the rules that govern the architecture of crystals: the subject of crystallography.

Often a crystalline body, for example a metal bar made of steel, does not consist of a single, macroscopic crystalline-particle. Instead the crystalline object is made of an assembly of crystalline domains. There is no systematic arrangement of the constituent domains over length-scales greater than the characteristic size of the individual domains (typically 1×10^{-6} metres). Materials of this type are described as polycrystalline. There is no long-range ordering within the assembly of small crystalline-domains whereas there is long range order within every individual crystalline-domain. This apparent paradox neatly illustrates the necessity for an appreciation of the length scales and time scales that are relevant to particular branches of modelling.

Some material properties of bulk polycrystalline-substances relate directly to the atomic-scale i.e. the microscopic crystalline-structure. This describes individual domains within a material (length scales of 1×10^{-11} to 1×10^{-8} metres). Modelling at the meso-scopic scale (length scales of 1×10^{-8} to 1×10^{-4} metres) is required to evaluate, for example, the over-all mechanical strength of a polycrystalline material and for this one might employ a model based on finite elements. In the case of modelling at the meso-scopic scale, clearly it is not practical (or necessarily correct in terms of the underlying physics) to employ methods that treat every individual atom constituting a material object over the extended volume implied by the length-scale hence, a purely molecular-modelling paradigm is not applicable over this greater length-scale. The emerging discipline of multi-scale modelling[6] seeks to bridge the gap between individual modelling-techniques that apply to different regions of length and time scales.

There are further states which are intermediate between the fully crystalline state, which has long-range order in three dimensions, and the liquid state. Where these states are characterised by a reduction in the long-range order they are known collectively as liquid crystals. Another similar phenomenon, manifested by plastic crystals, is the ability of some molecules, with a spherical or elliptical shape, to rotate within the confines of a crystal whilst maintaining their long-range ordering in the crystal. Solids in which the relative positions of all the constituent ions, atoms or molecules remain the same throughout time but which do not manifest any long range order are referred to as amorphous.

7.3 Crystallography and the specification of crystalline structures

The following summary of some important aspects from the field of crystallography is very brief and highly selective, for a complete description it is recommended to refer to an excellent treatise on the topic.[7] The most fundamentally important feature of the crystalline solid-state is its periodicity. In practical terms, periodicity means that a fully atomistic model of an extended region of a solid can be constructed; it requires the atomic coordinates to be known for just a small subset of all the atoms that go to make up the extended region. The mathematical operations required to generate the coordinates of the atoms are simple and best represented in terms of matrices and vectors. As far as the molecular modeller may be concerned, it is the job of appropriate modelling software to store the elements of the relevant matrices and vectors and apply correctly the appropriate mathematical manipulations to produce a model of a crystal structure.

Although this 'black box' situation may be desirable for the modeller it is nevertheless sensible to develop a basic understanding of an approach for classifying crystal structures, crystallography, on which such software relies. So, as will be explored below, to understand the physics and chemistry of the crystalline state of matter it is best to begin with an abstract mathematical concept called a *lattice*. Importantly however, this does not imply that a very extensive knowledge of the subject of crystallography is necessary even to begin exploring the application of molecular modelling to solids.

7.3.1 The lattice concept
Fundamentally a lattice is just a collection of points in space that are indistinguishable one from another. The simplest possible lattice can be described by defining a reference point, the origin of the lattice, and a line direction. A one-dimensional lattice satisfies the requirement that some distance exists, let us call it $|\mathbf{a}|$, that defines an infinite set of points, which are all indistinguishable, located along the line at distances from the origin which are integer multiples of the distance $|\mathbf{a}|$. Hence the vector $\mathbf{a}$ defines the repeating distance between the lattice points in the one-dimensional lattice.

The lattice concept can be extended to lattice points in a plane, yielding a two-dimensional lattice, by defining a second, non-parallel direction represented by a vector $\mathbf{b}$ so that the spacing between adjacent lattice points in the second dimension is $|\mathbf{b}|$. To obtain a complete description of the plane lattice so created, the angle, γ, between the rows of lattice points must be specified this is illustrated in Figure 7.1. When identical plane-lattices are stacked along a third direction in space, indicated by a vector **c**, that is not co-planar with **a** and **b** a three-dimensional lattice is formed. The connection between, on the one hand, the purely mathematical concept of a lattice and, on the other hand, the physical structure of a real crystal is the existence, in crystals, of a periodic array of points at which the structural environment is identical.

It is important, conceptually, to emphasise the difference between the abstract lattice and a physical crystal which the lattice helps to describe. Take the case of a two-dimensional lattice, the lattice points are not physical objects however, they are often represented in diagrams as circles (which could, for example, be cut out of a piece of paper and take on a physical existence) having their centres at the locations of the lattice points.[8] An implicit feature of a lattice is its symmetry.

The elements of symmetry which are important for two-dimensional lattices are axes of rotational symmetry and mirror lines (planes) and combinations of these through the lattice points. At this juncture there are two ways of proceeding; either we first consider an isolated lattice point in a two-dimensional lattice, and go through all the possible combinations of rotational-symmetry axes and mirror lines that could pass through that isolated point or, we first consider all the possible ways of arranging arrays of lattice points in space. Here, the latter approach is chosen and isolated lattice points and the definition of point groups will be considered latter.

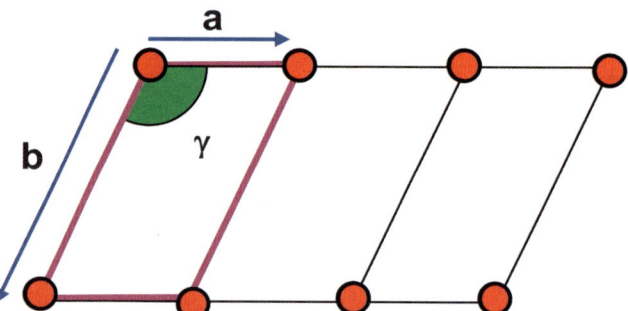

Figure 7.1 Arbitrary Plane Lattice with a Unit Cell Indicated (Magenta Lines)
The lattice points are shown as red circles, with the origin at the lattice point at the top left.
The vectors, **a**, **b**, and angle γ are the unit cell parameters for this two-dimensional lattice

7.3.2 The crystal lattice in two dimensions
Now that the concept of a lattice has been introduced, we can proceed by attempting a classification of the two-dimensional plane lattices. The classification can be made in terms of the groups of mathematical operations or transformations which, when applied to a lattice, map individual lattice points onto each other so that, over-all, the final configuration of the lattice is indistinguishable from the initial configuration of the lattice. The mathematical operations in question are all symmetry operations. The first operation to consider is rotation.

7.3.2.1 Axes of rotational symmetry normal to the plane

In the context of a two-dimensional lattice, the only rotation operations that are meaningful are ones that keep the lattice points within the plane. Hence the axis of rotation must be normal (perpendicular) to the plane of the lattice. Axes of rotational symmetry are classified by their order, n, where a rotation of 360/n degrees produces a final arrangement of lattice points which is indistinguishable from the initial arrangement. Figure 7.1 illustrates the most general form of plane lattice in which there are no restrictions on the spacings between adjacent lattice points along the axes a and b or the angle, γ, between the axes. This type of plane lattice is called oblique.

For an oblique plane-lattice, the greatest value of the integer n, describing the order of rotation, that applies is two. Hence, when an oblique plane-lattice is subjected to a rotation of 180° about an axis normal to the plane and passing, arbitrarily, through any of the lattice points, the configurations both before and after the rotation are indistinguishable. This type of rotation axis is referred to as a two-fold rotation axis or a rotation axis of order two. At the same time of course, a rotation of 360° also produces an indistinguishable final configuration of the lattice points but more than this, it produces an identical one. A rotation of 360° is called the identity operation for rotation, the positions of individual lattice points are invariant under the action of the identity operation. All groups must contain the identity operation which is usually denoted by the symbol, **I**. The two mathematical operations: identity operation and two-fold rotation define a group which characterises or classifies the oblique plane-lattice type.

Taking this approach one step further, consider the case where the parameters describing the plane lattice are restricted as follows: a = b and $\gamma = 90°$. Now we have a square type plane-lattice and the highest-order rotation axis that applies is a four-fold axis of rotation. Hence rotations of 90° produce indistinguishable arrangements of the lattice points in this case. Finally consider the case where a = b and $\gamma = 120°$, this produces a hexagonal arrangement of lattice points in the plane and rotations of 60° produce indistinguishable arrangements of the lattice points so there are six-fold axes of rotational symmetry normal to the plane and passing through every lattice point. Six-fold is the highest order of rotational symmetry that needs to be considered in the study of crystallography. In the context of plane lattices, five-fold rotational symmetry never applies but three-fold rotational symmetry is important when considering three-dimensional lattices. Hence the orders of rotational symmetry that are important in crystallography can be specified as $n \in \{1, 2, 3, 4, 6\}$.

7.3.2.2 In-plane rotational symmetry and lines of mirror symmetry

So far we have considered only axes of rotation normal to the plane lattice, and this has led to a classification of plane lattices into three types, oblique, square and hexagonal with the highest order of rotational symmetry being two, four and six respectively. This turns out to be insufficient to complete our classification scheme for plane lattices. We must consider the case where a $\neq$ b and $\gamma = 90°$ and the plane lattice is of rectangular type. As with the oblique plane-lattice type, the highest order of rotational symmetry is two-fold. However, there are now, additionally, two-fold axes of rotational symmetry in the plane and coincident with the rows of lattice points hence, three mutually perpendicular axes of two-fold rotational symmetry pass through every lattice point in a rectangular plane-lattice. The in-plane two-fold axes of rotational symmetry are also present in a square plane-lattice.

In conjunction with reading through the following paragraphs it is recommended to refer to Figure 7.2. It is conventional to view these in-plane rotation axes rather as mirror lines (the analogue of mirror planes in the case of a three-dimensional lattice). Hence every lattice point in a rectangular lattice has two perpendicular mirror lines that pass through and are, in turn, mutually perpendicular to the two-fold axis of rotational symmetry (Figure 7.2 b(ii)). Similarly in the square lattice, two perpendicular mirror lines pass through every lattice point and are, in addition, mutually perpendicular to the four-fold axis of rotational symmetry passing through the lattice point. For the square lattice there is an additional pair of perpendicular mirror lines which are rotated by an angle of 45° with respect to the first pair of mirror lines (Figure 7.2 d(ii)). In the case of the hexagonal lattice, there are three pairs of perpendicular mirror lines passing through every lattice point with the second and third pairs rotated by 30° and 60° respectively with respect to the first pair. Similarly all six mirror lines are, at the same time, perpendicular to the six-fold axis of rotational symmetry (Figure 7.2 e(ii)).

Up to this point, only axes of rotational symmetry and mirror lines which pass through the lattice points have been considered. However in certain cases, additional elements of symmetry are present that do not pass directly through the lattice points but are present by virtue of the symmetry of the entire array of lattice points. Hence in the case of the rectangular lattice, there are two-fold rotation axes located at the mid-points between adjacent lattice points along the lattice rows. Two perpendicular mirror lines also pass through these mid-points and are orientated parallel to the rows of lattice points. Similarly these additional symmetry elements are present in the case of the square lattice but there is also a four-fold axis of rotation at the mid-points of the diagonals joining the opposite corners of the squares of lattice points. In the case of the hexagonal lattice there are two-fold rotation axes at the mid-points between adjacent lattice points (including diagonally across the unit cell). Further there are three-fold rotation axes located one third and two thirds off the distance between adjacent lattice points along the longer diagonal of the unit cell.

7.3.2.3 Location of lattice points in primitive plane lattices

The four types of plane-lattice that have now been classified are referred to as being of primitive type. In a two-dimensional primitive lattice the locations of lattice points, with respect to an arbitrary origin, are given by a position vector, **r**, which must obey the condition specified in Equation 7.1 where h and k are integers only, and the vectors **a** and **b** specify the directions and magnitudes of the edges of the repeating unit of the plane-lattice known at the unit cell of the lattice.

$$\mathbf{r} = h\mathbf{a} + k\mathbf{b} \tag{7.1}$$

Equation 7.1 encapsulates mathematically the fundamental property of a lattice as described above.

7.3.2.4 Location of lattice points in centred plane lattices

In one of the cases we have considered so far, it is possible to identify a set of locations in the plane lattice which are identical in their symmetry to all the lattice points and yet, do not contain lattice points. Further, if lattice points are added at these locations it is not the case that a unit cell of the same type but smaller area is produced. This case is for the rectangular plane-lattice since it is possible to locate lattice points at locations that obey the condition specified in Equation 7.2.

$$\mathbf{r} = (\tfrac{1}{2} + h)\mathbf{a} + (\tfrac{1}{2} + k)\mathbf{b} \tag{7.2}$$

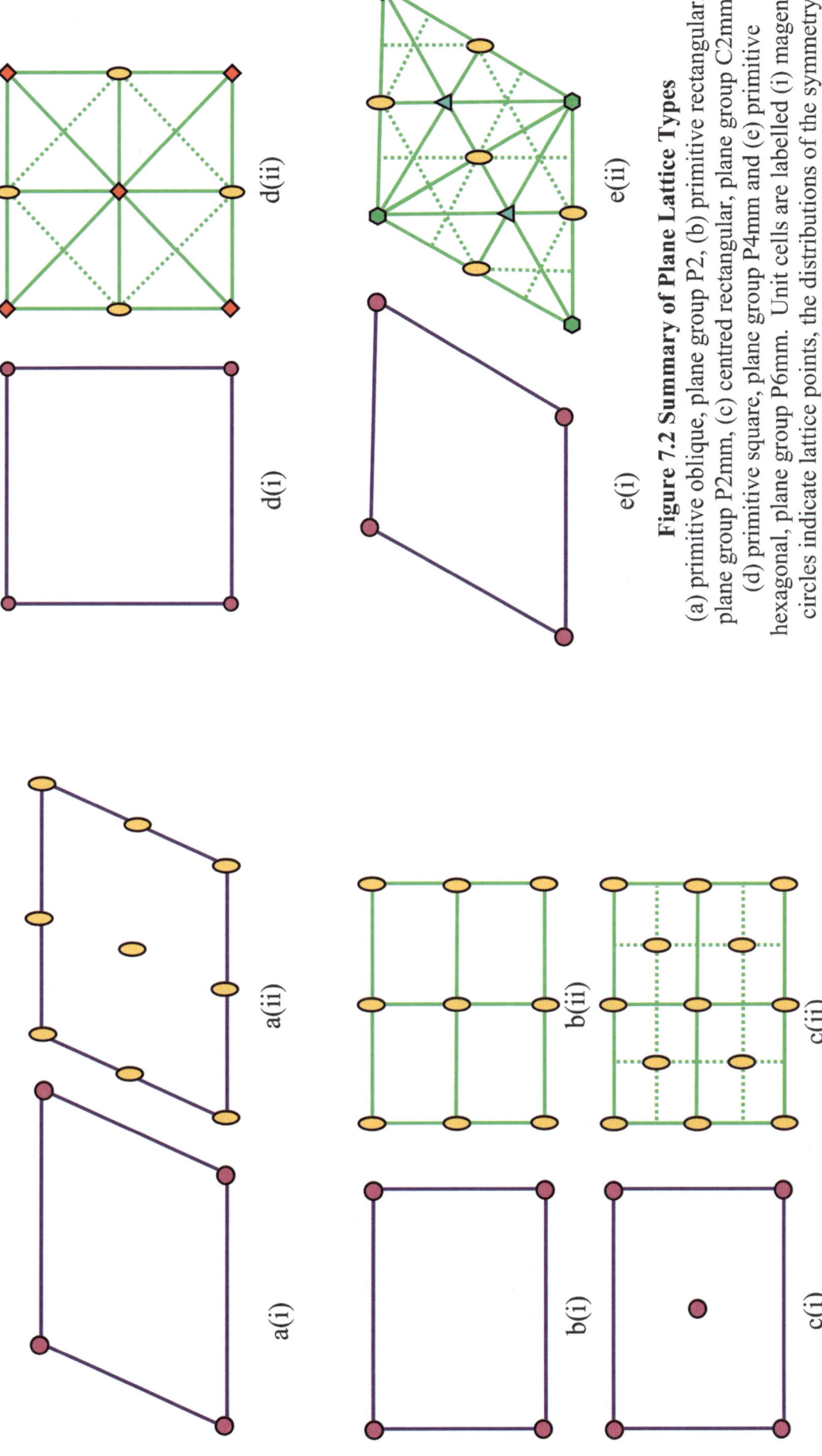

Figure 7.2 Summary of Plane Lattice Types
(a) primitive oblique, plane group P2, (b) primitive rectangular, plane group P2mm, (c) centred rectangular, plane group C2mm, (d) primitive square, plane group P4mm and (e) primitive hexagonal, plane group P6mm. Unit cells are labelled (i) magenta circles indicate lattice points, the distributions of the symmetry elements are labelled (ii). Green solid lines indicate mirror lines dotted lines indicate glides, two, three, four and six fold axes of rotational symmetry are indicated by orange ellipses, light blue triangles, red diamonds and dark green hexagons respectively.

The rectangular plane-lattice that is produced by adding these additional lattice points is no longer primitive but is centred. In fact, a primitive plane-lattice with smaller unit-cell area can be defined for the centred lattice but this is of the oblique rather than the rectangular type.

7.3.2.5 Crystallographic point groups in two-dimensions
Describing the five types of two-dimensional lattice has, at the same time, highlighted the maximum degree of symmetry consistent with each lattice type. Hence there are specific groups or combinations of symmetry elements passing through lattice points that are consistent with each of the five lattice types. These groups of symmetry elements are called point groups. The point group with the highest symmetry that is compatible with a specific lattice type defines the holosymmetric class for that lattice type. However, there may be further point groups (combinations of symmetry elements passing through the lattice points) and hence other classes with lower symmetry which are, nevertheless, compatible with the same lattice type.

For the case of two-dimensions the total number of possible point groups is ten. A point group symbol is used to denote a point group. Taking the primitive oblique lattice as an example, two of the ten possible point groups are compatible with this lattice type hence there are two classes for the primitive oblique lattice. The holosymmetric class is defined by the point group containing a single symmetry element, in addition to the identity operation, a two-fold rotation axis, and this group is denoted as **2**. The second, lower-symmetry class for the primitive oblique lattice is defined by the point group containing a single symmetry element, a one-fold rotation axis, and this is denoted as **1**.

Similarly for the primitive square lattice type, two of the ten point groups are compatible. The holosymmetric class is defined by the point group having a four-fold rotation axis normal to the plane and two perpendicular mirror lines in-plane denoted **4mm**. The second class associated with the primitive square lattice type is defined by the point group having the single symmetry element, a four-fold rotation axis denoted **4**.

Table 7.1 summarises the associations between the five lattice types and the ten point groups. Note that both the primitive and centred rectangular lattice types are associated with the same two point groups namely **2mm** and **m**.

Table 7.1 Summary of the Crystallographic Classification of the Planar Lattice

Planar Lattice Type	Associated Planar Points Groups	Associated Planar Groups (Space Groups)
Primitive Oblique	1, 2	p1, p2
Primitive Rectangular	m, 2mm	pm, p2mm, pg, p2mg, p2gg
Centred Rectangular	m, 2mm	cm, c2mm
Primitive Square	4, 4mm	p4, p4mm, p4gm
Primitive Hexagonal	3, 3m, 6, 6mm	p3, p3m1, p31m, p6, p6mm

7.3.2.6 Symmetry involving translations: space groups in two-dimensions
A complicating factor, not considered so far, arises from the act of combining lattice points, characterised by symmetry elements defined by the point group, into periodic arrays to form a lattice. This introduces further elements of symmetry which are not intrinsic to the point groups as they combine either a rotation or a reflection with a translation. For two-dimensions one new type of symmetry element, the glide line (plane), is introduced. The combination of a reflection through a mirror line with a

translation of the reflected image along the direction of that line defines a glide line. The consequence of observing symmetry elements that include the translation of an object following rotation or reflection is to require an expansion of the 'classification scheme' for crystals. A further set of groups, that take account of glide lines, can be derived such that any two-dimensional crystal structure is fully and correctly described by one of the groups. There are seventeen of these groups known as the plane groups (the equivalent set of groups for three-dimensional crystal structures are called the space groups).

7.3.2.7 Visualising plane (space) groups with INTERCHEM

We can use INTERCHEM to help illustrate the differences between the plane groups for some hypothetical, two-dimensional crystal structures. The first step is to select a simple molecule from which to construct the repeating patterns in-plane, here the tri-atomic (hence planar) molecule hypofluorous acid HOF is adopted. Using the sketch facility create a single molecule of HOF (click on *Build via SKETCH* (R9/C3) refer to Section 4.2.1). Select *Convert 2D->3D* (R4/C6) to create a molecular model of HOF and use *Copy 3D to A* (R1/C2) then *Exit Sketching* (R7/C10).

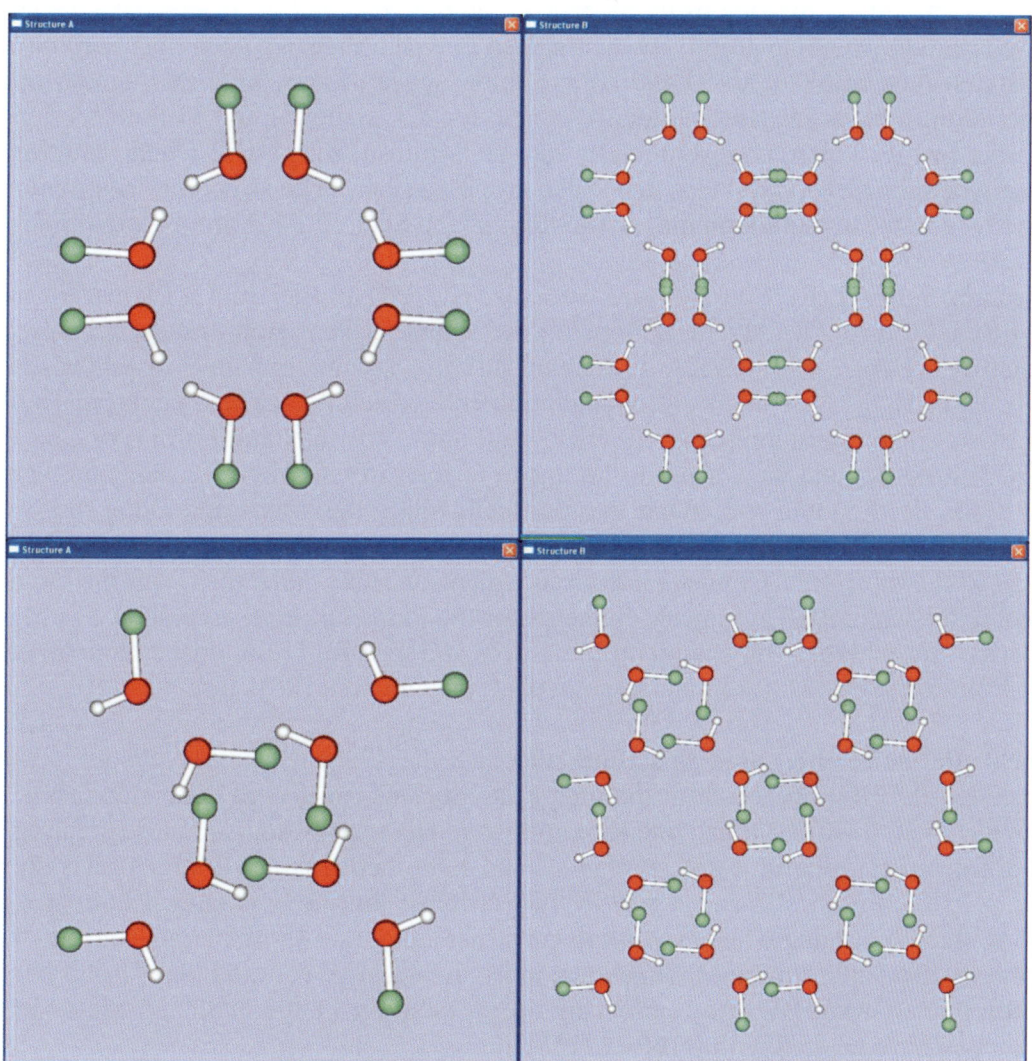

Figure 7.3 Hypothetical Two-dimensional Crystal Structures
Constructed in INTERCHEM for hypofluorous acid HOF in Plane Group 11 P4mm one unit cell (top left) two by two array of unit cells (top right) and alternatively in plane group 12 P4gm one unit cell (bottom left) and a two by two array of unit cells (bottom right).

We need to manipulate the molecular model of HOF by translating it in the plane. Use is made of the *Structure Merging* (R9/C4) facility accessed from the main menu. In structure merging first select *Merge B with A* (R1/C1) (make sure there is no model in the Structure B window before doing this). Now click *Move Host Left* (R3/C1) three times followed by *Move Host Up* (R4/C1) once (you can experiment by applying different shifts to the position of the molecule and/or in-plane rotations later). Click *Complete Merging* (R30/C1) and the translated molecule will be in the Structure A window.

Now we are going to record the coordinates of the atoms in HOF so that we can create a hypothetical two-dimensional crystal using this molecule as the building block or, more correctly, as the *asymmetric unit*. In the main menu click *Store Structure* (R2/C6) and select *XR Format* from the row of menu buttons that appears. A dialogue box will appear in which you are invited to supply the cell parameters for the unit cell of the crystal structure that will be created and to supply the unique number for the space group. Note this is designed for the more common situation of three-dimensional crystal structures hence prompting for three unit-cell edge lengths and three inter-axial angles (see Figure 7.4). Return 8.0 Angstrom units as the length of the a, b and c unit cell axes and 90° as the size of the inter-axial angles α, β and γ. In this case the space group is actually a plane group. To distinguish between the seventeen plane groups and the first seventeen of the two-hundred and thirty space groups, give the number of the plane group (from the International Tables for Crystallography Volume A)[9] as a negative integer. Here choose plane group 11 P4mm of the square planar lattice type hence type minus eleven (-11) as the space group number. The structural information is now saved in the crystallographic .xr file format (see Section 8.7.1 and Figure 8.1).

Now click *Select B* (R1/C2) and *Load Structure* (R2/C7) and select XR Format for input. INTERCHEM provides several options for the extent of the region of a crystal structure to display. Hence it is possible to display just the asymmetric unit, in this case one molecule of HOF, the contents of a single unit cell, which is eight molecules of HOF in this case or an array of unit cells. Referring to Figure 7.3, selecting *Form One Unit Cell* (R1/C3) produces the view seen in the image to the top left whereas, selecting *Nest of Unit Cells* (R1/C5) and specifying two unit cells along the x-axis and along the y-axis and one unit cell along the z-axis (not relevant here) produces the view seen in the image to the top right. Compare the arrangement of HOF molecules with the diagram showing the location of symmetry elements for the Plane Group P4mm shown in Figure 7.2 d(ii). See whether you can identify the two- and four-fold rotational symmetry axes, mirror lines and glide lines in this hypothetical, planar crystal structure.

Repeat the entire procedure this time selecting the Plane Group 12, P4gm. When translating the HOF molecule in the merge facility this time, apply one step to the left (Move Host Left (R3/C1)) and one step up (Move Host Up (R4/C1)) before creating the XR format file. One unit cell of this second hypothetical crystal structure is shown bottom left in Figure 7.3 and a two by two array of unit cells is shown bottom right. Test yourself by seeing whether you can construct a diagram showing the location of the symmetry elements, similar to that shown in Figure 7.2 d(ii) for the Plane Group P4mm, for the Plane Group P4gm by observing the disposition of the HOF molecules in the model you have created with INTERCHEM.

Accepting that neither of these hypothetical, planar crystal-structures have any validity as 'real' crystals in nature this exercise has, hopefully, illustrated how crystallography can be harnessed by molecular modellers to generate atomisitic models that display the packing of molecules in crystalline solids. Subsequently these models give insights into the forces operating between the atoms in different, but adjacent molecules (the inter-molecular forces) which govern the packing motifs adopted by molecules to fill space in crystalline solids. In the crystal structure observed experimentally,[10] the HOF molecules, although planar, do not all occupy the same plane. Hence we must move quickly on from two-dimensional crystals (with which we started since they are inherently less complex and given that crystallography in two-dimensions underpins that in three-dimensions) to crystals in three-dimensions.

7.3.3 The crystal lattice in three dimensions

The next step is to extend the concept of a lattice to three spatial dimensions. In doing this it will become apparent, hopefully, why the discussion up to this point has addressed only two-dimensional lattices. The action of creating a three-dimensional lattice can (in most instances but with one notable exception) be regarded as the stacking of plane lattices in a third spatial dimension in ways that preserve the elements of symmetry that are intrinsic to the plane lattices. Additional parameters are required to describe three-dimensional unit cells, an extra cell-edge direction and length characterised by a vector **c**, and three angles rather than one to describe the mutual orientation of three edges of the unit cell. The angle between cell edges **a** and **b** is gamma, cell edges **c** and **a** is beta and cell edges **b** and **c** is alpha (Figure 7.4).

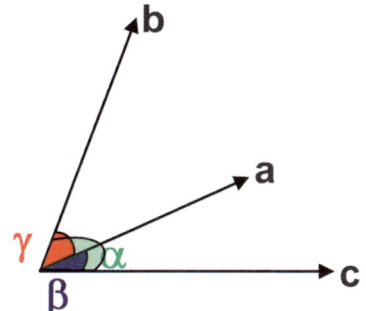

Figure 7.4 Unit Cell Parameters

7.3.3.1 Stacking plane lattices: tetragonal, orthorhombic, monoclinic, triclinic and hexagonal crystal systems

Consider first the possibility of stacking the square plane-lattice type along the plane-normal and selecting a stacking distance other than that corresponding to the edge length |**a**| of the unit cell of the two-dimensional lattice. This choice produces a tetragonal unit cell in which the cell parameters are related such that a = b, c is unrestricted and $\alpha = \beta = \gamma = 90°$. Stacking copies of a primitive rectangular plane-lattice along the plane normal produces an orthorhombic unit cell characterised by unrestricted values for the lengths of the unit cell edges and having $\alpha = \beta = \gamma = 90°$. Stacking copies of an oblique plane-lattice along the plane normal produces a monoclinic unit cell characterised by unrestricted values for the lengths of the unit cell edges and angle β whilst $\alpha = \gamma = 90°$. Alternatively, if an oblique plane-lattice is not stacked along the direction of the plane-normal but along an arbitrary direction, not coplanar with the plane-lattice, a triclinic unit cell is formed. Notice that the two-fold axis of rotational symmetry that characterised the holosymmetric class of the oblique plane-lattice is not present in the three-dimensional, triclinic unit-cell. There are no restrictions on the values of the cell parameters for a triclinic unit cell.

Stacking a hexagonal plane-lattice periodically along the plane normal (corresponding to the six-fold axis of rotational symmetry) produces a three-dimensional hexagonal unit cell. The repeat distance between adjacent planes in the stack need not correspond to the edge length of the unit net hence the c edge length of a hexagonal unit cell is unrestricted whereas a = b and $\gamma = 120°$ and $\alpha = \beta = 90°$. The rhombohedral type of cell is directly related to the hexagonal cell but it is not necessary to go into further details here.

7.3.3.2 Cubic crystal system

Now, finally, we come to a case in which the approach taken above is not entirely satisfactory. Consider a square plane-lattice, such planes can be stacked along the direction of the four-fold rotation axis, normal to the plane, at intervals of distance equal to the edge length of the unit net of the two-dimensional lattice. When this is done a cubical array of lattice points is generated. The cubical array preserves in its three-dimensional arrangement the four-fold rotational symmetry that was identified in the two-dimensional square plane-lattice. However, this is where the difficulty lies and it is quite a subtle problem. In all the other cases described above, the rotational symmetry axis of highest order from the plane lattice is also present in the lowest symmetry class of the three-dimensional lattice generated by stacking copies of that plane lattice. In the cubic (or isometric) crystal system this is not the case. It turns out that the point group of the lowest-symmetry class for the cubic system is **32**. Hence the description of symmetry in the cubic system should be based, more correctly, on the four, three-fold rotational symmetry elements that lie along the four body diagonals. These axes join the four-pairs of opposite corners of a cube and mutually intersect at the same angle as observed between vertices, through the body-centre, of a perfect tetrahedron namely 109.47°.

Figure 7.5(a) shows a primitive cubic cell in which one of the axes of three-fold rotational symmetry is picked out between the two lattice points shown as black circles. Two planes normal to this three-fold rotation axis are picked out with red and blue lines respectively. These planes, as depicted, are in the form of two equilateral triangles that have as their vertices the remaining six vertices of the cubic cell. The points of intersection of the planes with the three-fold rotation axis are at one third of the total distance along the body diagonal from the closer of the two vertices describing the axis.

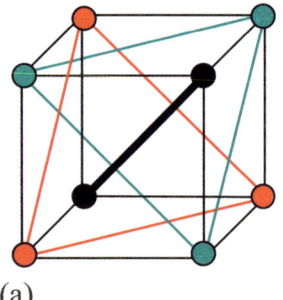
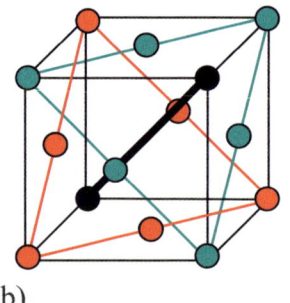
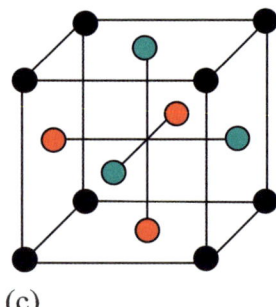

(a)　　　　　　　　　　(b)　　　　　　　　　　(c)

Figure 7.5 Cubic Crystal Systems
(a) view of primitive cubic cell showing one of the three-fold rotation axes and vertices interrelated by this axis in two sets coloured red and blue respectively, (b) shows the derivation of a face centred cubic (fcc) cell via addition of lattice points at the face centres, (c) another view of the fcc cell showing four-fold rotation axes through the body centre.

It can be seen how the action of the three-fold axis is to interchange the three lattice points in one case coloured red and in the other case coloured blue. If additional lattice points are added at the mid-points of the edges of the red and blue equilateral triangles then a face-centred cubic (fcc) lattice is produced this is illustrated in Figure 7.5(b). These additional lattice points at the centres of the faces are also interchanged, as required, by the action of the three-fold rotation axis.

There is, of course, no difference between the lattice points and the different colours are merely to guide the eye and, hopefully, make the symmetry elements easier to see. In Figure 7.5(c) three mutually perpendicular four-fold rotation axes are picked out intersecting at the body centre of the unit cell. There is no lattice point at the body centre of the fcc lattice. All the lattice points in the fcc lattice must lie on the planes indicated in Figure 7.5(b) and these planes do not pass through the body centre of the cell hence no lattice point can be located there.

7.3.3.3 The Bravais Lattices
The seven crystal systems have now been introduced and details of the cell parameters for unit cells describing these systems are summarised in Table 7.2. When seven cases are taken into account in which it is possible to identify additional set of points in the primitive unit cells (of the seven crystal classes), which are characterised by the same sets of symmetry operations, the fourteen Bravais lattices result. Hence in addition to the seven types of primitive cell there are another seven types of centred cell. The Bravais lattices are not illustrated here, but diagrams of these can be located in most physical chemistry texts together with more specialised texts dedicated to crystallography.

7.3.3.4 Inversion axes of symmetry
One example of a further element of point group symmetry which was not encountered in the case of the planar crystallographic point groups is the inversion axis. An inversion axis of symmetry combines an n-fold rotation with either a centre of inversion or a mirror plane. For example, a three-fold inversion axis (or inversion triad), denoted $\bar{3}$, combines three-fold rotation with inversion. An example of this element of symmetry can be seen in a molecule of ethane which is held in the anti-conformation. Which pairs of hydrogen atoms, bonded to different carbon atoms, are interrelated by the inversion triad which is coincident with the carbon-carbon covalent bond?

7.3.3.5 Point groups in three-dimensions
Evaluating all the possible combinations of symmetry elements in three-dimensions which are consistent with a single lattice point leads to thirty-two point groups c.f. ten point groups in two dimensions. Similar to the case in two-dimensions, each of the thirty-two point groups is consistent with one of the seven crystal systems. The allocation of point groups to crystal systems is summarised in Table 7.2.

7.3.3.6 Symmetry involving translations: space groups in three-dimensions
A further element of symmetry, the screw axis, combining a rotation with a translation along the direction of the rotation axis (impossible to have in a single plane) is present in 3-D crystal structures. As with glide planes (lines), screw axes are not part of the classification of the 32 point groups but must be considered in deriving the full set of 230 space groups just as glide lines had to be considered to derive the 17 plane groups. Hence all 3D crystal structures can be allocated to one of the 230 space groups.

A screw axis is denoted n_m where n is the order of the rotation, $n \in \{1, 2, 3, 4, 6\}$ as for pure rotation axes, and m is the so called 'pitch'. The pitch is the total number of unit-cell translations that have been applied after n rotations (i.e. a total rotation of 360° in n stages). So a 2_1 screw axis involves a rotation of 180° and translation of ½, (m/n), the lattice-repeat distance in the direction of the rotation axis. With higher order screw axes the sense of rotation is important. A 6_1 screw axis by convention has an anticlockwise sense, it is right handed. Imagine ascending a spiral staircase, if your right hand is always pointing to the outside, then the spiral is right handed. Similarly a 6_5 screw axis is left handed so the spirals produced by 6_1 and 6_5 screw axes are mirror images of each other.

A summary of the 230 space groups is shown in Table A2.2 in Appendix A2. For each entry, there is listed the Hermann-Mauguin symbol that summarises the symmetry properties of the group, the group's allocation to one of the seven crystal classes and a ranking of the frequency with which the group is encountered in crystal structures of organic and organometallic compounds. Finally those groups with appropriate symmetry properties (of which there are 65) that can be used by chiral molecules are marked with an asterisk.

Table 7.2
The Seven Crystal Systems, their corresponding Bravais Lattice and Symmetries

System	Bravais lattices	Axial lengths and angles	Characteristic (minimum) symmetry	Non-centrosymmetric point groups		Centrosymmetric point groups[a]
				Enantiomorphous[b]	Non-enantiomorphous[c]	Non-enantiomorphous[c]
Cubic	PIF	$a = b = c$; $\alpha = \beta = \gamma = 90°$	Four triads equally inclined at 109.47°	23, 432	$\bar{4}3m$	$m\bar{3}, m\bar{3}m$
Tetragonal	PI	$a = b \neq c$; $\alpha = \beta = \gamma = 90°$	One rotation tetrad or inversion tetrad	4^P, 422	$\bar{4}, 4mm^P, \bar{4}2m$	$4/m, 4/mmm$
Orthorhombic	PICF	$a \neq b \neq c$; $\alpha = \beta = \gamma = 90°$	Three diads equally inclined at 90°	222	$mm2^P$	mmm
Trigonal	R	$a = b = c$; $\alpha = \beta = \gamma \neq 90°$	One rotation triad or inversion triad (= triad + centre of symmetry)	3^P, 32	$3m^P$	$\bar{3}, \bar{3}m$
Hexagonal	P	$a = b \neq c$; $\alpha = \beta = 90°, \gamma = 120°$	One rotation hexad or inversion hexad (= triad + perp. mirror plane)	6^P, 622	$\bar{6}, 6mm^P, \bar{6}m2$	$6/m, 6/mmm$
Monoclinic	PC	$a \neq b \neq c$; $\alpha = \gamma = 90°, \neq \beta \geq 90°$	One rotation diad or inversion diad (= perp. mirror plane)	2^P	m^P	$2/m$
Triclinic	P	$a \neq b \neq c$; $\alpha \neq \beta \neq \gamma \neq 90°$	None	1^P		$\bar{1}$

[a] All the crystals which possess a centre of symmetry and/or a mirror plane are non-enantiomorphous.
[b] The eleven enantiomorphous point groups are those that do not possess a plane or a centre of symmetry. Hence enantiomorphous crystals can exists in right- or left-handed forms.
[c] Eleven of the twenty-one non-enantiomorphous point groups are centrosymmetric. Crystals that have a centre of symmetry do not exhibit certain properties, *e.g.* the piezoelectric effect.
The ten polar point (non-centrosymmetric) groups (indicated by the superscript P) possess a unique axis not related by symmetry. They are equally divided between the enantiomorphous point groups (1, 2, 3, 4, 6) and non-enantiomorphous point groups (*m*, *mm*/2, 3/*m*, 4*mm*, 6*mm*).
Trigonal crystals are divided into those that are represented by the hexagonal *P* lattice and those that are represented by the rhombohedral *R* lattice.

Reproduced (with permission of Dr. Christopher Hammond) from C. Hammond, *The Basics of Crystallography and Diffraction*, (*Third Edition*), OUP, Oxford, 2009.
Reproduced (with permission of Oxford University Press) from C. Hammond, *The Basics of Crystallography and Diffraction*, (*Third Edition*), OUP, Oxford, 2009

7.3.4 Examining crystal structures with INTERCHEM

In this section we will explore the crystal structure and solid-state chemistry of the molecule hexamethylenetetramine (or hexamine for short), $(CH_2)_6N_4$, by way of illustrating some of the facilities in INTERCHEM for examining crystal structures. First, however, a number of preliminary concepts are described. Recall that the smallest group of atoms which, together with the symmetry elements of the crystal lattice, allow the construction of a crystal structure is referred to as the *asymmetric unit*. When dealing with crystals of molecular materials, the asymmetric unit may be a subset of the atoms in a single molecule, for example in the case of urea, equivalent to one molecule, which is typically the case, or it may consist of more than one molecule and, moreover, more than one type of molecule. Hence the material in question may be a one-to-one salt where the asymmetric unit contains two molecular moieties having equal and opposite charge, for example L-Arginine acetate, or a cocrystal where the asymmetric unit contains two or more uncharged molecular species, for example resorcinol urea. In cases where the asymmetric unit consists of more than one molecule of a single type, the molecule adopts different molecular shapes, known as conformations, and the positions of the corresponding atoms in the different molecular conformations cannot be related to each other using the symmetry elements of the space group describing the crystal structure.

7.3.4.1 Space group symmetry operations

For every space group there is a well defined set of symmetry operators that construct the images of the asymmetric unit in a complete unit cell of the crystal structure. For those who wish to investigate this further, the mathematical details are provided in Appendix A1. However to give one example, the space group in the monoclinic system with the least symmetry is space group number 3, **P2**, with just a single two-fold rotation axis along the b-axis of the unit cell. Given the asymmetric unit is at a general position in the unit cell with fractional coordinates **x, y, z** (with respect to the **a, b** and **c** unit cell axes), other than the identity symmetry operation there is just one other symmetry operator required which produces the fractional coordinates **-x, y, -z**. Hence for the space group **P2**, for general positions in the unit cell, the number of asymmetric units present is equal to the total number of symmetry operations, i.e. two and is called the *site multiplicity*.[11]

7.3.4.2 Exploring the crystal structure of hexamine

Returning to hexamine, this is quite rare as an example of an organic, molecular material that crystallises in the cubic system. A recent search of the Cambridge Crystallographic Database revealed just 143 hits for cubic crystals containing solely the elements C, N, O and H. In the database there is for every one of the two hundred and thirty space groups at least one example of a crystal structure belonging to that group. The ranking of the space groups from the monoclinic space group 14, $P2_1/c$, which is the most populated, to the tetragonal space group 99, P4mm, the least populated is given in Table A2.2. The table also lists all the space group symbols in the Hermann-Mauguin notation and the crystal system to which each space group belongs.

Hexamine crystallises in a body centred cubic lattice in space group number 217, with space group symbol $I\bar{4}3m$, there are two complete molecules in the unit cell. The centre of coordinates of one molecule coincides with the origin of the unit cell and that of the second molecule coincides with the body centre of the unit cell (the point (½, ½, ½)). Volume A of the International Tables for Crystallography reports that the multiplicity of a general position in this space group is 48. Since the total number of nitrogen atoms in the unit cell is only 8 and the total number of carbon atoms only 12 this demonstrates that all the carbon and nitrogen atoms must lie on special positions of one sort or another. Hence if the molecular modelling software applies all 48 of the symmetry operations many multiple copies of atoms result which must be identified and deleted. INTERCHEM employs a routine for identifying duplicate atoms and reports how many are deleted when the model is first constructed.

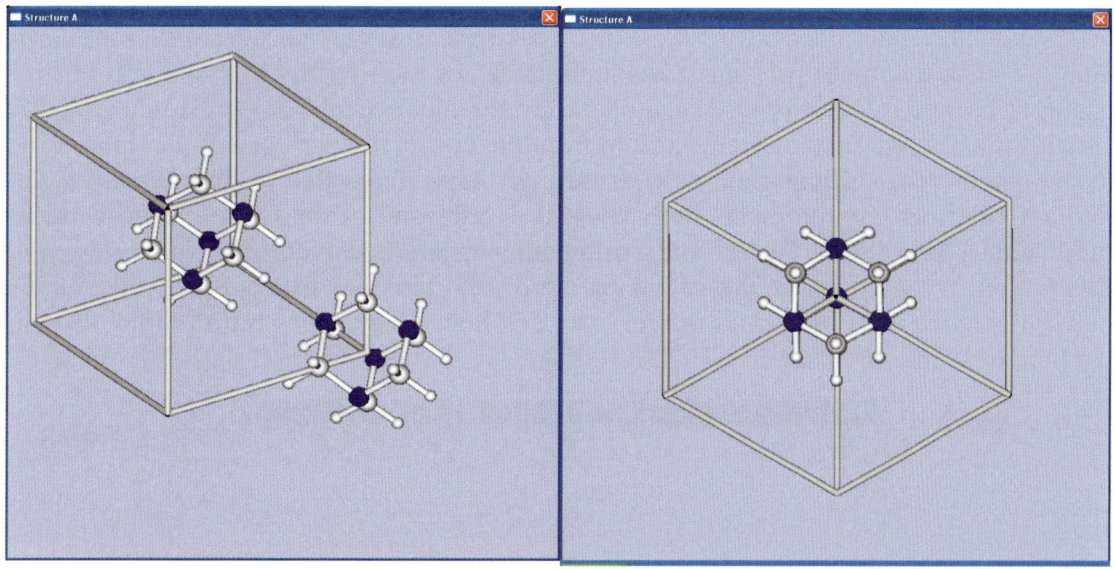

Figure 7.6 Views of the Unit Cell for the Hexamine Crystal Structure
View highlighting the inversion tetrads that coincide with the unit cell edges (left) and down a body diagonal of the cubic cell showing the action of a rotation triad (right).

The unit cell of hexamine as displayed in INTERCHEM is shown in Figure 7.6. On inspecting this figure the elements of symmetry can be identified readily. Looking first at the left image, the relative positions of the hydrogen atoms in methylene groups on opposite sides of a hexamine molecule are related by four-fold inversion axes that are coincident with the edges of the unit cell. The inversion axis combines a 90° rotation with a reflection in a horizontal mirror plane. Now looking at the view down the body diagonal of the cubic unit cell in the right image this coincides with a three-fold rotation axis. The nitrogen and carbon atoms that are interchanged by the three-fold rotation axis can be seen clearly. In this view the hexamine molecule at the corner of the unit cell completely obscures, eclipses, the molecule at the body centre of the unit cell.

The facility for adding the edges of the unit cell that appear in Figure 7.6 is provided within the *Build Structure* (R9/C1) routines accessed from the main menu. For example after loading hexamine into the Structure A from the CIF format file (see Section 8.7.5) called `hexamine.cif` and choosing the option to *Form One Unit Cell with Frame* the number of duplicate atom deletions reported is 1012. Now *Build Structure* (R9/C1) is selected and *Copy A to Base* (R1/C1) then *Extra Displays* (R8/C7) followed by *Wire Frame* (R7/C1). Eight dummy atoms are added to form the vertices of the unit cell. It will be noted that the cell edges are not displayed in other modes such as *Ball and Cylinder*, however, bonds can be formed between the appropriate dummy atoms manually using *Form Bond* (R5/C5) which does then enable the cell edges to appear when the display mode for the molecules is set to Ball and Cylinder.

7.3.4.3 Displaying Miller planes

Another useful facility is provided to overlay specific Miller planes on crystal structures. A Miller plane is usually denoted by three integers within round brackets, (h k l), and is the plane that intersects the **a**, **b**, and **c** axes of the crystallographic unit cell at the points a/h on **a**, b/k on **b** and c/l on **c**. If a plane is parallel to, for example, the **a** axis of the unit cell then the value of the integer h is zero and similarly planes perpendicular to the b axis have the general index (0 k 0). Since the surfaces of crystals are parallel to Miller planes with low indices (i.e. small absolute values of the integers h, k and l) it is very useful to be able to superimpose such planes on the model of the crystal structure. This enables the arrangement of atoms and molecules on specific surfaces of a crystal to be investigated. This is important as it is through its surfaces that an individual crystal interacts with its environment. For example, if a crystal of an active pharmaceutical ingredient (API) forms a particle, originating from a tablet, that is undergoing dissolution in the stomach, the nature of the crystal surfaces that are present governs how quickly the drug molecules are released and hence the concentration of the drug in the blood plasma.

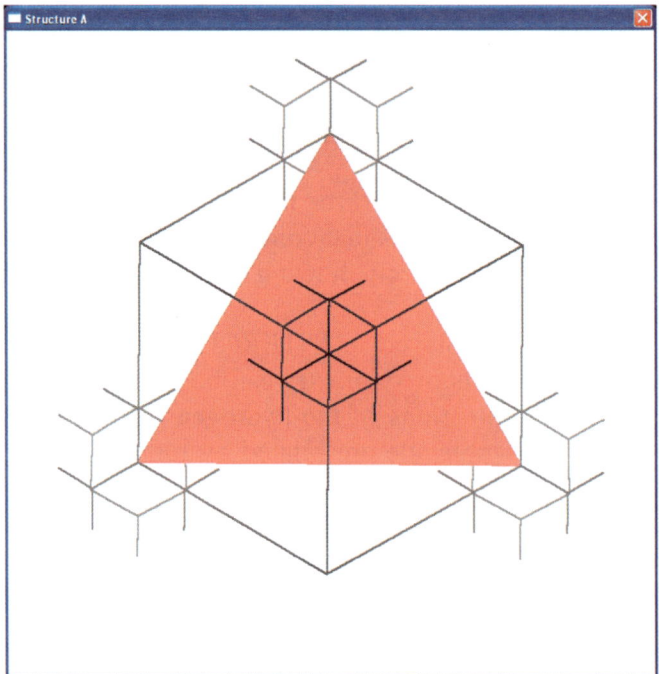

Figure 7.7 View of Miller plane (1 1 1) in a Unit Cell of Hexamine

To view a Miller plane again we use a facility in the Build Structure menu. This time select *Set Lattice Plane* (R10/C3) and enter the indices h, k and l for the plane (1 1 1) in the dialogue box that appears. Next click on *Unit Cell Displays* (R7/C9) and the plane, coloured red, should appear superimposed on the model of the unit cell including the cell edges. It is possible to select viewing directions for the crystal structure either perpendicular to this plane, by selecting the menu button *View Perp. to Plane* (R10/C4), or in the plane by selecting *View in the Plane* (R10/C5). The view looking along the normal to the (1 1 1) Miller plane in hexamine is shown in Figure 7.7.

7.3.4.4 Intermolecular hydrogen bonds in crystals

A further important aspect in modelling the solid state of molecular materials is the characterisation of the intermolecular interactions. The concept of molecular synthons and supramolecular crystal engineering is currently very fashionable in directing attempts to design and control molecular self-assembly in condensed phases[12, 13]. Supramolecular means simply an assembly or cluster of more than one molecule removed from a crystal lattice. The ultimate objective is to realise a genuinely molecule-up approach to the design of novel functional-materials. The basic concept is an extension of functional groups, as employed for the systematisation of synthetic organic chemistry, to non-covalent bond formation (otherwise known as inter-molecular interactions) observed in molecular solids. A discussion of the energetical aspects of this is left until later. However, here we touch on systematic approaches which have been applied to characterise the three-dimensional patterns of specific intermolecular interactions operating in the solid state. The majority of these interactions are hydrogen-bonding interactions of types (i) O-H...O, (ii) O-H...N, (iii) N-H...N and (iv) N-H...O.

Hydrogen bonding interactions can be identified by measuring the inter-atomic distance between the hydrogen-bond donor and acceptor atoms (e.g. in type (iv) the nitrogen atom is the donor atom and the oxygen is the acceptor atom). Measurements can be performed either from the main menu or Build Menu in INTERCHEM. Typically donor to acceptor atom distances are in a range from 2.5 to 3.0 Ångstroms. It is also useful to examine the size of the hydrogen-bond angle the ideal value being 180°.

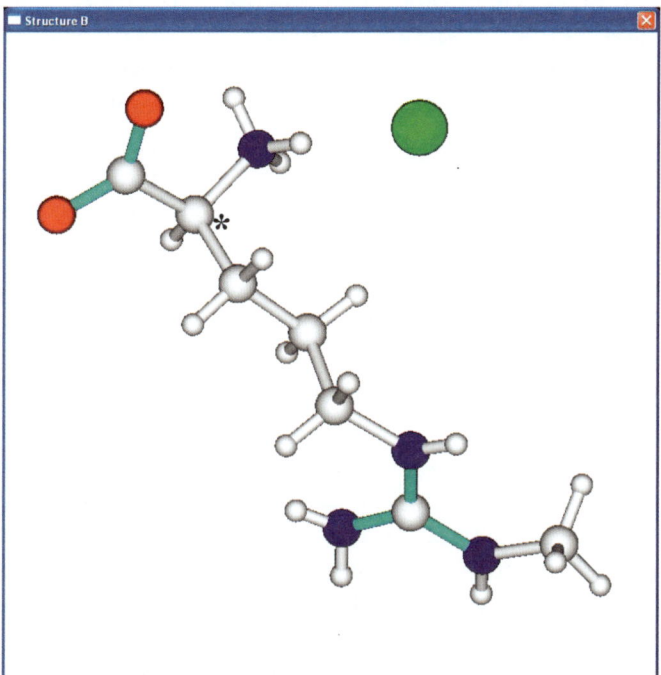

Figure 7.8 Molecular Structure of N^G-monomethyl-L-arginine Hydrochloride

Initially it is worthwhile to check the total numbers of hydrogen-bond donor atoms and acceptor atoms together with the total number of hydrogen atoms that can be donated in hydrogen bonds (usually considered to be those hydrogen atoms forming a covalent bond with either an oxygen or a nitrogen atom). From these numbers it is possible to assess whether it is plausible for a given molecule to employ all the atoms that have the potential to form hydrogen bonds in actually forming hydrogen bonds in a crystal structure. Even if the complete hydrogen-bonding potential could be satisfied from the point of view of a simple accounting exercise (matching donor and acceptor pairs), it may not be possible to realise this potential by creating a three-dimensional network of hydrogen bonds in a crystal structure. This is by reason of the constraints imposed by the particular shapes and sizes of the molecules that are being packed together to form a crystal. However, in terms of assessing the relative merits of one hydrogen-bonding arrangement over another, there is a tendency to favour configurations which more fully satisfy the hydrogen-bonding potential of the molecules. Consequently this concept can be harnessed in the so-called synthonic approach to designing the solid state.

7.3.4.5 Intermolecular hydrogen bonds: example of N^G-monomethyl-L-arginine hydrochloride polymorphs

Some of these concepts can be illustrated by examining two crystal structures of the material N^G-monomethyl-L-arginine Hydrochloride (L-NMMA.HCl) which is an example of a chloride salt which contains a molecular cation. The molecular structure is shown in Figure 7.8. This compound was synthesised for its possible application as a drug for the treatment of septic shock. L-NMMA.HCl exhibits polymorphism. This is the phenomenon where the same molecular species can adopt different packing arrangements to create distinct three-dimensional periodic structures. The crystal structures of two polymorphs are reported[14]: Form A is orthorhombic, space group $P2_12_12_1$, with four molecules per unit cell and Form D is monoclinic, space group $P2_1$, with two molecules per unit cell. In both polymorphs the asymmetric unit is a single formula unit of the salt (one molecular cation and one chloride anion). Note that the molecule is a single optical- enantiomer, (see Section 6.1) and so is associated with two space groups that have neither inversion symmetry nor mirror symmetry (point groups **222** and **2** for the orthorhombic and monoclinic polymorphs). The presence of either an inversion centre or a mirror plane would necessarily generate the mirror image of the molecule. Then the crystal structure would contain equal numbers of left and right handed molecules and be a racemate. In both crystal structures the hydrogen ion from the carboxylic acid group transfers to the alpha-amino group in the molecular ion.

In L-NMMA.HCl there are seven hydrogen atoms bonded to nitrogen atoms that can be accepted in hydrogen bonds. The four nitrogen atoms are the hydrogen-bond donor atoms they cannot act as acceptor atoms. There are three hydrogen-bond acceptor atoms: the two oxygen atoms and the chlorine atom. Both oxygen atoms are capable of accepting two hydrogen-bonds but this leaves a further three hydrogen atoms that could, potentially, be accepted in hydrogen bonds. The chlorine atom can also accept hydrogen-bonds. In both Form A and Form D the chlorine atom accepts three hydrogen bonds and both oxygen atoms accept two hydrogen bonds. Hence, the hydrogen bonding potential of the molecule is completely satisfied in both polymorphs. However, the exact pairings of the hydrogen-bond donor and acceptor atoms are different in Forms A and D. There are many possible combinations however, because the intermolecular hydrogen-bonds must generate a three-dimensional, periodic framework, only certain combinations of hydrogen-bond donor-acceptor pairs are compatible with the requirement to packing molecules to make a crystal.[15] Figure 7.9 shows the molecular packing in the polar, monoclinic Form D. The hydrogen bonds were added manually and are only shown where both the acceptor and donor atoms are present so called 'dangling bonds' are not shown.

Another interesting aspect of the two polymorphs of L-NMMA.HCl is that the conformation of the molecule in the asymmetric units is not the same. This relates to the relative positions of the N-methyl group in the two polymorphs and the fact that within the guanadinium group there is restricted rotation about the three carbon-nitrogen bonds due to their partial double-bond character. In Form A there is the E geometric isomer whereas in Form D there is the Z geometric isomer (see Section 6.3 for an explanation of geometric isomers). Structure files for the two polymorphs are provided so that the packing arrangements and intermolecular hydrogen bonds in the two polymorphs may be compared. The files containing the two crystal structures are provided on the software disc: SONSUK.cif (form A), SONSUK01.cif (form D)

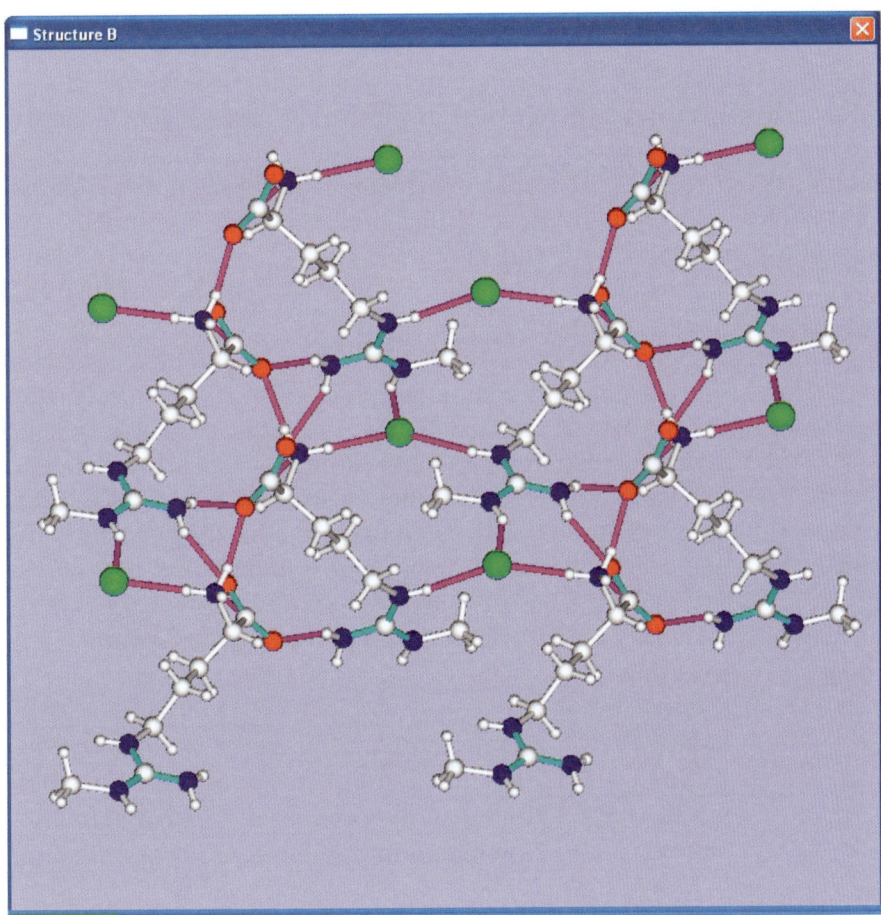

Figure 7.9 N^G-monomethyl-L-arginine Hydrochloride Form D Molecular Packing
View looking down the **a** axis (projection in the **b**, **c** plane) with a vertical alignment of the **b** axis; note the four carboxylic acid groups pointing towards the viewing direction and the four pointing away. Intermolecular hydrogen bonds are indicated with magenta cylinders. The dangling hydrogen bonds are not shown here.

7.3.4.6 Supramolecular engineering concepts applied to crystals

Over the last decade keen interest in the potential for exploiting cocrystals to improve the processability of APIs during secondary manufacturing, for example tablet formation, has been a major driver in promoting supramolecular-engineering approaches. Here the choice of the cocrystal former, CCF, which is the molecule that is combined with the API to form a cocrystal, can be directed by first principles approaches using molecular modelling. The categorisation of hydrogen bonding interactions becomes useful in a game of 'molecular Lego' in which the API molecule is matched up with various CCFs. The groups that can form hydrogen-bonding interactions in crystals of the separate API and CCF molecules are referred to as homo-synthons whereas the pairing of one group from the API with one group from the CCF defines a hetero-synthon. It is then possible to speculate whether certain homo-synthons or hetero-synthons are more or less likely to appear in the any cocrystal combining the API and CCF. This has led to drives to collect a significant library of crystal structures of particular APIs combined with very many CCFs. For instance recently, fifty crystal structures for cocrystals of carbamazepine have been analysed collectively.[16] The analysis seeks to infer rules that can allow the three-dimensional self-organisation of the two different types of molecules in crystals to be inferred directly from a knowledge of the synthons which are defined from the molecular structures.

7.3.5 Chirality and crystallography

In this section we touch briefly on the extent to which the assembly of chiral objects into three-dimensional crystal structures differs from the assembly of achiral objects. There are a number of aspects of molecular structure that can render a particular molecule chiral, taken to be a molecule than cannot be superimposed on its mirror image. Perhaps the most common aspect is the presence of an asymmetric centre for example, a carbon atom bonded to four different types of atom or groups of atoms. We have already discussed an example of a molecule with such an asymmetric centre, the N^G-monomethyl-L-arginine cation shown in Figure 7.8. The asymmetric centre is marked by an asterisk. Since the crystalline state contains one enantiomer only, the crystallographic space group must be enantiomorphous along with the corresponding point group (see Table 7.2). There are two symmetry elements which, when present, render a space group (point group) non-enantiomorphous, the mirror plane and the inversion centre. A truth table, Table 7.3, can be constructed that allows space groups and the corresponding point group to be allocated under the headings: enantiomorphous/ non-enantiomorphous and centrosymmetric/ non-centrosymmetric.

Table 7.3 Truth Table for Categorisation of Space Groups

Mirror Plane	Inversion Centre	Space Group is
No	No	Enantiomorphous & Non-centrosymmetric
Yes	No	Non-enantiomorphous & Non-centrosymmetric
No	Yes	Non-enantiomorphous & Centrosymmetric
Yes	Yes	Non-enantiomorphous & Centrosymmetric

There are twenty-one non-centrosymmetric point groups, eleven are enantiomorphous and ten are non-enantiomorphous. There is a further categorisation of the non-centrosymmetric point groups into ten polar and eleven non-polar point groups. Polar point groups possess a unique crystallographic-axis which is not related by symmetry to the other axes. There are five enantiomorphous and five non-enantiomorphous polar point-groups (see Table 7.2). In the case of the two polymorphs of L-NMMA.HCl only the monoclinic form is polar (point group **2**) the orthorhombic form (point group **222**) is merely enantiomorphous.

7.3.5.1 Benzophenone: enantiomers in the crystalline state by virtue of molecular conformation

The conformation that a molecule adopts can also render it an enantiomeric object. In the gaseous and liquid or solution states molecules can usually transform rapidly between different conformations whereas in solids this may not be possible due to the packing forces. A molecule of benzophenone is shown in Figure 7.10. The benzophenone molecule does not adopt a planar conformation in the two polymorphs which have been reported.[17] In Figure 7.10 the direction of rotation about the carbonyl-carbon to phenyl-carbon bonds is indicated by the blue, circular arrows. As a result of

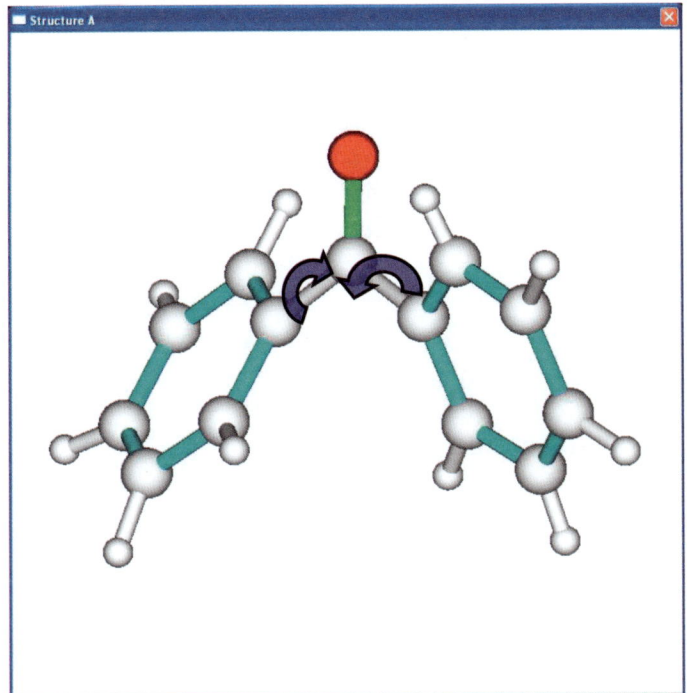

Figure 7.10: Molecular Structure of Benzophenone
Arrows indicate the direction of rotation of the phenyl groups away from co-planarity

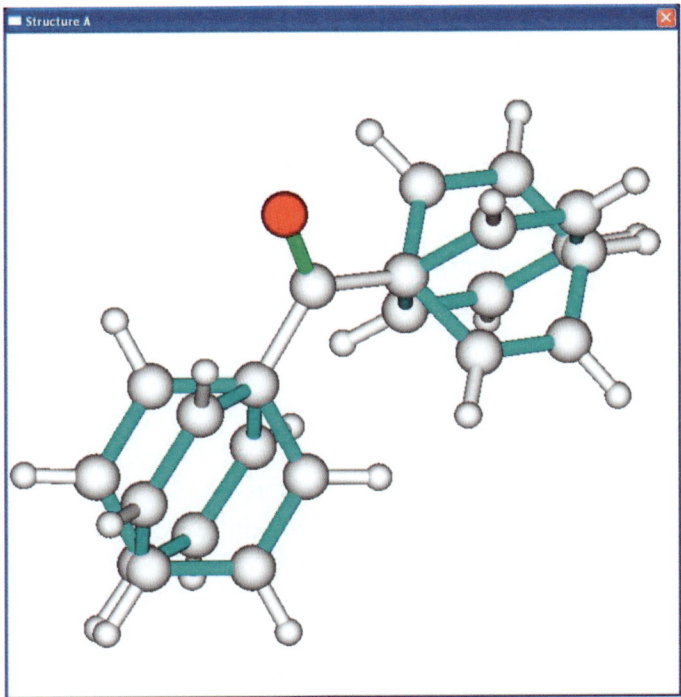

Figure 7.11 Overlay of Two Molecules of Benzophenone taken from the Unit Cell of the Metastable Polymorph.
Within the unit cell these molecules are related via inversion symmetry: this demonstrates that the molecule cannot be superimposed on its mirror image, it is enantiomeric.

the twist of the phenyl groups, when the molecules are packed together in crystals and can no longer change their shape, the molecule becomes a chiral object.

Benzophenone is of interest historically as it is one of the first substances to have been identified as polymorphic. The stable, or α-form, crystalline phase of benzophenone is orthorhombic in space group $P2_12_12_1$ which is an enantiomorphous space group. This indicates, since the benzophenone molecules are chiral, that the stable form is comprised of a single enantiomer i.e. all the molecules are identical. Hence crystals of the stable polymorph contain either all left handed or all right handed molecules. The metastable or β-form is monoclinic in space group C2/c which is non-enantiomorphous and centrosymmetric. Similarly, this indicates that both enantiomers of the benzophenone molecule are present in the metastable crystal. Hence in the metastable form, unlike the stable form, one half of the molecules should not be superimposable on the other half. INTERCHEM can be used to confirm this.

7.3.5.2 Exploring the crystal structures of benzophenone
Click *Select A* (R1/C1) and *Load Structure* (R2/C7) and select CIF format for input and load the metastable structure of benzophenone from the .CIF format file provided selecting *Asymmetric Unit Only*. Click *Centre Structure* (R3/C5) so that the molecule of benzophenone is displayed clearly. Now Click *Select B* (R1/C2) and load the same structure again but this time choose the option to *Form One Unit Cell*. Eight molecules of benzophenone should be displayed in model window B.

Select *Build Structure* (R9/C1) and *Copy B to Base* (R1/C2). Now display as a *Wire Frame* (R7/C1) with *Atom Numbers* (R8/C2) and check the sequence in which the molecules are numbered. Use *Delete Segment* (R5/C6) to remove all the molecules other than the molecule numbered second in sequence. Next click *Copy Base to B* (R2/C2) and *Exit Building* (R8/C10). Using the wire frame display mode and atom numbers manipulate the molecules in the Structure A and Structure B windows separately until they have, apparently, a similar orientation. Now use *25-Point fit B on A* (R6/C3) to check whether the molecules can be exactly superimposed by selecting pairs of corresponding atoms. It is necessary only to select three pairs of atoms (1, 1), (2, 2) and (3, 3). Repeat the entire procedure for the molecule numbered third in sequence. You should find that in one case the pair of molecules is exactly superimposed, and appears as a single molecule, and in the other case that the superimposed molecules appear as in Figure 7.11.

7.4 Origin of Cohesive Forces in Solids
We start this discussion of cohesive forces in solids by considering a purely ionic material composed of spherical ions, for example sodium chloride. Although real ions clearly occupy a finite volume, in certain instances the charge density around an ion has a spherical distribution and then the total charge on the ion can be treated as lying at a single point at the centre of the spherical distribution of charge density. Taking two such ions, i and j, with charges q_i and q_j having a centre-to-centre separation distance of r_{ij} the electrostatic force F_{ij} between the ions is given by Equation (7.3) Coulomb's Law.

$$F_{ij} = \frac{q_i q_j}{4\pi\varepsilon_0 r_{ij}^2} \qquad (7.3)$$

The potential energy, U_{ij}, of one ion in the presence of the other is given by Equation (7.4).

$$U_{ij} = \frac{q_i q_j}{4\pi\varepsilon_0 r_{ij}} \tag{7.4}$$

Since potential energy is an additive scalar-quantity, it is possible to perform a summation and calculate the potential energy possessed by any given ion in the presence of all the surrounding ions. This leads to an important concept for evaluating the internal energy of crystalline solids. We select a representative, central ion, i, within the crystalline structure under consideration, and evaluate where all the neighbouring ions are positioned with respect to the central ion. In principle this gives an infinite number of separation distances r_{ij} and associated charges q_j. The summation process to evaluate the over-all potential energy is expressed in Equation 7.5.

$$U = \sum_{j, j \neq i}^{\infty} \frac{q_i q_j}{4\pi\varepsilon_0 r_{ij}} \tag{7.5}$$

If the summation converges at some finite value of j then Equation (7.5) provides the means to evaluate the electrostatic potential-energy. However, this is where due care and attention is required from the practical point of view of a molecular modeller. You will note that in our example of sodium chloride, if we attempt to evaluate the potential energy by sitting on a central sodium ion and evaluating the separation distances to all the other ions in the limit, we should count an equal number of positively charged sodium and negatively charged chloride ions. This means that Equation (7.5) is composed of an infinite series of terms of alternating sign. Mathematically such a series is known to be *conditionally convergent*. The mathematician Bernhard Riemann (1826-1866) demonstrated that a conditionally convergent series can be made to converge to any value or to diverge depending on the order in which the terms in the series are evaluated.

The German physicist Erwin Madelung[18] (1881-1972) was first to explore such sums for simple ionic substances such as sodium chloride for which the crystal structure is illustrated in Figure 7.12. If the summation is performed in spherical shells centred on a central sodium ion then the first shell contains six nearest neighbours which are chloride ions at a separation distance, **a**, where **a** is the distance of closest approach between the centres of two unlike ions in a crystal of sodium chloride. The distance **a** is exactly half the edge length of the unit cell of sodium chloride.

Similarly, there are twelve second nearest neighbours, sodium ions, at a separation distance of √2a and eight third nearest neighbours, chloride ions, at a separation distance of √3a. If we denote the potential energy of the first three terms, in order of increasing separation distance, in the sum given by Equation 7.5 as $U_{11'}$, U_{11} and $U_{12'}$ respectively then the sum of these terms is given by Equation 7.6. Note that the second term in brackets on the right hand side of Equation 7.4 is a group of constants, the charge on an ion, q, the permittivity of free space ε_0, and the half lattice spacing, **a**. The first term in brackets depends on the lattice geometry and consists of the first three terms in a series of terms of alternating sign.

$$U_{11'} + U_{11} + U_{12'} = \left(-\frac{6}{\sqrt{1}} + \frac{12}{\sqrt{2}} - \frac{8}{\sqrt{3}}\right)\left(\frac{q^2}{4\pi\varepsilon_0 a}\right) \tag{7.6}$$

If the terms in the series are added together in the order prescribed then the infinite series converges to a value known as the Madelung constant M (an irrational number).

The value of this constant for sodium chloride is found to be 1.74756 to six significant figures. Given the value of the Madelung constant the electrostatic potential energy of a mole of sodium ions in the sodium chloride crystal lattice can be calculated according to Equation (7.7) where L denotes Avagadro's number and e the charge on an electron.

$$U_{electrostatic} = \frac{-LMe^2}{4\pi\varepsilon_0 a} \qquad (7.7)$$

Given that the unit-cell edge length in sodium chloride is 5.64 Angstroms it is left to the reader to confirm that the potential energy of the sodium ions (and therefore necessarily the chloride ions) due to the electrostatic force is -861 kJ/mol. However, even for such a simple structure a purely electrostatic description of the cohesive energy, such as the one described above, is incomplete. The reason for this can soon be understood if one considers the net force acting on the nearest neighbour sodium and chloride ions.

With the current model there is a net attractive force which should act to reduce the separation distance in other words, the model predicts that the oppositely charged ions should coalesce. What is missing is a description of the repulsive force at short separation distances due to the overlap of the electron density associated with the different ions. Hence a complete description of the potential energy includes a further term to represent the potential energy, $U_{repulsion}$, associated with these short range repulsions. The mathematical function most frequently adopted to describe short range repulsions in ionic materials is the Buckingham potential. The second term on the right hand side of Equation (7.8) is the Buckingham potential which has two system specific constants here denoted as A and B.

$$U_{total} = U_{electrostatic} + U_{repulsion} = \frac{-LMe^2}{4\pi\varepsilon_0 a} + A\exp\left(\frac{-a}{B}\right) \qquad (7.8)$$

Often for molecular crystals, short-range repulsion between two atoms i and j separated by a distance r_{ij} is described by the Lennard Jones 6-12 potential in which the repulsion term has the form given in Equation (7.9), where A_{ij} is a constant that depends on the chemical identity of the two atoms. Methods of deriving parameters for atom-atom potential energy functions are discussed later.

$$U_{repulsion} = \frac{A_{ij}}{r_{ij}^{12}} \qquad (7.9)$$

We know that even in the total absence of any net charge on two atoms there is still a force of attraction. This is the dispersion force and to understand its origins a quantum mechanical description is required however empirically, the potential of this force, U_{disp}, has a simple mathematical form. The potential energy of two atoms i and j separated by a distance r_{ij} due to the dispersion force is given by Equation (7.10) where B_{ij} is a constant that depends on the chemical identity of the two atoms.

$$U_{disp} = -\frac{B_{ij}}{r_{ij}^6} \qquad (7.10)$$

This brief discussion has prepared the groundwork for understanding how lattice energies are calculated as described in detail in Section 7.6.1.

Molecular Modelling and the Solid State of Materials

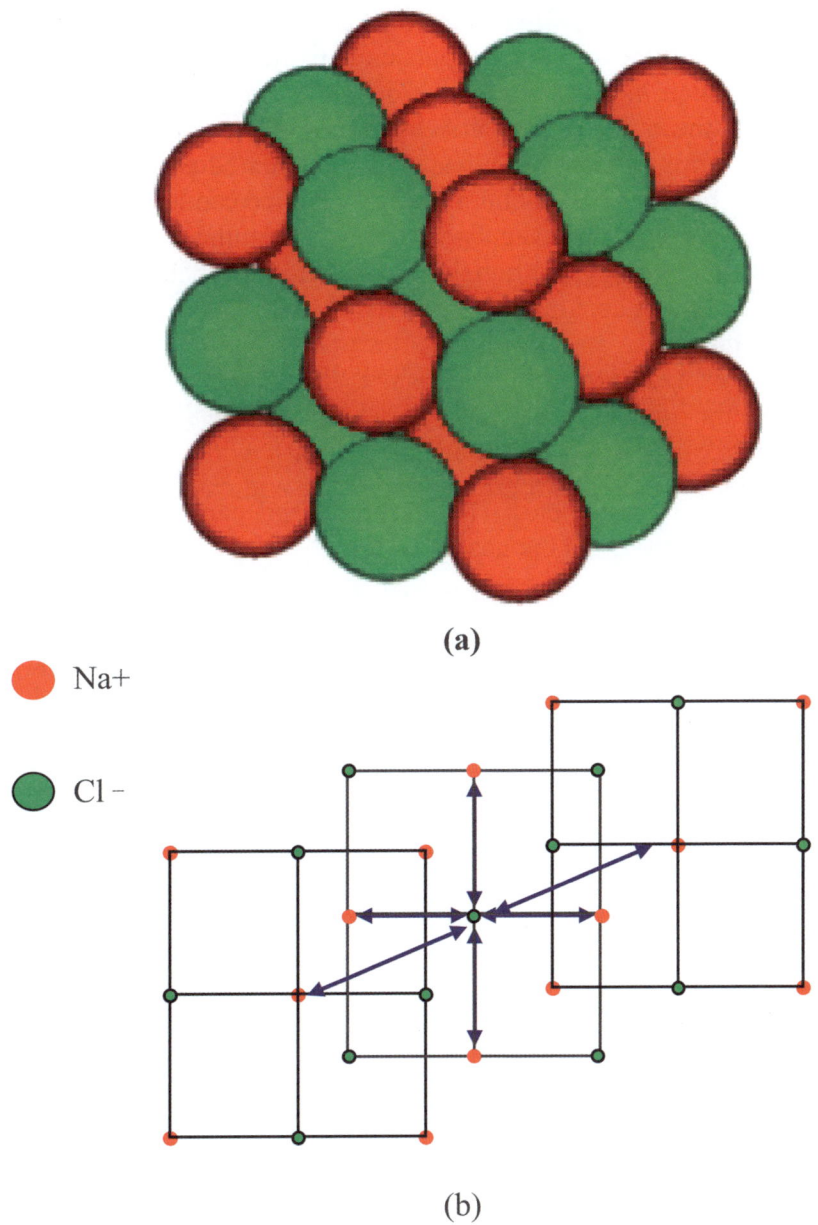

Figure 7.12
Model of Crystalline Sodium Chloride

Notes
(1) The top picture (a) shows one Unit Cell of a crystal of Sodium Chloride
(2) Red spheres represent sodium ions (Na+),
(3) Green spheres represent chloride ions (Cl−)
(4) The space group (No. 225) is Cubic ($\alpha = \beta = \gamma = 90°$, $a = b = c = 5.64$ Å).
(5) In the bottom picture (b), the blue arrows show the interactions of the central chloride ion with the six nearest neighbouring sodium ions.

7.5 Thermodynamics of crystalline solids and molecular modelling

If we consider a crystalline solid having all its constituent ions, atoms or molecules at rest and centred on their mean relative positions in space, it is possible to build a model of the potential of mean force and, thereby, to calculate the potential energy of the configuration. In terms of classical physics, the calculated potential energy represents the internal energy of a perfect crystal at the absolute zero of temperature. The energy calculated in this way does not take into account quantum mechanical effects such as zero-point vibrational energy (although a correction may be made for this if the calculation is of sufficient accuracy to merit it). Nevertheless, it is useful to be able to compare how strongly ions, atoms or molecules in different crystalline arrangements are bound together. For example, polymorphism is the manifestation of multiple packing arrangements of the same molecular species to create distinct three-dimensional periodic structures. It is often important to establish which of the polymorphs of a material is more (or most) stable thermodynamically and calculated lattice energies can provide a useful guide here.

As a starting point it is important to establish a working definition of lattice energy that can be employed in molecular modelling. The lattice energy can, to a first approximation, be regarded as the change in potential energy when the constituent ions, atoms or molecules of the crystal lattice, initially at rest and separated by infinite distances, are brought together to their respective locations at rest in the crystal lattice.

Notice the lattice energy, E_{Latt}, is actually an energy change or difference, ΔE, accompanying a transformation between two hypothetical states conducted reversibly at the absolute zero of temperature. Although both the initial state, a perfect gas at absolute zero, and the final state, a perfect crystal at absolute zero, cannot be realised exactly in any practical experiment, conceptually they are quite easy to understand as points of reference.

The real physical process to which this energy change most closely corresponds is sublimation, the formation of a vapour directly from a solid. Accordingly lattice energies calculated using molecular modelling approaches, E_{Latt}, are often compared with experimentally determined heats of sublimation, ΔH_{Sub}. To make a comparison between a calculated lattice energy and experimentally determined heat of sublimation at a finite temperature, T, for a material, a correction must be made for the relative change in enthalpy due to the difference between the heat capacities of the solid and gaseous states.[19] Hence, the correction is given by Equation (7.11).[20] The calculated lattice energy and sublimation enthalpy measured at temperature T are related by Equation (7.12).

$$\Delta H_{Sub}^{T=T} = \Delta H_{Sub}^{T=0} + \int_0^T (C_P^{gas} - C_P^{solid}) dT \qquad (7.11)$$
$$\text{where } C_P^{gas} - C_P^{solid} = (6.\tfrac{1}{2}R + R) - 6.R = -2R$$

$$\Delta H_{Sub}^{T=T} = -E_{Latt} - 2RT \text{ since } \Delta H_{Sub}^{T=0} = -E_{Latt} \qquad (7.12)$$

For some materials the heat capacities, at constant pressure, of the solid and gaseous states are known from experiment as a function of temperature. These experimental data allow the heat of sublimation at any temperature to be calculated directly.

The types of energy calculations we describe in this chapter are linked directly to changes in enthalpy (internal energy) for processes of interest e.g. sublimation. However, the conditions for equilibrium (temperature and pressure) and coexistence between different phases of a material are determined by changes in free energy. Along side free energy we have to consider entropy. The relationship between a change in free energy, ΔG, enthalpy, ΔH, and entropy, ΔS, at an absolute temperature T is given by equation 7.13.

$$\Delta G = \Delta H - T\Delta S \tag{7.13}$$

During a change of phase in a pure material, for example from solid to liquid, there is a precise temperature, for a specified pressure, at which the free energies of the solid and liquid states per mole are the same. This temperature is the melting or fusion temperature, T_f. Since $\Delta G_f = 0$ at T_f, the entropy of fusion, ΔS_f, is related to the enthalpy of fusion, ΔH_f, as $\Delta S_f = \Delta H_f / T_f$. To determine the value of T_f through molecular modelling we must calculate the free energies of the solid and liquid states of a material as a function of temperature. Tackling the problem requires us to consider the dynamics of molecules and this is beyond the scope of this chapter.

When comparing the stability of different solid forms, polymorphs, calculated lattice energies predict the relative stability when the absolute temperature is zero. For many pairs of polymorphs, their relative stability turns out to be independent of temperature. Such pairs are termed monotropes and the polymorphs have a monotropic relationship. Should there be a temperature at which the order of stability is reversed, the polymorphs are termed enantiotropes, they have an enantiotropic relationship. Again, predicting the temperature of transition between enantiotropes requires us to calculate the free energies of the polymorphs. However, doing so for solid phases is, perhaps, more straight forward than for liquids so this is a more tractable problem but, unfortunately, not one for which a push-button solution is readily available at this time. Now we consider practical methods for evaluating a lattice energy given the crystal structure is fully defined.

7.6. Lattice Energy Calculations
The information required about a crystal lattice to enable the calculation of the lattice energy is as follows.

- Unit cell parameters and space group,
- fractional coordinates of all the atoms in the crystallographic asymmetric unit,
- point atomic-charges and potential force-field parameters for the constituent atoms.

Given the connection between the enthalpy of sublimation and the lattice energy of a material, the identities of the species in the gas phase which result from the dissociation of the crystalline lattice must be specified clearly. Let us take a couple of examples. Firstly, sodium sulfate Na_2SO_4 here, typically, the crystal lattice is considered to dissociate into sodium cations and sulfate anions during sublimation but note that the covalent bonds between the sulfur atom and four oxygen atoms of the sulfate anions are regarded as remaining unchanged. Therefore in evaluating the lattice energy of sodium sulfate the sulfate ions are regarded as molecular units which persist in the gas phase. Secondly glycine $CH_2(NH_2)COOH$, this is an alpha amino acid and in the crystalline state the molecules are in the zwitterionic form (created via the transfer of a proton from the carboxylic acid moiety to the nitrogen atom of the alpha-amino moiety).

Hence there are two options for specifying the molecular species in the gas phase into which the glycine crystal-lattice is considered to dissociate: either the zwitterionic form or the non-zwitterionic form. Crucially, the value of the sublimation enthalpy (and by implication the lattice energy) depends on this choice differing by an amount equivalent to the enthalpy change associated with the proton transfer in the gas phase.

7.6.1 Worked example of lattice energy calculation for Sodium Sulfate

When molecular units are present in the crystal lattice it is necessary to differentiate between inter-molecular contributions and intra-molecular contributions to the energy. The intra-molecular energy contributions result from interactions between the constituent atoms of a single molecule. The inter-molecular energy contributions result from interactions between atoms in different molecules. For sodium sulfate the 'intermolecular' contributions to the energy result from interactions (i) between the atoms in different sulfate anions, (ii) between the atoms in sulfate anions and sodium cations and (iii) between sodium cations. So a connection between the method for evaluating the potential energy of a mole of sodium cations in crystalline sodium chloride, described in Section 7.4, and evaluating the lattice energy of crystalline sodium sulfate is starting to emerge.

To evaluate the lattice energy of sodium sulfate, the energy summation for the sodium cations in the sodium sulfate crystal lattice is performed in a completely analogous way to the energy summation for sodium cations in sodium chloride. However, for the terms in the energy summation associated with the sulfate anion, a separate pair-wise summation is performed for each of the five atoms of a single representative sulfate anion, interacting with all the neighbouring atoms (ions), but excluding pair-wise terms involving the other four atoms within the same sulfate ion. The exclusion of atoms within the same molecular unit from the pair-wise energy summation is the key difference when evaluating the lattice energy of molecular crystals compared with crystals such as sodium chloride. Commonly the lattice energy of sodium sulfate is expressed on the basis of the sublimation of one mole of formula units, Na_2SO_4, into two moles of sodium cations and one mole of sulfate anions.

7.6.1.1. Lattice energy calculation: obtaining required structural information

So the procedure for evaluating the lattice energy of sodium sulfate starts with a search of the Inorganic Crystal Structure Database (ICSD) to find the unit cell parameters, space group and fractional coordinates of the asymmetric unit. In fact sodium sulfate is a polymorphic material, and five different anhydrous polymorphs have been identified.[21] The forms commonly encountered at ambient temperature are Forms III and V (the latter form being the mineral thenardite). Hence the particular polymorph of interest must be identified. The second step is to use the crystallographic information to construct a model of the crystal lattice of sodium sulfate using, for example INTERCHEM. The distances between atoms in the crystal can be evaluated directly from the model of the crystal lattice. This leaves the question of the appropriate potential energy terms to employ in the pair-wise energy summation. One published atomistic potential for sodium sulfate[21] employs a Buckingham potential combined with dispersion energy and coulombic energy terms as given in Equation (7.14).

$$E_{lattice} = \frac{1}{2}\sum_{k=1}^{N}\sum_{j=1}^{n_2}\sum_{i=1}^{n_1} A_{ij}\exp\left(\frac{-r_{ij}}{B_{ij}}\right) - \frac{C_{ij}}{r_{ij}^6} + \frac{q_i q_j}{r_{ij}} \qquad (7.14)$$

Of the three summation terms in Equation (7.14), the innermost summation is over the n_1 atoms belonging to the 'molecule' taken as being central, the next summation is over the n_2 atoms in an adjacent molecule within the crystal lattice, the outermost summation is over N such pairs of molecules. The size of N is usually determined by imposing a maximum inter-atomic distance, r_{ij}, for the summation process. In practice the magnitude of this maximum inter-atomic distance depends on the rate of convergence of the lattice-energy sum with increasing separation distance. Whereas the first and second terms on the right hand side of Equation (7.14) rapidly become small with increasing r_{ij}, the coulombic energy term often makes a significant contribution to the energy to much greater separation distances. For organic, molecular materials, where the magnitudes of the partial atomic charges are usually less than for ionic materials, a typical limiting inter-atomic distance would be in the range of 20 to 30 Angstroms.

7.6.1.2. Lattice energy calculation: understanding the significance of the crystallography

Thenardite, which is selected to provide a fully worked-out example, crystallises with a face-centred orthorhombic lattice type, the space group is number 70 with the symbol **Fddd** (the letters **d** indicate diamond glides). The specification of the crystal structure[22] from the ICSD indicates that there are three independent atoms in the crystallographic asymmetric unit one each of sodium, sulfur and oxygen. The cell parameters are a = 5.858 Å, b = 12.299 Å, c = 9.814 Å and the unit cell volume is 707.08 Å^3. The fractional coordinates of the atoms in the asymmetric unit (as specified in a file in CIF format) are: Na 0.12500, 0.12500, 0.44180; S 0.12500, 0.12500, 0.12500; O 0.98070, 0.05820, 0.21480.

If the International Tables for Crystallography were available for the reader to consult, the entry for space group 70 in Volume A would make it possible to work out from the fractional coordinates that the oxygen atom is on a general position in the unit cell. But in the absence of these tables there may still be sufficient information contained in the CIF file. The symmetry operations needed to construct the unit cell are listed there and in this case there are 32 operations. From this we could deduce the number of general positions in the unit cell. If all 32 operations can be applied to the coordinates of the oxygen atom without, in the process, duplicating any coordinates then the oxygen is on a general position. Even if this is not the case the molecular modelling software should take care of the issue by removing duplicate atoms.

There are 32 general positions for the space group **Fddd** hence it can be concluded that there are 32 oxygen atoms in the unit cell and therefore 8 sulfate anions per unit cell. Simply to satisfy the stoichiometry of the compound, there must be 8 sulfur atoms and 16 sodium atoms per unit cell in addition to the 32 oxygen atoms. This implies that the sulfur atoms are on special positions in the unit cell with a site multiplicity of 8 (rather than a site multiplicity of 32 for a general position). Similarly the sodium atoms are on special positions with a site multiplicity of 16.[23]

The symmetry operations associated with space group 70 (for the second choice of unit cell origin) can be represented as follows:

(1) x, y, z (2) $\bar{x}$ + ¾, $\bar{y}$ + ¾, z (3) $\bar{x}$ + ¾, y, $\bar{z}$ + ¾ (4) x, $\bar{y}$ + ¾, $\bar{z}$ + ¾

(5) $\bar{x}$, $\bar{y}$, $\bar{z}$ (6) x + ¼, y + ¼, $\bar{z}$ (7) x + ¼, $\bar{y}$, z + ¼, (8) $\bar{x}$, y + ¼, z + ¼

A bar (macron) over a letter indicates that the negative value of the number represented by that letter should be taken. Hence if x y z represent the fractional coordinates of an atom in the asymmetric unit, then symmetry operation (1) leaves the fractional coordinates unchanged (the identity symmetry operation), symmetry operator (5) specifies that negating the values x y z produces the fractional coordinates of a symmetry copy of the atom. Similarly symmetry operator (2) specifies negating the value of the x fractional coordinate and adding ¾ to this, similarly for the y fractional coordinate, and leaving the z fractional coordinate unchanged produces the fractional coordinates of another symmetry copy of the atom.

So far only eight of the thirty-two general positions in the unit cell have been specified. The presence of additional lattice points, due to the face centring of the lattice, requires further translations to be considered. The International Tables show these additional translations as (0, 0, 0)+, (0, ½, ½)+, (½, 0, ½)+ and (½, ½, 0)+. These coordinates indicate the positions of lattice points. Hence, (0, 0, 0) is always present for any lattice type (primitive, body centred or face centred) and face centring adds lattice points at, (0, ½, ½), (½, 0, ½) and (½, ½, 0). So to generate the fractional coordinates of the second set of eight general positions, we take symmetry operations (1) to (8) and add ½ to the y coordinate and ½ to the z coordinate in every case. To generate the fractional coordinates of the third set of eight general positions, we take symmetry operations (1) to (8) and add ½ to the x coordinate and ½ to the z coordinate in every case and similarly for the fourth set producing the thirty-two general positions.

Although, strictly speaking, there are just three unique atoms in the asymmetric unit for thenardite, this is not the most convenient way to think about the crystal structure from the point of view of molecular modelling. If, for example, we want to calculate the lattice energy of the crystal structure it is preferable to maintain the sulfate ions as molecular units. Hence, it is better to specify the fractional coordinates of one sulfur and four oxygen atoms for one of the sulfate anions in the unit cell.[24] A structure file to represent the thenardite in .xr format is shown in Figure 7.13. Note a .xr format file containing the coordinates of all 56 atoms in one unit cell is also provided for the convenience of displaying the crystal structure of thenardite with INTERCHEM. Two styles of representation of the crystal structure of thenardite are illustrated in Figure 7.14: on the left, balls and cylinders are adopted, on the right a space-filling model where the ionic radii have been used to scale the spheres representing the atoms.

For the lattice energy summation to converge to a constant value, the sulfate anions and sodium cations in all the unit cells that surround the 'central' unit-cell must also be included in the energy summation.[25] It is apparent, therefore, that a manual calculation of all the pair-wise terms quickly becomes impracticable. Molecular modelling software packages often provide a facility for automatically evaluating all the terms in such an energy summation and hence for determining a lattice energy.

7.6.1.3. Lattice energy calculations: selecting a potential for the calculation

Returning to Equation 7.14, we have now looked in detail at how a model of the crystal structure of thenardite is constructed that allows all the relevant inter-atomic distances r_{ij} to be calculated. The set of point atomic-charges, q_i etc, used in a lattice energy calculation is an integral component of the atomistic force-field. Appropriate atomic charges should be assigned as part of the process of deriving the other parameters in the potential for example in the present case A_{ij} and B_{ij} associated with the Buckingham potential terms and C_{ij} associated with the dispersion potential term. The double indices for these potential parameters indicate that there is a separate parameter value for each unique combination of atom types.

Hence there are parameters defined for interactions of an oxygen atom with (i) another oxygen atom, (i) a sulfur atom and (iii) a sodium atom. All the oxygen atoms in a sulfate ion are taken as being equivalent in terms of the potential energy parameters. In the current example the values of the parameters are zero apart from two cases: interactions between sodium and oxygen atoms and between oxygen atoms in different sulfate anions. The values of the parameters and the atomic point charges, taken from a published potential for sodium sulfate,[21] are given in Table 7.4.

7.6.1.4. Lattice energy calculations: using an Excel workbook for the calculation

Having defined the parameters and point charges of the potential to be employed, the lattice energy summation can now be carried out. To demonstrate a practical method for carrying out the calculation a Microsoft Excel Workbook complete with a short Visual Basic macro is provided (`lattice_energy.xls`), see Figures 7.15 and 7.16. The fractional coordinates of all the atoms in one unit cell of sodium sulfate are entered in columns C to E of the worksheet. Also entered are the atom number, column A, atom name, column B, atomic charge, column F, and atom-type number, column G. For the atom type numbering, one is for oxygen atoms, two is for sulfur atoms and three is for sodium atoms. The cell edge parameters a, b and c are entered in Cells H3, I3 and J3 respectively. The atoms are grouped as eight formula units containing one sulfate and two sodium ions. (In the forgoing and following descriptions, the word 'cell' is used in two contexts: when used with an initial capital letter 'Cell' refers to the Excel spreadsheet, when used without an initial capital letter 'cell' refers to the crystallography context).

The number of unit cells constructed in the lattice energy calculation can be varied by changing the parameter in cell K5. For example if a value 1 is entered in cell K5 a nest of 27 unit cells is used (step one unit cell in both positive and negative sense along x, y and z Cartesian directions). To execute the calculation a Visual Basic procedure is run through the Macro dialogue box illustrated in Figure 7.15. The results of the calculation appear in a table on the worksheet. The interested reader can explore the effect of changing the number of unit cells on the value of the calculated lattice energy. When the lattice energy no longer appears to change with an increase in the number of cells the lattice energy sum is said to show convergence. Again for the interested reader the Visual Basic code can be displayed which reveals the details of the calculation. The calculated lattice energy is -1918 kJ/ mol of formula units of Na_2SO_4. This calculation assumes that the internal energy of the sulfate ions does not change when the sulfate passes from the crystal to the gas phase. The calculated lattice-energy is similar to two values reported in the literature for sodium sulfate of -1827 kJ/mol[26] and -1856 kJ/mol[27] respectively.

```
                       A,B,C =    5.858    12.299    9.814
   ALPHA,BETA,GAMMA =  90.000   90.000   90.000      SPGR =    70
    6    0

    1  S1      0.12500    0.62500    0.62500    2    3    4    5    0    0    0    0    0.000
    2  O1     -0.01930    0.55820    0.71480    1    0    0    0    0    0    0    0    0.000
    3  O1A    -0.01930    0.69180    0.53520    1    0    0    0    0    0    0    0    0.000
    4  O1B     0.26930    0.55820    0.53520    1    0    0    0    0    0    0    0    0.000
    5  O1C     0.26930    0.69180    0.71480    1    0    0    0    0    0    0    0    0.000
    6  NA1     0.12500    0.62500    0.94180    0    0    0    0    0    0    0    0    0.000
    7  NA1A    0.12500    0.62500    0.30820    0    0    0    0    0    0    0    0    0.000
```

Figure 7.13 Format of Crystallographic .XR File Describing a Complete Sulfate Anion

For the full specification of the format see Section 8.7.1 and Figure 8.1 in Chapter 8

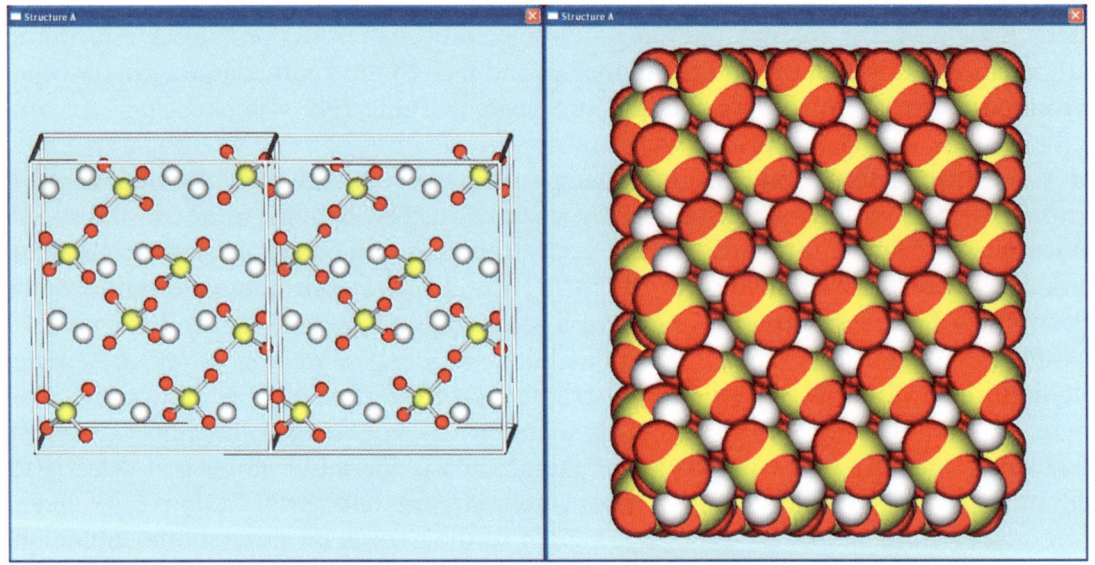

Figure 7.14 Crystal Structure of Sodium Sulfate Phase V (Thenardite).
Left two unit cells viewed down the crystallographic b axis, right similar view showing space filling (ionic radii used for sphere radii).

Table 7.4 Atomic Point Charges and Potential Parameters for Sodium Sulfate

Atom type or interaction	Atomic point charge	A parameter [eV]	B parameter [Å]	C parameter [eV Å^6]
Na	+1.0	-	-	-
S	+2.180	-	-	-
O	-1.045	-	-	-
Na-O	-	716.3	0.2955	0.0
O-O	-	16372.0	0.2130	3.47

7.6.1.5. Lattice energy calculation: application of intramolecular energy terms

Note an underlying assumption of the calculation is that the molecular geometry of the sulfate anion (sulfur-oxygen covalent bond length and oxygen-sulfur-oxygen bond angle) is the same in both the crystal and the gas phase. To account for any difference in energy, due to a change in the geometry of the sulfate ions on going from the crystalline state to the gaseous state, it is necessary to use additional mathematical functions to describe the intramolecular energy of a sulfate ion. For example a harmonic potential may be used to describe the increase in potential energy associated with a decrease or increase in the length of a sulfur-oxygen bond with respect to its equilibrium value. Similarly a harmonic potential can be used to describe an energy penalty associated with deviations of the oxygen-sulfur-oxygen bond angle from the value of 109.5° expected for a perfect tetrahedron. In the published potential, described above, a Morse potential is used to describe the sulfur-oxygen covalent bonds. The form of the Morse potential is shown in Figure 7.16.

Load in INTERCHEM the XR format structure file provided, named `sodium_sulfateX.DAT`, for sodium sulfate phase V and select to display only the asymmetric unit. Now measure the length of the sulfur-oxygen covalent bonds. You will discover that the length of these bonds is 1.472 Å this is in comparison to a reference, equilibrium bond length of 1.600 Å adopted in the Morse potential. Now measure the six oxygen-sulfur-oxygen bond angles. You will discover that the angles divide into three groups of two depending on bond angle. The values of the bond angle in each pair are 112.14°, 109.89° and 106.43° respectively compared with an equilibrium bond angle of 109.5 ° adopted in the harmonic potential.

An additional Visual Basic procedure is provided to calculate the increase in potential energy of the sulfate ion in the crystal compared with the reference state (gas phase) due to changes to the sulfur-oxygen bond-length and oxygen-sulfur-oxygen bond angles. The procedure is called *Sulfate_Ion_Energy*. The procedure populates the Cells of the table shown in Figure 7.16 with the calculated values. It can be seen that the deviations in the bond angles from the reference angle of 109.5° leads to a relatively small increase in energy. Contrast this with the deviation in the bond length from the reference value which results in an energy increase equivalent to roughly 15% of the bond dissociation energy. However, the Morse potential is 'hard-walled' for compression of bonds, (the rate of increase in energy with reduction in bond length is rapid), hence the energy change predicted is strongly dependent on the equilibrium bond-length used. This illustrates the point that great care must be taken over the selection of the parameters in atomistic potentials when quantitative accuracy is required. The parameter values are less important if it is only necessary to represent molecular structures qualitatively.

7.6.2 Lattice energy calculations for organic molecular materials

In some ways the procedure for calculating the lattice energy of an inorganic crystal, such as sodium sulfate, presents more technical issues than lattice energy calculations for molecular, organic materials such as benzophenone. The stable form of benzophenone has a melting point of 321 K whereas, the metastable form has a melting point in the range 297-299 K. Note the melting point of the metastable form is significantly lower than that of the stable form. Single crystals of the metastable form may undergo a spontaneous phase transformation to the stable form.[28]

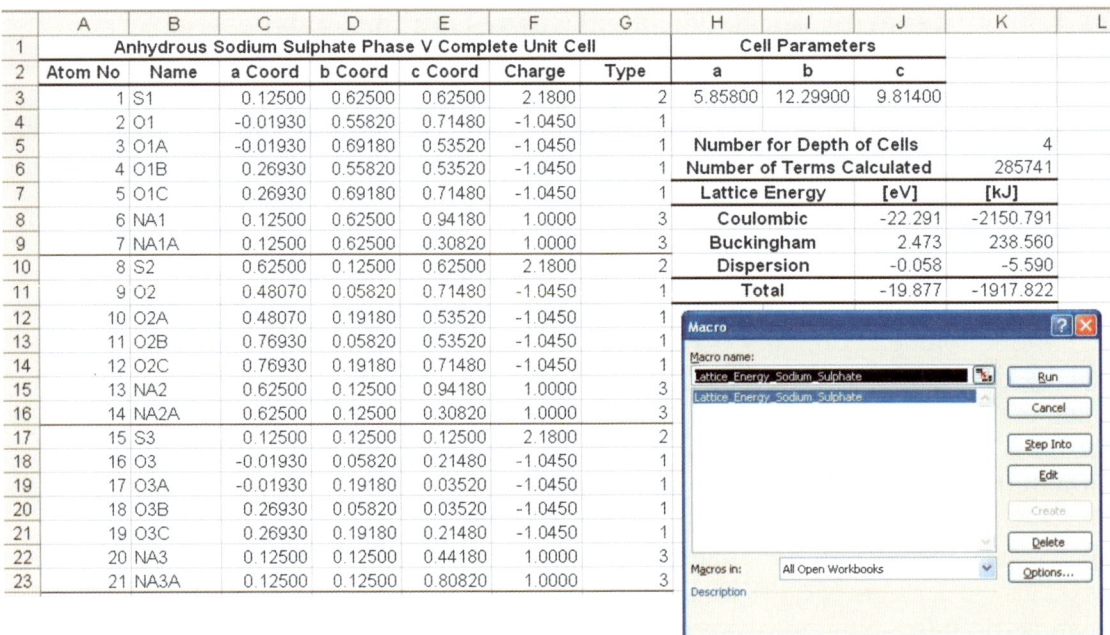

Figure 7.15
Excel Workbook used to Calculate Lattice Energy for Sodium Sulfate

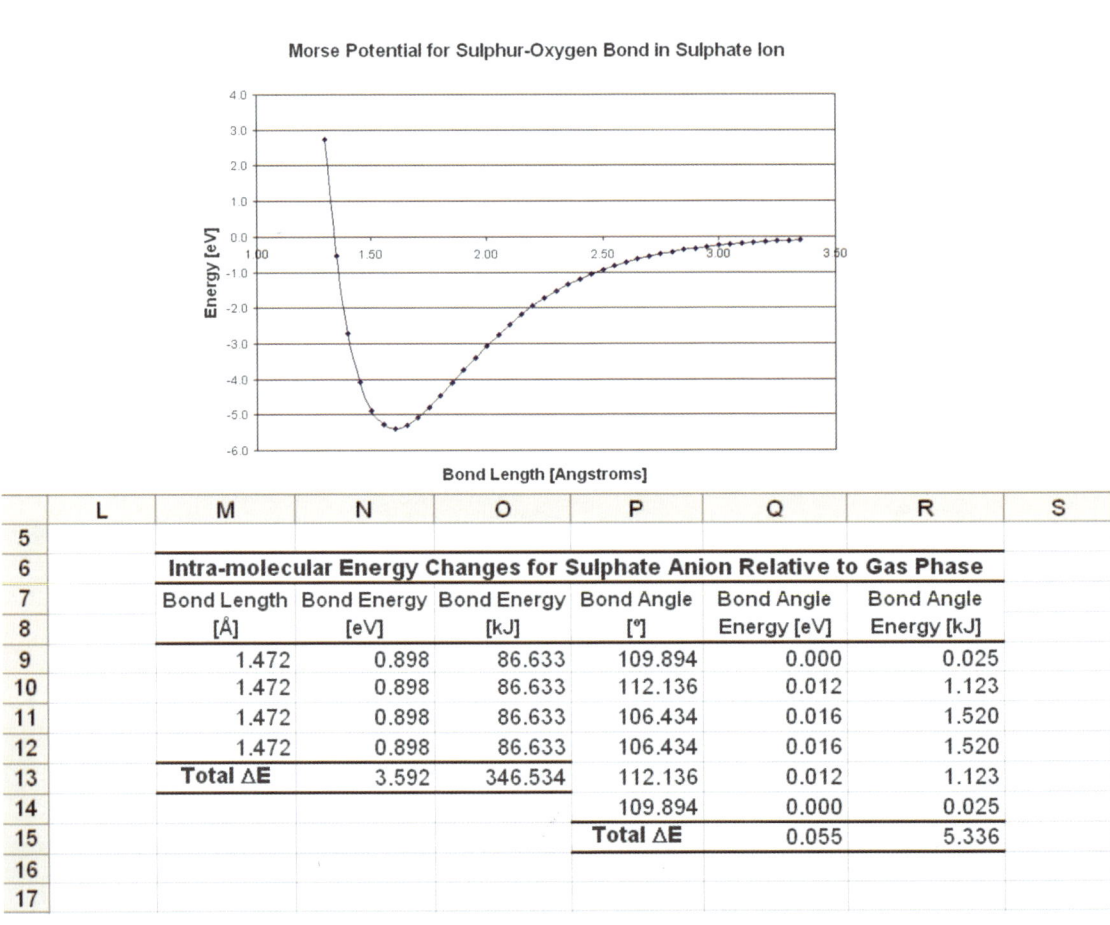

Figure 7.16 Functional form of the Morse Potential (top) and View of an Excel Workbook Presenting Intra-molecular Energy Contributions to the Energy of the Sulfate Anion (bottom)

Having already discussed some aspects of the crystallography of benzophenone, it is of interest to calculate the lattice energy of the stable and metastable forms to try and rationalise the relative stability of the two polymorphs. In the case of organic materials there are many published potentials that can be used for lattice energy calculations. Some of these potentials are examples of generic potentials such as the Dreiding potential.[29] These contain parameters for both intra-molecular and inter-molecular terms and can treat a very broad range of compounds. Alternatively there are potentials for treating groups of closely related compounds which, for example, may only provide parameters for intermolecular atom-atom potentials an example is the set of parameters from Scheraga *et al.* derived, in the first instance, for calculating the lattice energy of amino acids.[30]

In the case of some published calculations for benzophenone[17] the Scheraga potential was used and atomic charges were obtained using MOPAC. The calculated lattice energy is -85.9 kJ/mol for the stable phase and -83.0 kJ/mol for the metastable phase. The basis for both these calculations is one mole of benzophenone molecules. The stable phase is calculated to have a greater lattice energy and hence greater stability than the metastable phase as expected. However, these energies reflect the process of sublimation whereas melting is the transition from the solid to the liquid phase. Hence the calculated lattice energies do not indicate why there is a difference of 24 Kelvin between the melting points. Other more advanced forms of molecular modelling, such as molecular dynamics simulations, could be employed to try to evaluate the energy change associated with melting, (the enthalpy of fusion with sign reversed) and address the reason for the difference in the melting points.

7.7 The shapes of crystals

On examining single crystals of a material we observe, in general, that their shape is not spherical. Crystals are terminated by clearly defined sets of faces referred to as forms. Frequently the crystal surfaces that comprise different forms vary in both their physical and chemical properties. This variation of properties on a face-specific basis has important consequences from a scientific and technological perspective. For example, the precise hue of crystals of a pigment depends on the particular surfaces that are present on the crystals together with the crystal size and shape. Pigments are functional materials which are manufactured and sold on the basis of colour and not simply chemical purity. Consequently, manufacturing processes for pigments must be capable of delivering crystalline particles with minimal batch-to-batch variation in shape and size. But this begs a question: how is crystal shape and size related to parameters such as temperature and pH that may be subject to variability during manufacturing? The need to answer this question is one reason for seeking methods to predict the shape of crystals using molecular modelling.

The shape of a crystal depends on the relative rates of growth perpendicular to each surface during the crystal growth process. Every surface on a crystal can be specified by a Miller plane and its associated Miller index. The Miller plane that describes a crystal surface is the plane (or family of planes) exactly parallel to that surface. Therefore through the definition of the Miller plane a direct link is made between the crystal surface observed at the bulk, macroscopic scale and the atomic structure of the crystal surface at the nano-scale.

There are two terms that are often used interchangeably when referring to crystal shape: crystal *morphology* and *crystal habit*. However these terms are distinct; crystal morphology refers to the particular combination of forms that is present on a crystal; crystal habit refers to the relative surface areas of these forms on a crystal and hence the over-all shape of the crystal. Hence, two crystals which have the same morphology may have different shapes and therefore a different habit. Similarly two crystals may have the same habit but different combinations of forms and hence different morphologies. For materials of interest we need to know which forms are likely to appear on crystals of that material, the morphology, and the relative rates of growth of the different forms, the habit.

7.7.1 Bravais Friedel Donnay Harker (BFDH) approach for crystal habit prediction

To predict crystal shape from first principles a theory is required stating how the relative rate of growth perpendicular to each crystal surface relates to properties of the Miller plane describing the surface. Firstly Bravais[31] and later Friedel[32], Donnay and Harker[33] proposed that the rate of growth of a crystal surface is inversely proportional to the spacing between consecutive Miller planes in the family of Miller planes described by the Miller Index. This is the basis of so called Bravais Friedel Donnay Harker or BFDH approach for predicting crystal shape

Taking the example of a material that crystallises in a primitive cubic lattice, there is a straightforward mathematical relationship, Equation 7.15, relating the inter-planar spacing, d_{hkl}, for a set of reciprocal lattice planes with a Miller index (h k l) and the lattice parameter, a.

$$d_{hkl} = a\sqrt{\frac{1}{h^2 + k^2 + l^2}} \qquad (7.15)$$

In this particular case the relative rate of crystal growth perpendicular to a crystal face (hkl) is given by the ratio (d_{hkl}/d_{100}). Hence the relative distances from the centre of a crystal to each face along the direction perpendicular to the face are also given by the ratio (d_{hkl}/d_{100}).

Once the relative distances from the centre of a crystal to each face have been defined, a model of the crystal shape can be constructed. Referring to Table 7.5 a form (denoted by curly brackets) is a set of crystal faces which are all equivalent due to the symmetry of the crystal. The multiplicity is the number of faces in the form. For example in a cubic crystal the form {100} has a multiplicity of six containing the individual faces (100), (010), (001), ($\bar{1}$00), (0$\bar{1}$0) and (00$\bar{1}$). In a cubic crystal a property of the set of faces that comprise the form {100} is their ability to describe the surfaces of a polyhedron which completely encloses a central point. In fact the faces collectively describe a cube. The faces of form {111} can similarly form a polyhedron which encloses a central point symmetrically giving an octahedron.

The twelve faces of the form {110}, which are all equivalent in a cubic crystal, come in three sets of four {110}, {101} and {011}. The four faces of form {110} are parallel to the crystallographic c axis, the four faces of form {101} are parallel to the b axis and the four faces of form {011} are parallel to the a axis. None of these sets of four faces can, taken individually, define a closed polyhedron however in combination they can. The polyhedron created represents the {110} form for a cubic crystal. It is called a rhombic dodecahedron and has twelve faces, with the same surface area, and fourteen vertices shown in Figure 7.17.

Table 7.5 Parameters for BFDH Habit Calculation for Cubic Crystals

Primitive Cubic Lattice			
Form/ Miller Index	Multiplicity of Form	Interplanar Spacing d_{hkl}	Relative Growth Rate (d_{hkl}/d_{100})
{100}	6	a	1.00
{110}	12	$a\sqrt{(1/2)}$	1.41
{111}	8	$a\sqrt{(1/3)}$	1.73
{200}	6	$a/2$	2.00
{210}	12	$a\sqrt{(1/5)}$	2.24
Body Centred Cubic Lattice			
Form/ Miller Index	Multiplicity of Form	Interplanar Spacing d_{hkl}	Relative Growth Rate (d_{hkl}/d_{110})
{110}	12	$a\sqrt{(1/2)}$	1.00
{200}	6	$a/2$	1.41
{211}	8	$a\sqrt{(1/6)}$	1.73
{222}	8	$a\sqrt{(1/12)}$	2.50

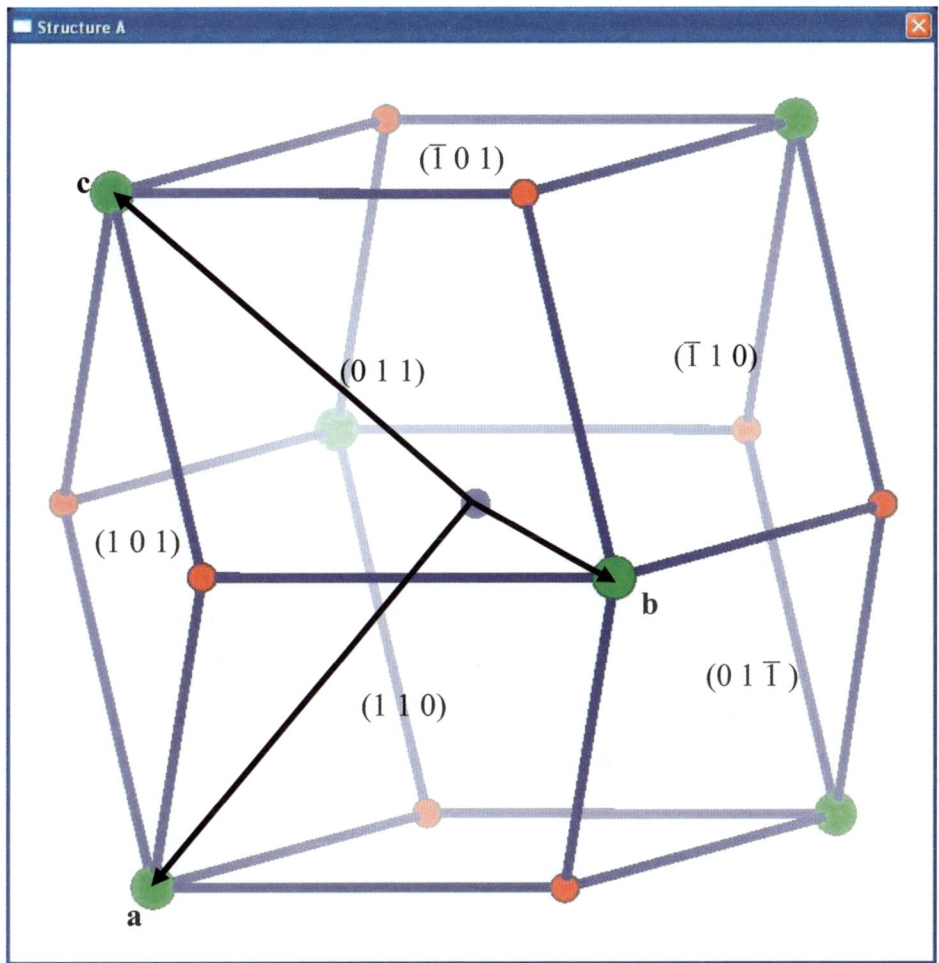

Figure 7.17 Habit Predicted by BFDH Approach for Body Centred Cubic Crystal

For the sake of argument, suppose that the morphology of a primitive cubic crystal is comprised of forms {100}, {111} and {110}. A model of the crystal habit is constructed by placing a cube, an octahedron and a rhombic dodecahedron so that their centres coincide at a single point. The respective distances from the central point to the faces of the cube, octahedron and rhombic dodecahedron, along vectors normal to the faces, are adjustable parameters call them λ_1, λ_2, and λ_3. The BFDH approach defines the ratios (λ_1/λ_2) etc. Once a value is selected for λ_1, this fixes the values of λ_2 and λ_3. The crystal habit is derived by creating faces out of the overlapping polyhedra so as to form the polyhedral shape with the smallest possible volume that encloses the central point. The procedure is essentially the same as that for combining the faces of the forms {110}, {101} and {011} in a cubic crystal but now the centre to face distances are not equal.

For a primitive cubic crystal, the directions normal to the faces of the form {111} coincide with the body diagonals of the cube. Hence we can calculate the maximum ratio for the centre-to-face distances for the {100} and {111} forms for which the {111} form is still visible in the crystal habit. The maximum ratio is $1:\sqrt{3}$ which is exactly the ratio given by the BFDH approach. A similar procedure reveals that none of the other forms are predicted to appear either. So the BFDH approach predicts that the morphology of a primitive cubic crystal is the single form {100} and that the habit is a cube.

In an X-ray diffraction profile for a cubic lattice containing additional lattice points, i.e. a centred lattice, there are certain families of reciprocal lattice planes that do not appear to reflect X-rays in accordance with the Bragg equation. Such reflections are extinguished as a result of total destructive interference of the X-rays. These missing reflections are referred to as systematic absences. What results is a set of relationships between the values of the indices h, k and l that must be fulfilled for a reflection to be observed in an X-ray diffraction profile. Donnay and Harker proposed that these extinctions should be taken into account when defining the external surfaces of crystals. For a body centred cubic lattice the extinctions are governed by the formula $h + k + l = 2n$ where n is a positive integer. Only Miller planes with indices satisfying this relationship are considered as crystal surfaces. So for a body centred cubic lattice rather than a primitive cubic lattice we do not include the sets of crystal faces associated with the forms {100}, {111} or {210} when considering the forms that could define the external shape of a crystal of this type.

Looking at Table 7.5 we see that for a body-centred cubic lattice the {110} form has the smallest centre to face distance. Again the BFDH approach predicts a morphology consisting of a single form, in this case the {110} form and the predicted crystal habit is a regular rhombic dodecahedron illustrated in Figure 7.17.

7.7.2 Potential deficiencies in the BFDH approach

The BFDH approach is the simplest theoretical approach for predicting the shape of crystals. However, it does not take account of inter-molecular forces in predicting the relative growth rates of surfaces and so the BFDH method may not always be accurate in predicting crystal shape. But why should the intermolecular forces be important? Crystals grow via a process of molecular self-organisation that take place at the crystal surfaces. Since the molecules within the crystal-bulk have a precise orientation and periodic placement, which is not present in the melt or solution, the growth of crystals must be a process which creates order.

Molecules in motion react to their environment through the action of intermolecular forces. Therefore, the rate at which molecules can self-organise into layers at the surfaces of growing crystals depends on the strength of the intermolecular forces acting between molecules within the layers. Random molecular motion (kinetic energy), measured by temperature, always tends to disrupt ordering. Consequently, stronger intermolecular forces are better able to overcome this randomising effect and self organisation is quicker. Since the molecular orientation varies with the particular surface so do the intermolecular forces and so different crystal surfaces manifest different rates of growth.

7.7.3 Attachment energy approach for crystal habit prediction

A more sophisticated approach for predicting crystal shape that does take into account inter-atomic interactions is the attachment energy approach.[34, 35] To describe this it is necessary to define a few terms.

- A **slice** is a slab of crystal bounded by two adjacent planes (hkl) in the crystal lattice. Hence the slice thickness is d_{hkl}. The crystal is envisaged as growing by the addition of crystal slices onto an existing crystal surface (hkl).
- The **slice energy** is the summation of all the atom-atom interaction energies between the atoms in a 'central' molecule in the slice and all the atoms in all the other molecules contained within the same slice.
- The **attachment energy** is the summation of all the atom-atom interaction energies between the atoms in a central molecule in the slice and all the atoms in all the other molecules outside the same slice. The sum of the slice and attachment energy is the lattice energy.

The attachment energy can also be interpreted as the energy released when a slab of crystal of thickness d_{hkl} is incorporated into the crystal lattice. As with the BFDH approach, the presence of centring and/or elements of space group symmetry, screw axes and glide planes, may reduce the thickness of the growth slab on a particular surface of a crystal. For example in the monoclinic space group $P2_1$ the growth slab on the (010) surface is specified by adjacent (020) planes and not adjacent (010) planes. Hence the correct slice thickness is d_{020} and not d_{010} (i.e. the slice thickness is halved).

The attachment energy model for predicting crystal shape states that the rate of growth perpendicular to a crystal surface (hkl) is proportional to the attachment energy for that surface. Hence the relative rates of growth perpendicular to the growing surfaces that enclose a crystal can be calculated from the attachment energies for the surfaces. Knowing the relative rates of growth allows the relative centre to face distances of all the crystal surfaces to be defined and this gives a model of the shape of the crystal.

When it comes to confronting the attachment-energy model of the crystal-shape with the crystal shape observed experimentally, there is often a good level of general agreement. However, this is provided that the particular growth environment does not play a major role. For example, if the interaction of the solvent from which the crystal is grown is not the same on all the growing surfaces then there will be an influence on the shape of the crystal. Since such solvent interactions are not accounted for in the attachment energy model, the implicit assumption is that these interactions are the same on all the crystal surfaces and therefore do not affect the relative growth rates or the crystal shape (although these interactions may affect the absolute growth rates of the crystal surfaces).

$$\gamma_{hkl} = \frac{ZE_{att}d_{hkl}}{2V_{cell}N_a} \qquad (7.16)$$

The surface energy in a vacuum, γ_{hkl}, of a specific surface of a crystal (hkl) can be estimated from the attachment energy according to the approximate formula given as Equation (7.16). In the equation Z is the number of asymmetric units per unit cell, d_{hkl} is the inter-planar spacing, V_{cell} is the unit cell volume, N_a is Avagadro's number and E_{att} is the attachment energy of the slice bounded by adjacent planes (hkl).

Whereas the attachment energy approach reflects the habit of crystals which are grown and then separated from their growth environment, surface energies define the shape of crystals which have had a sufficiently long time to come to equilibrium with their growth environment, the equilibrium habit. The equilibrium habit is the polyhedral shape which, for a given volume, minimises the total surface energy of the crystal.

Surface energies are also important in assessing the powder-flow behaviour of assemblies of crystalline particles for example, in pharmaceutical tablet manufacture. There is a tendency for particles to agglomerate on the surfaces with the highest energy to reduce the total surface area which is exposed and so reduce the surface energy over all. The mechanical properties of the agglomerates are determined by the strengths and distribution of the inter-particle contacts. Surface energies calculated through molecular modelling and then used in contact-mechanics models is an example of employing a multiscale modelling approach to predict material properties.

The interrelationship between the surfaces in the crystal habit and the underlying atomic structure is important to understanding how solid particles interact with their environment. Hence it is useful to be able to create molecular models of nano-sized crystals which have the combination of surfaces identified from a habit calculation. This allows not only the surface structure to be understood but also the structure along the edges and at the corners of crystals which becomes increasingly important when the particles are small. Figure 7.18 gives an indication of how a model of a facetted nano-particle of calcite can be derived from the crystal structure and displayed with INTERCHEM. This model can then be used as the basis for performing calculations, for example to explore the adsorption of water molecules on specific surfaces, and to investigate the formation of clusters of calcium-carbonate units on the crystal surfaces, the process of surface nucleation. Hence, for the future molecular modelling has the potential to provide a more complete understanding of processes of particle growth and surface catalysis.

7.8 Envoi

In this chapter we have described some key concepts which underpin the molecular modelling of crystalline solids. The aim has been to provide sufficient information so that those less familiar with molecular modelling have confidence in harnessing the possibilities and applying them in their areas of interest. Hopefully we have indicated the wide range of scientific and technological areas that can benefit from the careful application of molecular modelling of the solid state. There are many stimulating challenges yet to be completely solved, and we hope to see molecular modelling transition from an 'optional extra', employed by a few specialists, to a mainstay in any research strategy for materials design and accessible to many from the desk-top.

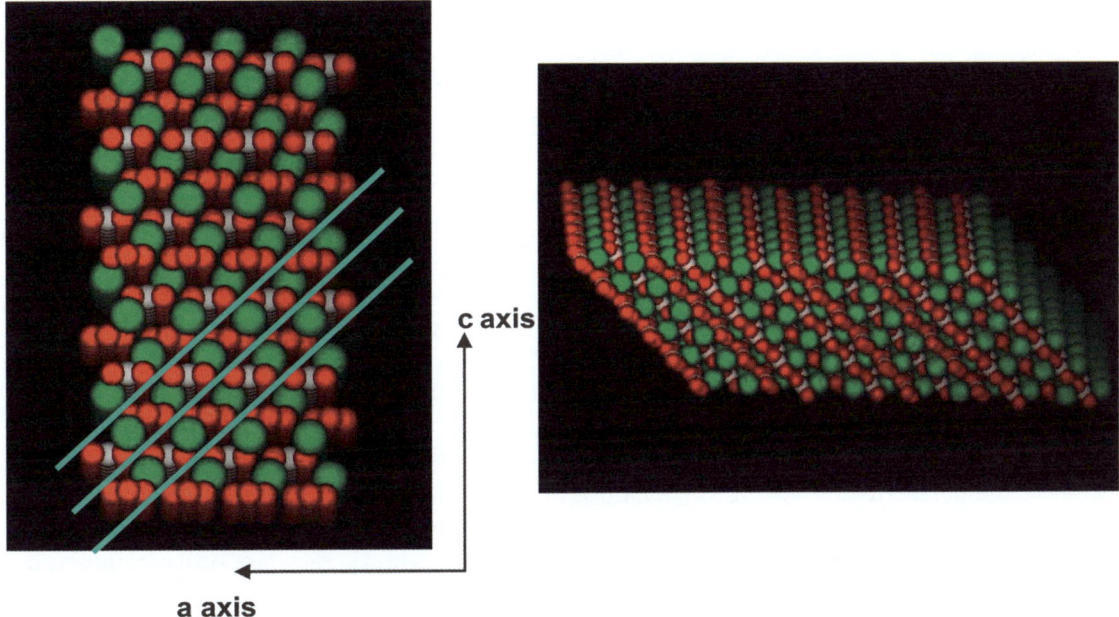

Figure 7.18 View in INTERCHEM of the Crystal Structure (left) and Derived Nano-particle of Calcite CaCO$_3$ (right)

Left crystal structure: calcium ions are represented as green spheres, oxygen atoms as red spheres and carbon atoms as grey spheres. The **b** crystallographic axis is towards viewing direction. Directions of the **a** and **c** crystallographic axes are indicated. The blue lines indicate three consecutive (104) Miller planes which are cleavage planes in the calcite structure. Right image of a nano crystal of calcite terminated by (104) surfaces (top and bottom) and displaying one possible termination of the surface (001) to the right of the image (a layer of calcium ions).

References and Endnotes. Chapter 7

[1] S. Lee and C. Huff in P.H. Stahl and C.G. Wermuth (Eds.) *Handbook of Pharmaceutical Salts, Properties, Selection and Use*, Wiley-VCH, Weinheim, 2008, Chapter 8, p192.

[2] J. Maddox, *Nature*, 1988, **335**, 201.

[3] S. M. Woodley and C. R. A. Catlow, *Nature Materials*, 2008, **7**, 937–946.

[4] M. O'Keeffe, *Phys. Chem. Chem. Phys.*, 2010, **12**, 8580.

[5] In reality it turns out that using molecular modelling to estimate energy changes associated with transitions from the solid state to the liquid state (and from there determining melting points) is challenging. The difficulties arise (i) because temperature must be taken into account, and free energy rather than internal energy is important in the study of phase transformations and (ii) from having to model the liquid state in which atoms, ions or molecules are subject to long range forces and are changing their relative positions continuously. A branch of molecular modelling known as molecular dynamics is required to model the liquid state and take account of temperature, but we will not attempt to describe this here. For information about molecular dynamics see for example A. R. Leach, *Molecular Modelling Principles and Applications (2nd Edition)*, Prentice Hall, Harlow England, 2001. For those familiar with molecular dynamics: examples of modelling melting transitions see J. M. Polson and D. Frenkel, *J. Chem. Phys.*, 1999, **111**, 1501 on the melting of n-octane and J. Anwar, D. Frenkel and M. G. Noro, *J. Chem. Phys.*, 2003, **118**, 728 on the melting of sodium chloride.

[6] J. A. Elliott, *International Materials Reviews*, 2011, **56**, 207.

[7] C. Hammond, *The Basics of Crystallography and Diffraction* (3rd Edition), Oxford University Press, Oxford, 2009.

[8] Representing the lattice concept by using circles in diagrams is sensible because the collection of circles manifests the same symmetry as the lattice itself but it is, nevertheless, an arbitrary choice. Hence we could select any two-dimensional object to mark the location of the lattice points for example the letter 'P'. When employing an arbitrary shape, here that of the letter 'P', to illustrate an array of lattice points, the choice of the exact locations of the lattice points with respect to the physical P shapes is also arbitrary. Most importantly looking at the arrays of 'P's, unlike the arrays of circles, they do not share the same symmetry elements as the arrays of lattice points themselves.

[9] T. Hahn (Ed.) *International Tables for Crystallography Volume A. Space Group Symmetry, Fifth Edition*. Kluwer Academic Publishers, Dordrecht, 2002.

[10] W. Poll, G. Pawelke, D. Mootz and E. H. Appelman, *Angew. Chem., Int. Ed.*, 1988, **27** 392.

[11] Another parameter known as the site symmetry is always equal to the identity operator (i.e. the subgroup of the full space group containing just the identity operation) for general positions in the unit cell. However, not all positions in the unit cell are general positions for example, the lines with fractional coordinates **0**, **y**, **0**; **0**, **y**, ½; ½, **y**, **0** and ½, **y**, ½ all coincide with two-fold rotation axes and so the fractional coordinates of such points are invariant under the action of the rotation. So instead of having the original and one image of an atom with coordinates e.g. **0**, **y**, **0** a duplicate image is formed. The site symmetry of these lines in the space group **P2** is **2** which is a subgroup identical to the full space group. Importantly, dividing the multiplicity of a general position, here two, by the number of symmetry elements in the subgroup that defines the site symmetry of a special position, again two, gives the number of images of the atoms on special positions that will be produced, in this case one. The above may appear very esoteric however it is of significance in real molecular modelling applications. If atoms are located very close to special positions in a particular unit cell a symptom may be the appearance of 'ghost'

images of atoms which then appear to overlap. Often modelling software will check for atoms having identical sets of coordinates as a way of remedying this situation, however the success or otherwise of this approach depends on the distance tolerance used to decide whether atoms are duplicate copies.

[12] G. R. Desiraju, *Crystal Engineering: The Design of Organic Solids*, Elsevier, Amsterdam, 1989.

[13] G. R. Desiraju, *Angew. Chem. Int. Ed. Engl.*, 1995, **34**, 2311.

[14] S. Dharmayat, R. B. Hammond, C. Kilner, X. Lai, R. A. Palmer, B. S. Potter, C. M. Rayner, K. J. Roberts, *Organic Process Research & Development*, 2008, **12**, 860.

[15] An approach known as graph-set analysis can be used to characterize hydrogen-bonding networks in crystals. This involves identifying the repeating motifs manifested by inter-molecular hydrogen bonds and assigning a graph type to characterize them. Hence hydrogen bonds can form infinite chains, labelled (C), cycles labelled (R), non-cyclic dimers or finite hydrogen-bonded sets labelled (D) or may be intra-molecular labelled (S). A superscript is added to the letter that denotes the total number of hydrogen-bond acceptor atoms in the motif and a subscript that denotes the total number of hydrogen-bond donor atoms. Finally a number is appended in parentheses which denotes the total number of atoms involved in the motif. The hydrogen bonded dimer in a carboxylic acid, such as benzoic acid (see Chapter 4, Section 4.3.1), belongs to the graph set $R_2^2(8)$. For further information see e.g. M. C. Etter, J. C. MacDonald, and J. Bernstein, *Acta Crystallogr.*, 1990, **B46**, 256 and J. Bernstein, R. E. Davis, L. Shimoni, and N. Chang, *Angew. Chem. Int. Ed. Engl.*, 1995, **34**, 1555.

The issue of whether chlorine should be regarded as a hydrogen-bond acceptor atom is contentious. Whereas water has a melting point of 273 K and a boiling point of 373 K, hydrogen chloride has a melting point of 162 K and a boiling point of 188 K and lithium chloride has a melting point of 883 K. So the inter-atomic forces in condensed phases of hydrogen chloride are clearly not ionic (as in lithium chloride) and, at the same time, the significantly lower melting and boiling points compared to water suggest an absence of hydrogen bonding. However, some theoretical studies (I. R. McDonald, S. F. O'Shea, D. G. Bounds and M. L. Klein, *J. Chem. Phys.*, 1980, **72**, 5710) have suggested that weak hydrogen bonds are present in liquid hydrogen chloride. Nevertheless solid-state chemists have different criteria on which they judge the ability of chlorine atoms and chloride ions to accept hydrogen bonds in molecular crystals. One approach has been to search the Crystal Structure Database to find average separation distances between halogen atoms and adjacent hydrogen atoms which are bonded to hydrogen-bond donor atoms (oxygen and nitrogen). These mean distances are then divided by the sums of the van der Waals radii of the atoms (for example chlorine and hydrogen). When the ratio obtained is less than one this is taken to indicate an interaction which cannot be assigned to dispersion forces alone but indicates a degree of hydrogen bonding see, for example, L. Brammer, E. A. Bruton and P. Sherwood, *Crystal Growth and Design*, 2001, **1**, 277.

[16] S. L. Childs, P. A. Wood, N. Rodríaguez-Hornedo, L. S. Reddy, and K. I. Hardcastle, *Crystal Growth and Design*, 2009, **9**, 1869.

[17] H. Kutzke, H. Klapper, R. B. Hammond, and K. J. Roberts, *Acta Crystallogr.*, 2000, **B56**, 486.

[18] E. Madelung, *Physikalische Zeitschrift*, 1918, **19**, 524-532.

[19] Applying the principle of equipartition of energy, individual degrees of translational and rotational freedom contribute ½R to the heat capacity at constant volume, C_v, whereas degrees of vibrational freedom contribute R to C_v where R is the ideal gas constant. The solid state has six more degrees of vibrational freedom per constituent molecule than the gaseous state whereas, the gas has three more degrees of translational and three more degrees of rotational freedom per constituent molecule (assuming the general case of a polyatomic molecule). Hence the difference

in the heat capacities at constant volume between the gaseous and solid states is -3RT. Finally, since most experimental determinations of the heat of sublimation are carried out at constant pressure rather than constant volume, it is more appropriate to consider the difference in the heat capacities at constant pressure. The heat capacities of a solid at constant volume and constant pressure are the same to a good approximation. The heat capacities at constant volume and pressure for an ideal gas are related by $C_{p,gas} = C_{v,gas} + R$.

[20] A. I. M. Rae and R. Mason, *Proc. Roy. Soc. A*, 1968, **304**, 487-499.

[21] T. G. Cooper and N. H. de Leeuw, *J. Cryst. Growth*, 2006, **294**, 137.

[22] S. E. Rasmussen, J. E. Jorgensen, and B. Lundtoft, *Journal of Applied Crystallography*, 1996, **29**, 42.

[23] The fractional coordinates of the sulfur atom are ⅛, ⅛, ⅛, the *International Tables, Volume A*[9] indicate that the second choice for the position of the origin of the unit cell has been selected and that the site symmetry is 222. The fractional coordinates of the sodium atom are ⅛, ⅛, z (where z represents any fractional coordinate) so sodium atoms are positioned on two-fold rotation axes and the site multiplicity for these positions is 16.

[24] A subset of eight of the thirty-two symmetry operations is applied to generate the fractional coordinates of the eight sulfate anions in the unit cell. We use symmetry operations (1) and (5) in conjunction with the additional translations associated with the lattice points. The positions of the sixteen sodium cations in the unit cell are produced by applying symmetry operations (1), (3), (5) and (7) in conjunction with the additional translations associated with the lattice points.

[25] In evaluating the lattice energy, for a single unit cell of thenardite there are N x n_2 x n_1 atom-atom pair-wise terms in the summation. For (i) interactions between one of the sulfate anions and all the other sulfate ions in a single unit cell N = 7, n_2 = 5 and n_1 = 5 producing 175 atom-atom terms. For (ii) interactions between one of the sulfate anions and all the sodium cations N = 16, n_2 = 1 and n_1 = 5 producing 80 terms. For (iii) interactions between one of the sodium cations and all the other sodium cations N = 15, n_2 = 1 and n_1 = 1 producing 15 terms. Since the unit cell contains 56 atoms, the total number of pair-wise interaction terms, excluding self-interaction terms, is 56 x 55 = 3080. From this number we subtract the 5 x 4 = 20 intramolecular terms associated with every sulfate anion, a total of 8 x 20 =160 terms per unit cell. This leaves 2920 terms, however, in this number every pair-wise interaction is counted twice so there are really 1460 terms. For interactions between sulfate anions there are 8 sulfate anions from which to select the 'central' sulfate for the purposes of the energy summation (the innermost summation term in Equation 7.11). This suggests a total of 8 x 175 = 1400 terms however, all interactions are doubly counted here so actually there are 700 terms. Similarly for interactions between sodium cations, there are 16 sodium cations from which to select the central sodium hence suggesting a total of 16 x 15 = 240 terms but again, these interactions are doubly counted. However, for the cross terms between a central sulfate anion and the sodium cations there is no double counting, there are 8 ways of selecting the central sulfate anion and hence 8 x 80 = 640 terms. So the total number of terms in the energy summation for the central unit-cell, allowing for double counting, is 700 + 120 + 640 = 1460 terms as required.

[26] H. D. B. Jenkins, *Mol. Phys.*, 1975, **30**, 1843.

[27] M. F. C. Ladd and W. H. Lee, *J. Inorg. Nucl. Chem.*, 1961, **21**, 216.

[28] The transformation can be initiated through mechanical stress or contact with a crystal of the stable phase. It is termed a reconstructive phase transition as it destroys any single crystal in which it occurs. What is left is a polycrystalline aggregate of the stable phase, which, nevertheless, retains the shape of the former single crystal. Once a transition event is initiated, a transition front moves through the crystal at a rate of a few millimetres per second. In advance of the front the crystal is still transparent whereas behind it has become turbid.

[29] S. L. Mayo, B. D. Olafson, and W. A. Goddard III, *J. Phys. Chem.*, 1990, **94**, 8897.

[30] F. A. Momany, L. M. Carruthers, R. F. McGuire, and H. A. Scheraga, *J. Phys. Chem.*, 1974, **16**, 1595.

[31] A. Bravais, *Etudes Cristallographiques*, Gauthier Villars, Paris 1866.

[32] G. Friedel, *Bull. Soc. Franc. Mineral.*, 1907, **30** 326.

[33] J. D. H. Donnay and D. Harker, *Am. Mineral.*, 1937, **22**, 463.

[34] P. Hartman and P. Bennema, *J. Cryst. Growth*, 1980, **49**, 145.

[35] G. Clydesdale, R. Docherty, and K. J. Roberts, *Computer Physics Communications*, 1991, **64**, 311.

Chapter 8
The Sources of Archived 3D Chemical Structure Information

8.1 Introduction
Since in molecular modelling we are frequently interested in the three-dimensional (3D) shape and disposition of the structures, it is appropriate to review how such 3D data are obtained.

There are two ways that we can acquire chemical structural information: we can get data from archival sources, or we can generate (*i.e.* build) structures by using computer software. This software will usually be included in the same software package that can be used to display and analyse the structures. In preceding chapters of this book you saw how the Interprobe software provides these tools. Methods for generating 3D structures from lower dimensional data is also dealt with in these chapters. This chapter is concerned with obtaining and using archived data.

Archived data can be categorised according to how it was generated (source), how it has been stored (storage format), and how it is made available (presentation format). The storage format is probably of less interest to a user who will be more concerned with source and reliability of the data, and the practicalities of coping with the presentation format.[1] However, before all this, we need to review some of the organizations that hold and provide access to the data.

8.2 Structures of small organic molecules from X-ray crystallography
In 1965 the Cambridge Crystallographic Data Centre (CCDC) began compilation of the Cambridge Structural Database (CSD)[2] of organic and metal-organic crystal structures of 'small molecules', (up to about 1,000 atoms including hydrogen atoms). The database now (mid-2011) contains crystallographic, chemical and bibliographic data for more than 570,000 compounds. Neutron and powder diffraction studies are included. Until the mid-1990s, all data entering the CSD were re-keyboarded from the primary literature and associated supplementary deposition documents. With the advent of the crystallographic information file (CIF, see section 8.7.5) the CCDC began to receive increasing volumes of data electronically, reaching close to 100% from around 2000. In the late 1980s, key journals in the area began to require authors to provide crystal structure information to the CCDC, initially as printed documents but now by electronic transmission. Currently, more than 100 major journals collaborate with the CCDC to ensure that the CSD continues to be comprehensive. The CCDC operates a 'request-a-structure' service which returns original deposited CIFs, or CIFs constructed from earlier CSD entries, usually by email.[2b]

Originally grant-funded within the University of Cambridge, the CCDC is now an independent non-profit institution with UK charitable status and retains close links with the University. Because it receives no public funding of any sort, the CCDC licences the CSD to both academia and industry on a cost-recovery basis. The CSD is supplied to academics in around 70 countries worldwide, usually through a local National (or Regional) Centre affiliated to the CCDC, but with some smaller countries being supplied directly from Cambridge. As part of its charitable remit, the CCDC provides discounts of up to 100% for countries in the developing world. Within the UK, the CSD (and other crystallographic databases) are hosted by the

Chemical Database Service (CDS)[3] at the Daresbury Laboratory. The CDS provides free database search services to registered UK academics.

8.3 Structures of inorganic compounds and metals

There are databases concerned with purely inorganic compounds,[4] and with metals, alloys and minerals (CRYSTMET).[5] These are also made available by Daresbury Laboratory and are capable of being searched by the same tools as used for the CCDC files. A separate database of zeolite structures is available at the International Zeolite Association (IZA-SC).[6]

8.4 The Protein Data Bank

The Brookhaven National Laboratory in New York State housed one of the first synchrotron radiation sources used for crystallography, and it was natural that, in 1971, it became the host for a depositary of macromolecular structures. This was called the Protein Data Bank (PDB). Starting with just seven structures, it has grown, so that on 21st June 2011 there were 73,951 structures. While the PDB started at the Brookhaven Laboratory[7] (and was frequently known as the 'Brookhaven' database), there was a major change in organization in 1999, so that now it is managed by a consortium called the Research Collaboratory for Structural Bioinformatics; the components of this are: Rutgers, - The State University of New Jersey; San Diego Supercomputer Center, University of California, San Diego; and the BioMagResBank (Wisconsin).[8] Now the databank includes data on protein and nucleic acid structures, and the methods used for structure determination include Nuclear Magnetic Resonance Spectrometry (NMR) and Electron Microscopy.

The PDB has had funding from the United States Government, and had made the data available for the cost of the distribution media. Recently the decision was made to make data available only by electronic transfer. Now anyone may download it without charge. (A domestic broadband connection will allow transfer of the structural data in the most compact format in about 28 hours).[9]

In parallel with the PDB, nucleic acids structures are available from the Nucleic Acid Database (NDB) from Rutgers University.[10] New structural data for nucleic acids is now deposited here rather than in the PDB.

For comprehensive articles on the crystallographic databases of biological macromolecules see [11, 12]

8.5 ZINC

ZINC is a free database of commercially available compounds for virtual drug screening.[13] It is hosted at the Department of Pharmaceutical Chemistry, University of California, San Francisco and is compiled from the catalogues of various companies who supply samples of 'drug-like' or 'lead-like' compounds for screening. By removing duplicates and compounds that are unsuitable drug candidates (because, for example, they are reactive or insoluble), the total number of compounds is around 2.3 million. The catalogues of the companies ('Diversity Suppliers')[14] usually have 2D structures, and the compilers of the database have converted these into 3D structures through intermediary SMILES strings. (SMILES is discussed in Chapter 4).

8.6 Interprobe Chemical Services 3D database

The 4.6 million compounds in this database have been derived from the 2D structure catalogues of diversity suppliers. In contrast to the ZINC database there has been no attempt to eliminate duplicates, and the offerings of each of the suppliers have been kept separate. The data is available from Interprobe and a copy is supplied with this book. Methods for searching it are described later in Chapter 10.

8.7 File formats

In molecular modelling, perhaps nothing causes as much misunderstanding, contention, and controversy as the format of data files (except perhaps the choice of computer programming language!). Some of this arises because the aims of the experimentalists and the modellers are different. A crystallographer will want to record the results, and also explain them. A typical PDB file will have data lines that contain atom coordinates, but there will be comment lines, and also indications that perhaps the structure is disordered, or that part of the structure is missing. A modelling program (and a modeller!) must be able to cope with these eventualities.

Turn to any paper dealing with synthetic organic chemistry of thirty years ago (*i.e.* before the fruits of molecular modelling in the form of coloured pictures became obligatory), and the text will be interspersed with structural diagrams (in 2D) that show the *bonds* between atoms. There may be indications of stereochemistry, but the atoms themselves are often not shown explicitly, and (when dealing with organic compounds) are assumed to be carbon by default. It turns out that the minimum requirement for representing a structure in three dimensions is the *positions* of the *atoms*. Provided that the identities of the atoms are known and that their positions show that the atoms are within bonding distance, a modelling program should be able to (and usually will) infer where the bonds are and their nature (single, double, etc), and display the structure. After all, this is just the information from an experiment that is provided to an X-ray crystallographer. It is true however that the tasks of a modelling program are made easier if some bonding information is provided.

While most formats are intended to be read by a computer, some are capable of being understood by the human eye as well. One problem encountered in some formats is that a good initial design has been inelegantly modified to accommodate extra features. We do not attempt to rate the formats detailed below as 'good' or 'bad'; we do attempt to place them in historical context, and comment on instances where the format has been adopted for inappropriate use.

In many molecular modelling software packages, the user may not need to be concerned about the details of the file formats. This will apply for most of the time, but there will be occasions when things just do not work; an error message appears to say that a format is not recognized. Knowing about formats, and how to correct a file using an editor can mean the difference between failure plus frustration, and success!

8.7.1 XR format

This is so called because the computer file names are suffixed '.XR' (e.g. xxxxxxx.XR). This is one of the optional formats offered by the CrystalWeb utility at Daresbury Laboratory for output and downloading data from the CCDC. A typical file is shown in Figure 8.1.

The format can be used to specify either fractional coordinates (as in this example) or Cartesian coordinates. When dealing with Cartesian coordinates, the '0' on line 3 is replaced by '1', the unit cell dimensions (A,B,C) are set to 1.0, the angles (ALPHA, BETA, GAMMA) are set to 90.0 degrees, and the space group is set to 1. Modelling software should be able to handle the conversion of the fractional coordinates into orthogonal Cartesian coordinates, making use of these cell dimensions and angles.

The space group number can run from 1 to 230. It is followed by the Hermann-Mauguin symbol (without any regard to the 'case' of the characters or subscripting; i.e. $P2_12_12_1$ is shown as P212121, and $P4_2mc$ is shown as P42MC). Because some space groups have alternative settings, the interpretation of XR files can be problematical. It is better to rely on CIF files in these cases (See paragraph 8.7.5).

The example shows the X-Ray crystal structure of the salt of an alkaloid and is typical of results of its time (1970) in not showing the positions of hydrogen atoms. Connections between pairs of atoms are shown, but the nature of the bonds (single, double, etc) is not shown. Note that connections are shown twice (once for each atom in a pair). The limit of eight connections per atom does present problems when dealing with organo-metallic compounds, for example *ferrocene,* which requires ten connections for the iron atom.

8.7.2 TRIPOS MOL2 format

This is a very popular format, devised by the firm Tripos for their Sybyl modelling program. It is a second generation format (replacing an older MOL format) and consequently has many of the inelegant features of the earlier format removed. It is a 'tagged' format; by this we mean that each section of a structure file is introduced by a 'Record Type Indicator' (RTI). The complete specification has been published on the internet,[15] and while the format is proprietary, its wider use has been encouraged by Tripos. There are over 30 RTIs, most of which are used only by the Tripos software. The example shown in Figure 8.2 only has five RTIs. Of these, only three are needed for specifying a structure (in this case the same molecule that is represented in Figure 8.1). Each RTI begins with the string of characters @<TRIPOS>.

The essential ones are:

```
@<TRIPOS>MOLECULE
@<TRIPOS>ATOM
@<TRIPOS>BOND.
```

Lines beginning with the hash character (#) are treated as comments and ignored. Empty lines are also ignored.

Line 1 A,B,C are the unit cell dimensions in Ångstrom Units
Line 2 ALPHA,BETA,GAMMA are the cell angles in degrees
Line 2 SPGR is the space group number and space group symbol
Line 3 "34" specifies the number of atoms
Line 3 "0" signifies that data is in fractional coordinates
          "1" would indicate Ångstrom coordinates
Line 4 "40" signifies the type of compound (alkaloid in this case)

Lines 5 onwards specify the atoms
    Atom 1 is the bromine (ion), which has no connections
    Atoms 2 to 26 are carbon atoms
    Columns 3 to 5 are the fractional coordinates (a,b,c)
    Columns 6 (to 13) are the connections to other atoms
    Atom 34 is (unconnected) Oxygen (H$_2$O of crystallization)

```
REFERENCE STRUCTURE = 12542    A,B,C =   11.370   10.960   12.410
   ALPHA,BETA,GAMMA =  90.000 120.300   90.000      SPGR =   4 P21
   34    0 CODEN=LUNARB10 SYMOPS=50028
   40       RFAC=13.1 ERRFLAG=0 (C-C)ESD=3
    1 BR1      0.09970    0.00000    0.08840
    2 C1       0.26480    0.19620   -0.40170    13    23    31
    3 C10      0.53250    0.10800    0.01710     4    26
    4 C11      0.41560    0.07260   -0.08640     3     5
    5 C12      0.38870    0.12900   -0.19620     4    24    31
    6 C13      0.39770    0.39070   -0.34250     7    23
    7 C14      0.50400    0.46970   -0.31490     6     8
    8 C15      0.49030    0.60340   -0.31250     7    27    32
    9 C16      0.61370    0.79750   -0.25730    10    27
   10 C17      0.64290    0.84560   -0.12540     9    11
   11 C18      0.78020    0.80650   -0.02280    10    12
   12 C19      0.81560    0.86310    0.10810    11    28
   13 C2       0.22690    0.12320   -0.52140     2    20
   14 C20      0.96680    0.68930    0.23520    15    28
   15 C21      1.11450    0.65050    0.27800    14    16
   16 C22      1.13330    0.51270    0.31960    15    29
   17 C23      0.95490    0.35810    0.24910    18    29    33
   18 C24      0.84170    0.30660    0.13240    17    19
   19 C25      0.74380    0.23810    0.12980    18    26
   20 C3       0.33560    0.03680   -0.49970    13    21    30
   21 C4       0.47980    0.07200   -0.41790    20    22
   22 C5       0.49430    0.21400   -0.39720    21    23
   23 C6       0.41270    0.25590   -0.33600    22     2    24     6
   24 C7       0.46870    0.21090   -0.20610     5    23    25
   25 C8       0.58650    0.24880   -0.10200    24    26
   26 C9       0.61950    0.19110    0.01240    19     3    25
   27 N30      0.60610    0.66200   -0.26000     8     9
   28 N31      0.95480    0.82840    0.20230    12    14
   29 N32      1.03240    0.43850    0.22230    16    17
   30 O26      0.30360   -0.06790   -0.55350    20
   31 O27      0.27070    0.10310   -0.31380     2     5
   32 O28      0.37980    0.65630   -0.35410     8
   33 O29      0.97490    0.33780    0.35390    17
   34 O33     -0.10480    0.15960    0.44760
```

Figure 8.1
**Listing of the XR File for the Crystal Structure
of Lunarine Hydrobromide**

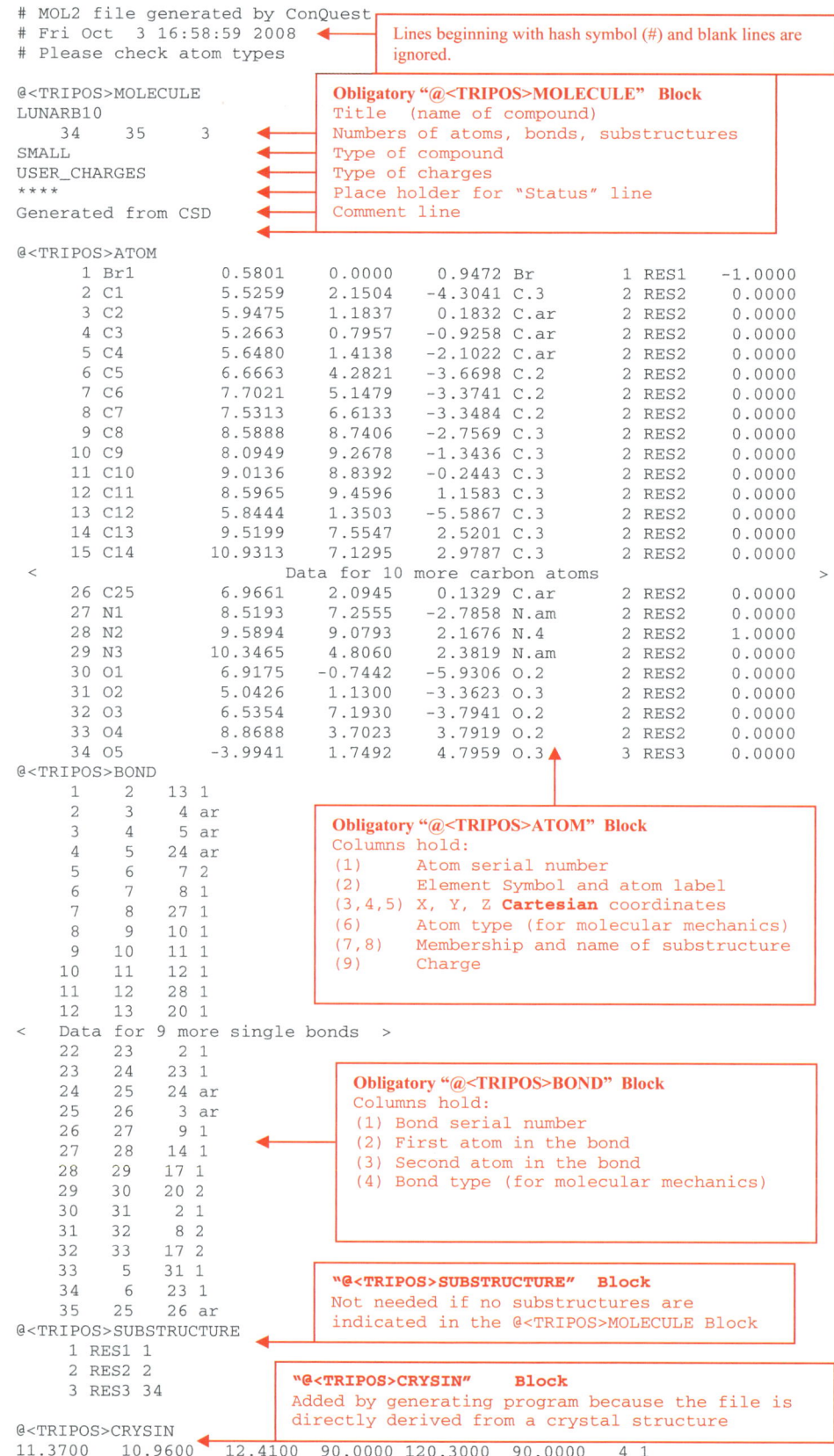

**Figure 8.2
Listing of Tripos MOL2 File for
Crystal Structure of Lunarine Hydrobromide**

8.7.3 PDB format

The Protein Data Bank has had its own format from its first days. However, it has recently been substantially revised, and this has meant that the data files have been rebuilt ('remediated' in PDB speak). The revised format is described in a document that is available from the PDB.[16] A substantial part of a PDB structure file is concerned with describing the experiments that relate to it, the affiliation of the experimenters, and the relevant references in the scientific literature. Each record in a data file is exactly 80 characters long, and the first six characters define its role. In effect, each record begins with a tag. Examples of the bookkeeping records are HEADER, TITLE, REMARK, AUTHOR, and MASTER. The types of records that define a structure are ATOM, HETATM, and CONECT, (note that the tag strings that define the record type can have fewer than six characters, and in that case are padded out with spaces). The main purpose of PDB files is to hold structures of proteins and nucleic acids, and these are mainly polymers of the 20 standard amino acids and 8 standard nucleotide monomers (4 each for DNA and RNA). The internal connections of these monomers are well defined. Thus given a set of ATOM records for a sequence of amino acid residues, it is easy to construct the table of their connections. HETATM records are needed to define structures of non-polymeric molecules (ligands, cofactors, solvent molecules including water, metal atoms etc). The CONECT records are necessary for joining atoms in ligands etc., and for defining disulfide bonds in the protein chains. HETATM and CONECT records are also used to define the structures of any unusual amino acids or nucleotides. Examples of various types of record are shown in Figures 8.3 and 8.4. The large size of typical PDB files means that it is not possible to show a complete file.

All the atom coordinates in a PDB file (ATOM and HETATM records) are in Ångstrom units in orthogonal Cartesian form. For some purposes it is necessary to be able to recover the fractional coordinates based on the unit cell dimensions. Records for doing this are provided (CRYST1, ORIGXn, SCALEn, n = 1,2,3). These records are obligatory even when the data comes from non-crystallographic experiments.

In protein crystals the structures are frequently disordered. This can mean that in parts of a proteins chain alternative sets of coordinates may be given for some atoms, or (in extreme cases) that whole sections of a protein chain may be missing. The complete amino acid *sequence* of the protein will usually be known, and the PDB file will contain this and a reference to the paper when that is recorded. If there are coordinates that are missing for certain residues, then there *may* be discontinuities in the listing of the residue numbers. (Considerable latitude was offered to submitters of data in the past, and this latitude often causes problems in interpretation).

Nuclear magnetic resonance (NMR) experiments designed to determine the structures of proteins (and other biomolecules) will normally be run on solutions. These experiments can thus complement the results from crystallography. The way that the raw results are processed leads to a set of 'models', in each of which the coordinates are consistent with the data, together with an 'average' structure. Each model is delimited by MODEL and ENDMDL records.

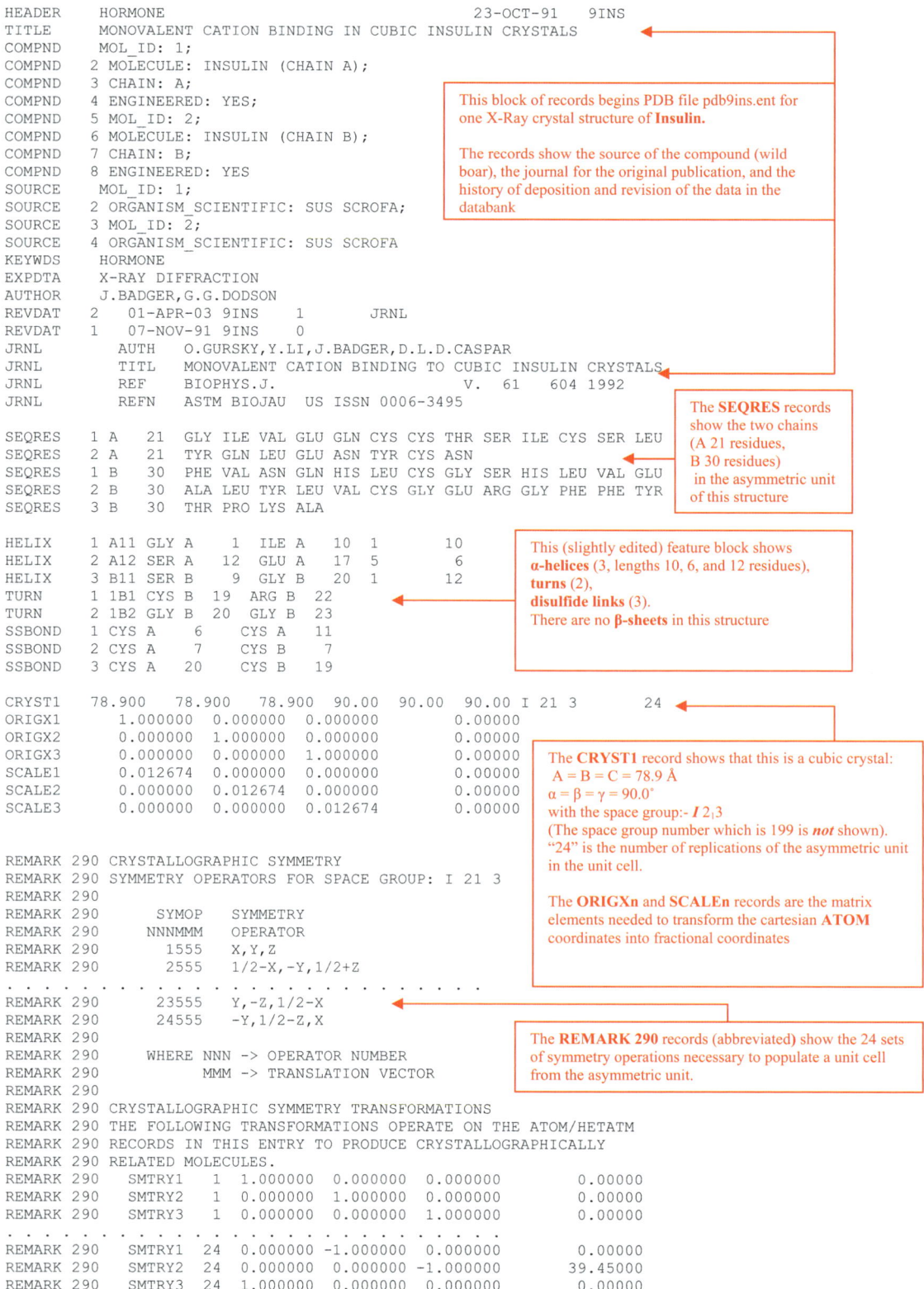

Figure 8.3 (to be read with Figure 8.4)
The Descriptive Parts of the PDB File (9INS) for the Crystal Structure of Insulin

The Sources of Archived 3D Chemical Structure Information 197

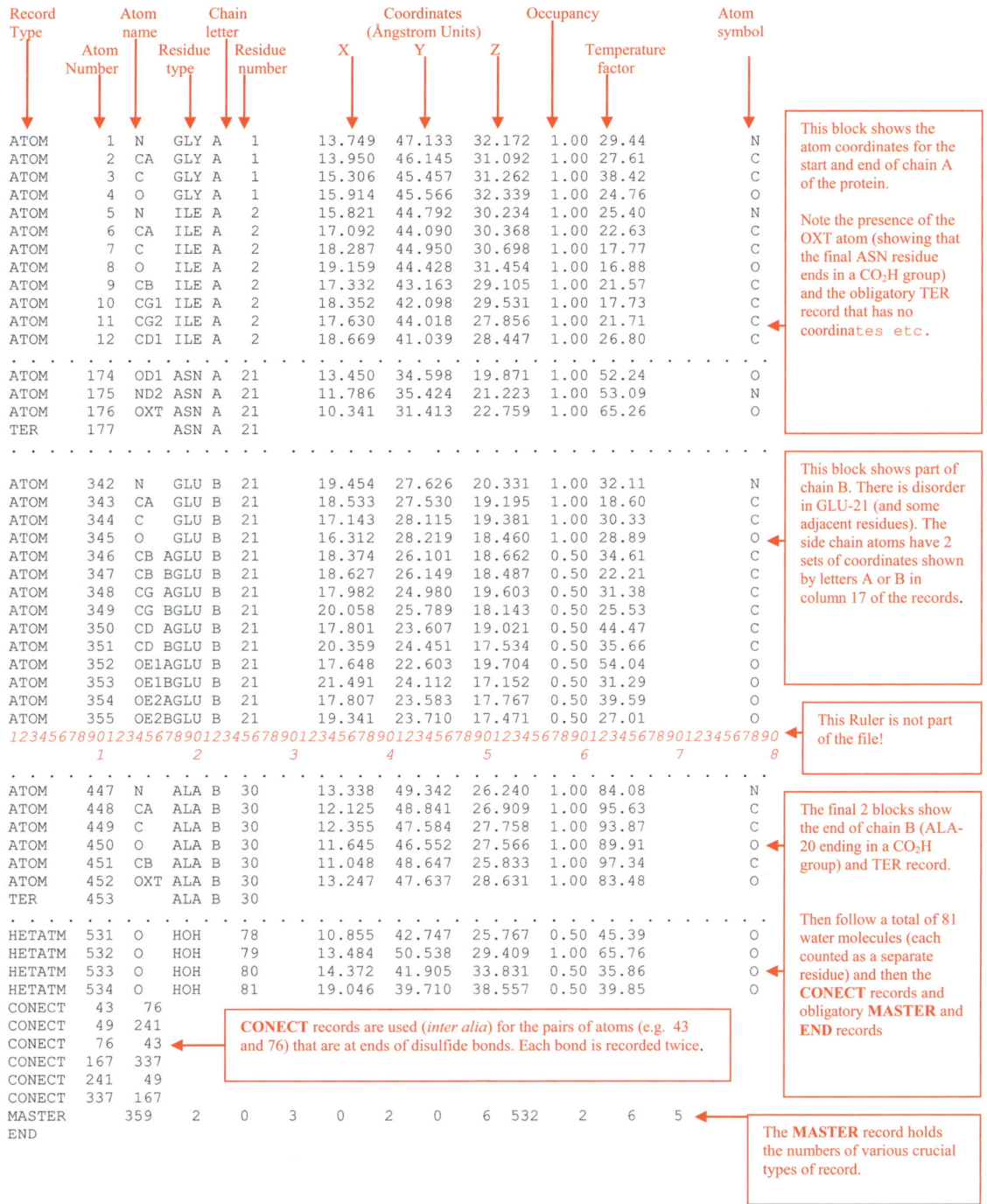

**Figure 8.4 (to be read with Figure 8.3)
The Atom Data and Connectivity Data Parts of the File (9INS) for the
Crystal Structure of Insulin**

In the current version of the Protein Data Bank there are over 73,000 files, and in compressed form they require 10.0 Gbytes for storage. A typical file will be named (for example) `pdb4ins.ent.gz`. When uncompressed the resulting text file will be `pdb4ins.ent`. This file name adheres to the '8.3' filename format [eight characters (maximum) followed by a decimal point and then a three character suffix]. This type of name is widely accepted by current operating systems (Windows, Linux, Unix, MacOS, although often more characters are allowed before the decimal point). The PDB uses the four characters immediately before the dot to identify the file; the two central characters (in this case 'in') are used to name the folder (subdirectory) that holds this file and others similarly named. In the early days, the last three characters of the four were assigned in a way that indicated the nature of the molecule. Thus, in this case, '4ins' implies a fourth version of an insulin structure. The rapid growth of the PDB means that now the four characters are assigned arbitrarily. With the restriction that the first character of the four is only assigned the digits 1 to 9, and the remaining three are either alphabetical or digits, there are a total of 419,904 possible filenames or codes. The four-character codes are used in the literature to identify PDB structures; and are used by the PDB itself to allow users to access the files. The codes are treated as case insensitive. The PDB guarantees that the four-character code is unique and will never be reused, even if a structure is deleted from the databank. The PDB format is also used by the Nucleic Acid Database. The PDB is supported by the European Bioinformatics Group (EBI) and there is a parallel mirror site for online access.[17]

The PDB organization now also uses the mmCIF format devised by crystallographers, but this does not appear to have found favour with users (See section 8.7.5).

8.7.4 MDL format
Molecular Design Limited (MDL) was a company set up in 1978 by W.Todd Wipke and Stuart Marson to apply the ideas used in computer aided engineering design to chemistry. The company was acquired by Robert Maxwell's Pergamon Press, and when that organization was broken up, it passed into the hands of Elsevier Ltd. The MDL facilities were briefly owned by Symyx Technologies Inc., but that company has been taken over by Accelrys Inc.[18] Over the years, MDL produced many computer programs and needed to have defined file formats to facilitate transfer of data between them. The popularity of the programs caused the formats to be adopted widely. The company acknowledged this in setting out the details in a paper.[19] One particular file type, variously called in the paper MOLfile, SDfile, and now often referred to as MDL format, is widely used by Diversity Suppliers for their computer readable catalogues. The format is capable of handling 3D coordinate information but when used in a catalogue the data is more frequently two dimensional (the third coordinates are set to 0.0). There are four sections to each structure definition: *Header Block, Connection Table (coordinates and connections), Data items (introduced by tags)*, and a *Delimiter*. When used for 2D data any stereo-chemical information is listed in a non standard way, that is best avoided. The ending of each structure by the delimiter ($$$$) facilitates the construction of multi-structure files. The 3- letter suffixes, (in either upper or lower case), SDF and MDL are often used for these files. The detailed specification is shown in Figure 8.5.

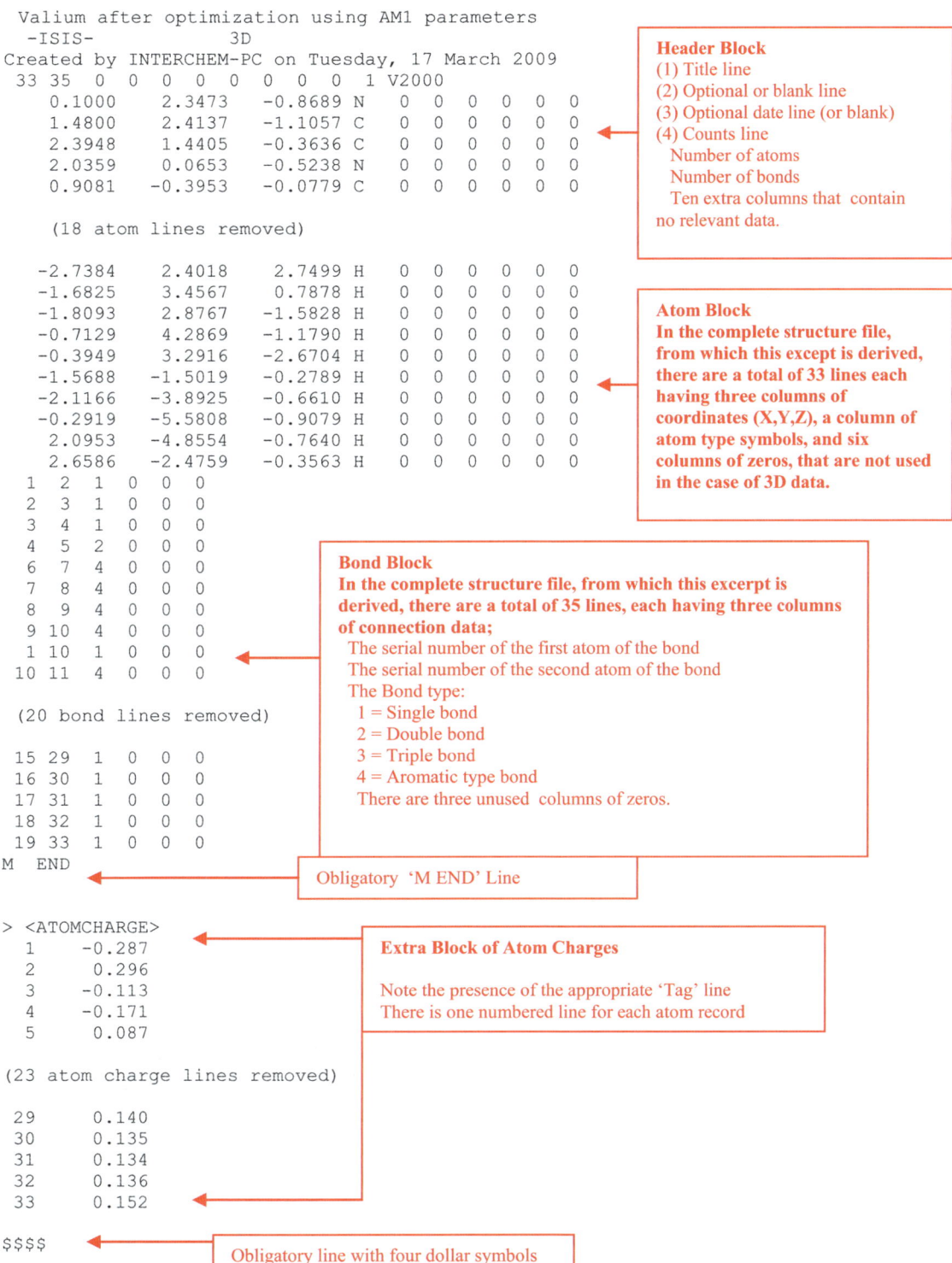

Figure 8.5
Listing of the Molecular Design Limited SD File for Diazepam

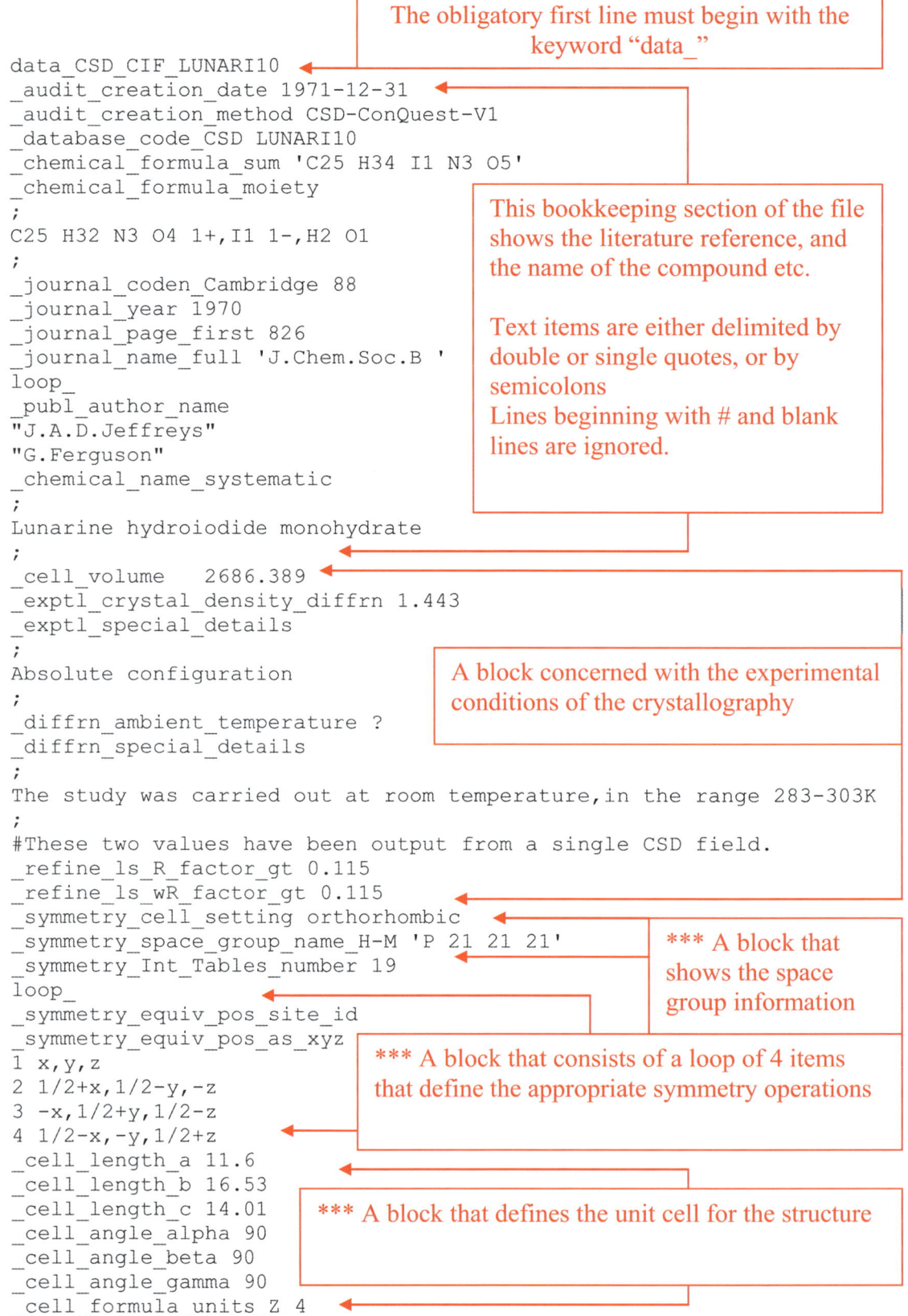

Figure 8.6
Crystallographic Information File (Part 1, to be read with Figure 8.7)
The CIF of Lunarine Hydroiodide

```
character suloop_
_atom_type_symbol
_atom_type_radius_bond
C 0.68
I 1.33
N 0.68
O 0.68
loop_
_atom_site_label
_atom_site_type_symbol
_atom_site_fract_x
_atom_site_fract_y
_atom_site_fract_z
C1  C  0.12753 -0.20180 -0.19820
C2  C  0.33689 -0.03828  0.00948
C3  C  0.30310 -0.04668 -0.08810
C4  C  0.22283 -0.10479 -0.10880
C5  C  0.11740 -0.29935 -0.05830
C6  C  0.06650 -0.32655  0.02156
C7  C  0.10348 -0.40731  0.06305
C8  C  0.07967 -0.50137  0.19619
C9  C  0.04488 -0.49507  0.30176
C10 C  0.14744 -0.45373  0.35513
C11 C  0.10955 -0.42753  0.46123
C12 C  0.02383 -0.19063 -0.26555
C13 C  0.26986 -0.33205  0.49702
C14 C  0.21285 -0.25803  0.53900
C15 C  0.26463 -0.17958  0.51350
C16 C  0.31443 -0.13101  0.35035
C17 C  0.27685 -0.12407  0.24949
C18 C  0.32128 -0.08236  0.17865
C19 C -0.04669 -0.12178 -0.23298
C20 C -0.07174 -0.10992 -0.12911
C21 C -0.04009 -0.18860 -0.07292
C22 C  0.08508 -0.21576 -0.09145
C23 C  0.17537 -0.15478 -0.04170
C24 C  0.20024 -0.14800  0.05267
C25 C  0.28619 -0.08937  0.07677
I1  I  0.02323  0.26091  0.21659
N1  N  0.07071 -0.41937  0.15276
N2  N  0.24292 -0.16729  0.40622
N3  N  0.21294 -0.40912  0.51814
O1  O  0.15043 -0.45924  0.01034
O2  O  0.41023 -0.10017  0.37407
O3  O  0.18572 -0.12489 -0.20335
O4  O -0.08867 -0.07485 -0.29199
O5  O  0.11880 -0.41015 -0.29250
#END
```

A (looped) block that defines the bonding radii for the four elements that occur in this structure, This information is not used when the CIF is interpreted.

*** This block defines the crystal coordinates of the atoms in the structure.

It starts with the keyword "loop_" Then follow definitions of the contents of the "columns" (in this case five) of data.

Note that there is no symbol that marks the end of a loop.

Note that the columns are not well aligned; in fact the displaying of the data in columns is not required. The coordinate data would satisfy the rules for CIFs by being presented as continuous characters, the only requirement is that data items are separated by "white space" (*i.e.* spaces or new lines), and that the total number of data items is a multiple of the number of column definitions.

In this CIF, the only sections that are needed to define a valid structure are the initial "data_" line plus those blocks marked ***

Figure 8.7
Crystallographic Information File (Part 2, to be read with Figure 8.6)
The CIF of Lunarine Hydroiodide

8.7.5 Crystallographic Information Files

In the last twenty years there have been moves to standardize the ways that scientific information is stored and distributed. One outcome was the specification of the **S**elf-defining **T**ext **A**rchive and **R**etrieval (STAR) file format.[20] Derived from this are the **M**olecular **I**nformation **F**ile (MIF) and **C**rystallographic **I**nformation **F**ile (CIF) formats. All of these formats are described in detail in Volume G of International Tables for Crystallography.[21] Most of the over 600 pages in that volume deal with CIF format; consequently the description presented here will be relatively brief!

The key feature of the STAR approach is that files shall be readable by a wide variety of software tools. They are tagged files, and the tags themselves are defined in Dictionaries. (This is the meaning of the words 'self-defining'). A second feature is that it is not necessary that (for example) crystal data shall precede the atom coordinates. (*i.e.* the data are unordered). A CIF makes use of loop structures wherever possible to list coordinates, *etc.*, and (strangely) lists of authors' names and affiliations (perhaps a new meaning for the phrase 'in the loop'!). An example of a CIF is shown in Figures 8.7 and 8.8 Bonding data is usually inferred from the distances between the atoms of a structure, either from data supplied in the file, or calculated from the atom coordinates. A brief glance at a CIF may give the impression that there are no rules; this is deceptive – there are rules, but their aim is to make the files easy for computers (but not necessarily *homo sapiens*) to interpret. One problem is that there are alternative ways or recording some data items; a computer program can be made to handle this, but a casual human reader is easily perplexed.

Data items in a CIF are defined by character strings beginning with an underscore character, and using underscores instead of spaces. These underscore characters can either precede the use of an individual data item, or occur in the preambles to 'loops'. The dictionaries mentioned previously are the authorities that validate the use of strings. CIFs for small molecule crystals use the CORE dictionary (coreCIF).

From a crystallographer's point of view the CIF format is appealing, because the computers attached to diffractometers can be programmed to (almost) write papers that are acceptable to journal publishers! A variant of CIF format has been introduced by the Protein Data Bank that makes use of a special dictionary (mmCIF). This has yet to find favour with most users of the PDB

8.7.6 INTERCHEM D format

The final format that we discuss is the one that is used in the INTERCHEM program that accompanies this book. The name of the *format* is derived from the construction of the *file name*. With an '8.3' name, the last of the eight characters before the full stop is an upper case D. The first seven characters can then be chosen to have some relevance to the compound being characterized. The three-character suffix *was* then reserved for a project code or a user's three initials. However over the years, reasons of computer security have demanded that different computer users have separate user accounts, so now the suffix is invariably `'DAT'`. The foregoing discussion might appear to be largely irrelevant, but its inclusion emphasises the point that continuing development of software often has unexpected effects.

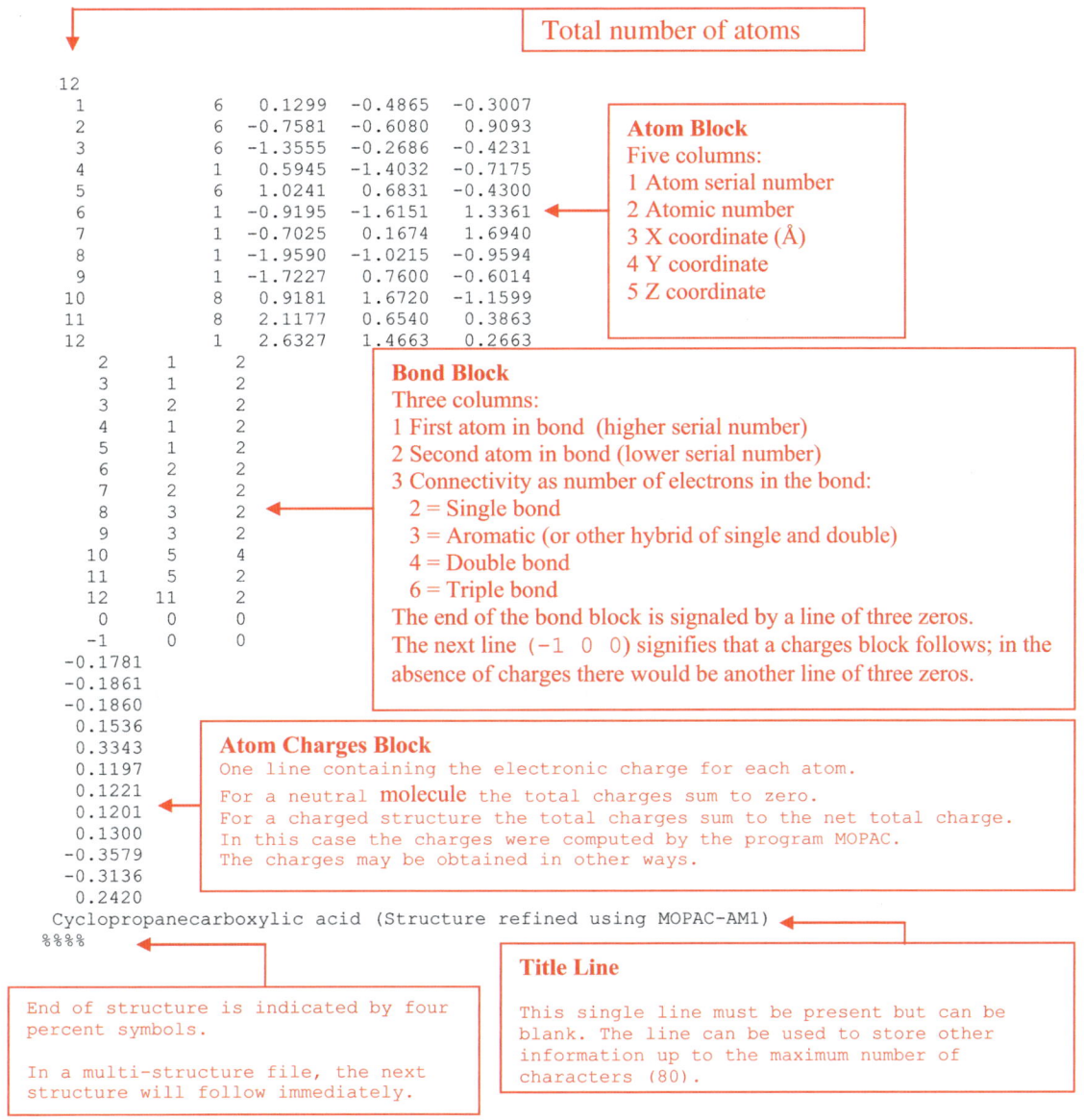

**Figure 8.8
Listing of the INTERCHEM 'D' File for
Cyclopropanecarboxylic Acid**

The details of a simple example of INTERCHEM D format are shown in Figure 8.8. While the format is *fixed* for *output from* the program INTERCHEM, *input to* the program is *free*. Thus, the program will accept a data file in which the exact spacing between columns varies. The only requirement is that 'white space' exists between columns and that each data line is terminated by a 'new line' sequence. We have examples of variants of the INTERCHEM D format in earlier chapters when we discuss the workings of the software

8.7.7 Multi-structure files

When it is necessary to hold sets of structures, it is convenient to store them in a single file. The ends of individual structures in such files need to be marked. For SDF files this is done by having a set of four dollar symbols ($$$$) as the first characters on a separate line. For INTERCHEM D files and MOL2 files ampersand characters (&&&&) and hash characters (####) respectively serve the same purpose. All of these terminators are guaranteed not to occur within the structure parts of the files.

8.8 General comments about files of 3D coordinates

All files of 3D molecular structures have essentially 3 columns of atomic coordinates; how these are used and interpreted is determined by the computer program that uses them, and this depends ultimately on the human beings involved in making measurements (crystallographers) and writing programs (programmers); it is essential that these two groups shall be in agreement.

8.8.1 Right-handed coordinate systems

The convention adopted by crystallographers and by the writers of this book is to use a *right-handed coordinate system*. Adapting this system to a computer screen, this has the positive X direction going from left to right, the positive Y direction going from bottom to top, and the positive Z direction is directed towards the viewer. To some extent this is counter-intuitive, since it is usual to associate large Z-coordinate values with a large distance *from* the viewer. This conflict can be resolved by rotating the *whole coordinate system about the X-Axis*. The direction of this axis is not changed, but the Z-Axis is now vertical, and the positive direction of the Y-axis is directed away from the viewer. This operation does not change the handedness of the coordinate system.

The *right-handed coordinate system* is made memorable by the rule that the extended digits on the *right hand* are pointing thus: thumb-> X-axis, first finger->Y-axis, middle-finger>Z-axis.

The choice of the right-handed 3D-coordinate system is not adopted by all users of computer graphics systems. Indeed two of the books, that were influential in the development era of computer graphics, disagreed in the conventions that they used, That written by W. M. Newman and R. F. Sproul[22] adopted a *left-handed* eye coordinate system, while that written by J. D. Foley and A. van Dam[23] used a *right-handed* system. These two books are worth consulting for the details of algorithms that are used in 3D display systems. Several other books written from a more mathematical stance (and useful sources of the necessary geometry and algebra), are also worth consulting. However, the book by Angell[24] adopts a left-handed system, while others by Barrett and Mackay,[25] by Harrington,[26] and by Gasson[27] use the right-handed system. You have been warned! At the time when these books were written many of the algorithms were being implemented in software. Now display hardware has advanced to the stage where the hardware does nearly all the work.

There is a further minor conflict of conventions, that concerns the 2D coordinate systems used by the display screens. When graphics (*i.e.* pictures) are being drawn, the usual conventions of coordinate geometry are used (origin at the bottom left of the screen, x-coordinate horizontal running left to right, y-coordinate vertical running bottom to top). However when text is being displayed, the origin is at the top left hand corner of the screen, and the implied coordinate system follows the conventions of writing used in European Language books.

8.8.2 The rules of engagement

Handling 3D data on a computer is governed by the geometric system being used. Some of the confusion that this generates can be alleviated by remembering a few rules; these are stated here without proofs and further justification.

(a) When data is read into a computer, unless the three columns of coordinate data are marked otherwise, they are interpreted so that the columns reading from left to right on the page are X, Y, Z orthogonal Ångstrom coordinates.

(b) If, when a 3D coordinate data set is read by the computer, one of the columns has the sign of its values changed, this corresponds to forming a mirror image; thus if the sign of the values in the X-column is changed, this corresponds to a reflection in the YZ plane.

(c) If data in two of the columns have their signs changed, there is no change in chirality, but a different view is formed.

(d) If data in all three columns are subjected to a change of sign, then a mirror image structure is formed by inversion about the origin.

(e) Swapping the roles of two columns (for example X, Y, Z → Y, X, Z) implies a change in chirality.

(f) Cyclic permutation of the columns (for example X, Y, Z → Z, X, Y) does not change the chirality of the structure.

(g) When processing crystallographic data, it is usual to process fractional cell coordinates (a, b, c) so that a → X, b → Y, c → Z; implying that the X Cartesian axis coincides with the fraction a axis, and that in the axis pairs b, Y and c, Z the two axes are separated by acute angles.

References and Endnotes. Chapter 8

[1] In this book the word *format* is used in several ways:
   (1) To label the type of file being used;
   (2) To define the way characters are laid out in the file; we term this *microformat;*
   (3) In the context of the programming language *Fortran*. We frequently use the *Fortran* conventions.

[2] F. H. Allen, *Acta Cryst.* 2002, **B58**, 380;
[2a] Web address (Home page)   http://www.ccdc.cam.ac.uk ;
[2b] Web address (Requests)   http://www.ccdc.cam.ac.uk/products/csd/request;
[2c] Web address (Teaching)   http://www/ccdc.cam.ac.uk/free_services/teaching;
(Web addresses accessed 27 July 2011).

[3] Web address http://www.cds.dl.ac.uk (accessed 27 July 2011); D. A. Fletcher, R. F. McMeeking, and D. Parkin, *J. Chem. Inf. Comput, Sci.*, 1996, **36**, 746.

[4] E. Fluck, *J. Res.Natl. Stand. Technol.*, 1996, **101**, 217.
Web address http://www.fiz-karlsruhe.de (Accessed 27 July 2011).

[5] G. H. Wood, J. R. Rodgers, S. R. Gough, and P Villars, *J. Res.Natl. Stand. Technol.*, 1996, **101**, 205. Web address http://www.tothcanada.com. (Accessed 27 July 2011)

[6] http://www.iza-structure.org/databases (Accessed 27 July 2011)

[7] F. C. Bernstein, T. F. Koetzle, G. J. Williams, E. E. Meyer, M. D. Brice, J. R. Rodgers, O Kennard, T. Shimanouchi, and M. Tasumi, *J. Mol. Biol.*, 1977, **112**, 535. (This was the general reference to the PDB while it was based at Brookhaven, which was required in a publication that made use of the data).

[8] H. M. Berman, J. Westbrook, Z. Feng, G. Gilliland, T. N. Bhat, H. Weissig, and I. N. Shindyalov, *Nucleic Acids Res.,* 2000, **28**, 235. (This is the general reference to the PDB after the transfer from Brookhaven, which is now required in a publication that makes use of the data).

[9] Web address http://www.rcsb.org/pdb/home (Accessed 27 July 2011)

[10] Web address http://www.ndbserver.rutgers.edu (Accessed 27 July 2011)

[11] *Structural Bioinformatics,* ed. P. H. Bourne and H. Weissig, Wiley-Liss, Hoboken, 2003, p 199.

[12] Crystallography of Biological Macromolecules, International Tables for Crystallography, Volume F. ed. M. G. Rossmann and E. Arnold. Kluwer, Dordrecht, 2001.

[13] J. J. Irwin and B. K. Shoichet, *J. Chem. Inf. Model.*, 2005, **45**, 177. ZINC means ZINC Is Not Commercial|!

[14] The Directory of Molecular Diversity Suppliers, Wendy Warr, Research and Markets, Dublin, 2004.

[15] Web address http://tripos.com/data/support/mol2.pdf (accessed 27 July 2011)

[16] Web address: http://www.wwpdb.org/docs.html (accessed 27 July 2011)

[17] Web address: http://www.ebi.ac.uk/pdbe (accessed 27 July 2011).

[18] Web address: http://accelrys.com (accessed 27 July 2011).

[19] A Dalby, J. G. Nourse, W. D. Hounshell, A. K. I. Gushurst, D. L. Grier, B. A. Leland, and J. Laufer, *J. Chem. Inf. Comput, Sci.*, 1992, **32**, 244.

This paper is mentioned since its publication was aimed at defining the format and other related ones that were already in widespread use. However, the paper was badly written and printed. In places it is almost unreadable because of the small size of the printing

However the following web addresse explains the format adequately
http://www.agnostic.inf.ethz.ch/MDL-SD-format.html (accessed 27 July 2011)

[20] S. R. Hall, *J. Chem. Inf. Comput. Sci.,* 1991, **31**, 326.

[21] *Definition and Exchange of Crystallographic Data, International Tables for Crystallography, Volume G.* ed. S. Hall and B. McMahon. Springer, Dordrecht, 2005.

[22] W. M. Newman and R. F. Sproull, *Principles of Interactive Computer Graphics, Second Edition*, McGraw-Hill, New York, 1979

[23] J. D. Foley and A. van Dam, *Fundamentals of Interactive Computer Graphics*, Addison - Wesley, Reading, Mass., 1982.

[24] I. O. Angell, *A Practical Guide to Computer Graphics*, Macmillan, London, 1981

[25] A. N. Barrett and A. L. Mackay, *Spatial Structure and the Microcomputer*, Macmillan, London, 1987.

[26] S. Harrington, *Computer Graphics, A Programming Approach*, McGraw-Hill, New York, 1987

[27] P. C. Gasson, *Geometry of Spatial Forms*, Ellis Horwood, Chichester, 1983.

Chapter 9
Molecular Modelling and Medicinal Chemistry

9.1 The need for new (legal) drugs, and how it is met
In today's society, among the chief concerns of people, one stands out - personal and family health and safety. This is true in both developed countries and those in the third world. It does not always feature among the prime concerns of those who govern these people, but it should do! Where healthcare *is* of serious concern to governments, the cost of providing it will rank high in their budgets. The magnitude of this cost is due to various factors. One factor is the high salaries of medical practitioners and administrators. The high cost of the drugs that are used for treatment of diseases is another. Can molecular modelling help to reduce this latter cost, and if so how? To answer these questions, we need to digress a little into the history of drug discovery.

9.1.1 Recent history
From the beginning of the Twentieth Century, advances in medicine and the introduction of new drugs were slowly responsible for tackling the scourges of many diseases, consigning them to distant memory, with the growing expectation in the populace that all the remaining diseases and afflictions would similarly disappear.

However, in 1961, it was realised that the drug *thalidomide* (1a, 1b), widely prescribed to alleviate the problems of morning sickness in pregnancy, could have severe teratogenic effects on the foetus.[1] The drug was harmless to the mothers, but many babies were born with deformities of limbs and other organs. A large number of these babies have survived into adulthood, and despite the severe handicaps, have lived normal lives, to the extent that this is possible.

(1a, *R-enantiomer*)   (1b, *S-enantiomer*)

The thalidomide tragedy has had a profound effect on the way new drugs were licensed for use. More extensive testing on animals is now required, particularly looking for teratogenic effects, before testing on human subjects is allowed. One requirement before a new drug can be licensed, is that it shall have clear advantages over drugs already being used for the same purpose.

Thalidomide has not disappeared from the pharmacy shelves completely; it has had a revival as a sedative, which was its original intended use. Thalidomide is chiral (there are two mirror image forms), and it was found following the original debacle, by testing on animals, that the harmful effects on the foetus are confined to one of the isomers. This discovery has led to the requirement, that, before any new chiral drug is licensed for use, either as a racemate (combination of both enantiomers) or as separate (resolved) enantiomers, **both** forms shall have been tested separately. It is also a requirement for drugs administered as solids (for example as tablets) that any polymorphic crystalline forms (see Chapter 7) are tested separately, since they may differ in their solubilities.

9.1.2 Economics of drug development

The bringing to market of a new drug was always expensive, but the new regulations have increased the cost so that now the total can range from $500 million to $2 billion (U.S. dollars). In the context of official reports, new drugs are referred to as 'New Molecular Entities' (NME) or 'New Chemical Entities' (NCE). In recent years (2001-2010) the average number of NMEs approved *per annum* has fallen (from around 45 in 1990) to around 23. Assuming an average figure of one billion dollars for each, the total worldwide yearly expenditure on drug development is around 23 billion dollars. This figure is almost certainly an underestimate,[2] since it takes no account of the abandoned projects.

9.1.3 The plight of pharma

To these development costs must be added the costs of manufacturing a drug from possibly expensive intermediates. To stay in business and to be profitable, a pharmaceutical firm has to recover its expenditure on a new drug from sales of drugs already in its portfolio and from any new ones, during the years remaining from their patent cover. Patents normally last for 20 years, but since they are usually taken out before any clinical trials are started, there are only about 13 years of protection left, when a drug is finally approved. After this time manufacture and sale of a drug can be done freely by so-called *generic* manufacturers, who, not having had to bear the research and development costs, can sell the drug more cheaply.[3]

These facts account for the high price of new drugs. Two consequences of this are the seemingly unfavourable reputation that many pharma companies have among the general populace, and the reluctance of public funded services, such as the United Kingdom National Health Service (NHS), to authorise the use of new drugs.

9.2 What makes a chemical compound a drug?

The answer to this question is at the heart of medicinal chemistry; it is what research in the pharmaceutical industry is all about, and what drives it. This book is not intended to be a textbook on medicinal chemistry (there are already many excellent books for this).[4,5,6] However, to understand why molecular modelling is essential to drug design, and how it fulfils its tasks, the basics of how drugs operate needs to be dealt with.

A compound may be considered to be a drug if its meets some of the following criteria:
(1)   It is capable of being administered to a human (or animal) patient
(2)   It is relatively non-toxic
(3)   It acts upon, and modifies a biological process somewhere in the patient, to the ultimate benefit of the patient.

The first criterion may seem obvious, but the ease of administering a drug can be an important factor in determining its success or failure. In the second criterion the important word is 'relatively'; a balance has to be struck. In choosing and then testing a potential drug much effort will be expended in optimising the therapeutic index (therapeutic ratio). The final criterion is, perhaps, the most controversial. Questions arise when a patient cannot personally assess the benefit, or when the benefit, consciously assessed, is seen to be only transient. The word 'somewhere' is inserted because the 'biological process' may be in an infective agent in the patient!

9.2.1 The way that drugs work – a basic classification

In most cases, a drug exerts its action by being bound to an *active site* in a large biomolecule. This will be in most cases an *enzyme,* (a protein that acts a a catalyst in a biological process), but in some cases can be a nucleic acid. The interaction between the drug and the biomolecule is a **chemical** interaction, may be only transient.

If it is the aim of the drug to treat a bacterial infection, then the active site is most likely to be in an enzyme associated with the bacterium, and the drug would be classified as an *inhibitor* or an *antagonist*. If the drug actively promotes an effect it would be classified as an *agonist*. There are other descriptive names used depending on circumstances. If the drug actively binds to the active site, then the word *ligand* is frequently used.

9.2.2 How modern drugs are designed

The process of drug design is perhaps unique in the fields of scientific endeavour, in the large number of scientific disciplines involved: mathematicians, statisticians, computer scientists, physicists, chemists (of all sorts!), pharmacologists, pharmacists, biologists, and clinicians will all be found working together in a modern pharmaceutical firm. It might be that some of the disciplines are outsourced; (stages in the design process being assigned to separate divisions of a company, or even separate companies). However, to be successful the whole process depends on cooperation. It is in the ambit of chemistry, that we place molecular modelling.

9.2.3 Some definitions

In looking at books and papers dealing with drug design, you will frequently encounter other descriptive terms, and it is well to have these (roughly) defined before we start in earnest. For the purposes of this introduction, the following will suffice, but for more precise definitions refer to those in the review by Valler and Green.[7]

Hit is used to describe a molecule or structure that has been obtained from some search process, either experimentally based, or by using a computer, perhaps by comparison with some already known structure.

Lead is used for a structure, possibly derived from a hit, that possesses some of the desired properties of a final drug. Again a more restricted definition:

Drug in this context means the final target drug.

Thus the process sequence:

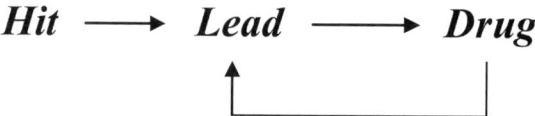

is a summary of the drug design process. The arrow in the loop indicates that an established drug can be an inspiration for further development.

The individual steps in this overall process may involve different techniques and scientific disciplines. These would be described as *in vitro* (literally in glass) when they involve chemical synthesis, chemical analysis, or pharmacological testing using non-living materials; *in vivo* when living organisms are involved, or *in silico* (in silicon!) when a computer is used.

The term **Drug-Like** acknowledges that an experienced medicinal chemist can recognize a chemical structure as being that of a likely lead. The presence of certain functional groups (and the absence of others), as well as the overall shape of a structure contributes to this.

The term **Drugability** is applied to an active site that may be amenable to control by a drug.

The **Therapeutic Index** or **Therapeutic Ratio** is initially determined from testing a drug on animals, and because of the variability of such tests it is usually defined as the ratio of LD_{50} (the dose that results in the death of 50% of the animals) and ED_{50} (the dose that produces maximal beneficial effect in 50% of the animals). Thus a drug with a high therapeutic ratio is safer than one with a low value

The Rule-of-Five. [8] This rule (often known as Lipinski's rule, or ROF) states that a chemical entity is **unlikely** (>90%) to be a good candidate for a drug for oral administration (*i.e.* by mouth as a tablet etc.) if two or more of the following criteria are met:
(1) It has molecular weight greater than 500
(2) The Clog P value is greater than 5 [see Chapter 4 (4.5.10)].
(3) There are more than five hydrogen bond donors sites (OH and NH groups)
(4) There are more than 10 hydrogen acceptors (oxygen and nitrogen atoms)

Note that the criteria should only be applied to the structure that is intended to interact with a receptor site; the formation of a salt would certainly increase the molecular weight, and conjugation with a sugar to form a glycoside would alter the hydrogen bonding character. Such changes to a structure do not count if they are only there to ease the administration of the drug. In fact they amount to formation of a ***pro-drug***.

Other factors have been shown to be significant; the number of rotatable bonds (*i.e.* single bonds that join separate parts of a structure, that allow existence of multiple conformers), and what is called *polar surface area*.[9]

9.2.4 Natural products as leads

Until recently *Materia Medica* was an important component of a medical student's training; recognition that from historical times extracts obtained from plant materials (and elsewhere) have been used in medicine for disease treatment. While some of these extracts were relatively easy to separate into their active components, it is only recently that modern chemical separation techniques (chromatography in its various forms), and analytical methods (Nuclear Magnetic Resonance [NMR], and Mass Spectrometry [MS]) have allowed a large scale routine searching of these sources.

That natural products are still an inspiration for drug designers is emphasised in the review by Harvey.[10] The total number of known natural products is estimated to be 153,000,[11, 12] and of these ~22,000 are of marine origin.[12, 13] A large proportion of marine natural products are of animal origin, whereas most non-marine sources are from plants or bacteria. Of the estimated ~250,000 plant species, less than 10% have been tested for biological activity.[14]

Significantly, many established drugs derived from natural products break the rule-of-five criteria, and these criteria tend to be relaxed (or ignored) in searching for new natural product drugs.

9.2.5 How molecular modelling helps in drug discovery

For the reasons given in the previous paragraphs the drug design process is very inefficient. The economic pressure on pharma is such that the major firms have started divesting themselves of the early stages of drug discovery.[16, 17] These have been taken up by specialist firms that combine molecular modelling with experimental tools such as nuclear magnetic resonance (nmr).[86]

9.2.6 The drug design pathway

Despite the vast amount of knowledge and insight that has been assembled in the last 100 years, about how diseases are propagated and how drugs can combat them, the chances that any one 'chemical entity' will be successful as a drug is pitifully small. One estimate of the attrition rate is that starting with 10,000 structures submitted to High Throughput Screening (HTS),[15] less than ten might reach clinical testing,[16] and only one of these might become an approved drug.

There is also pressure to continually reappraise all of the techniques (laboratory and computer based) in the light of recent successes or failures. There is, however, no doubt that computation and modelling will still have a role.[18]

9.2.7 High throughput screening

Since this is a method that uses real chemical samples and either chemical or biochemical assay methods, it is expensive in terms of both materials and labour. The latter aspect has been addressed by using robots for the sample preparation and analysis, and the use of computer for recording the results and for statistical analysis. It is mentioned here, since there is a long-established (1975) program of the United States National Cancer Institute, for testing compounds for anti-cancer activity.[19] While the emphasis of this screening program is on cancer therapy, the *structures* of compounds submitted are made available for other screening uses. (The actual samples are not distributed however).

9.3 Molecular Modelling in Drug Design.

In the design of drugs molecular modelling has hitherto come to the rescue in a variety of ways. We shall consider three of them:

(1) *Virtual High Throughput Screening (VHTS).* This is sometimes prefixed by '*In silico*' to emphasise that it is using a computer. This is often classified as one aspect of **Ligand Based Design** to emphasise that it can be used in situations where the site of action of a drug is not known.

(2) *Docking.* In contrast this is classified as **Structure Based Design** as is only feasible if the site of action of the drug is known (typically an active site of an enzyme).

(3) *Filtering Methods.* These are adjunct techniques applicable to the first two methods. In practice a design task will use all of these tools that are appropriate.

9.3.1 Virtual high throughput screening

The difference between *Virtual High Throughput Screening* (VHTS) and *High Throughput Screening* is that the virtual variety no longer incurs the expense of using real chemicals and assays; instead the screening takes place in a computer, the chemicals are replaced by a database of their structures, and the assays are replaced by parameters, calculated in subsidiary programs, that act as filters. To be used successfully it is necessary to have some information about a causal agent of the disease or condition, or some intermediary in its propagation. As a minimum, this could be the knowledge that a folk remedy could treat a disease. This would, however, require extensive investigation of the folk remedy to identify the active ingredient(s)! A more likely scenario would be that a known drug would do the job. The significant point is that details of how the known drug functions are not needed.

With the structure of the drug (or the active components of the folk remedy) to hand, the next step is to search the database(s) for structures that resemble that of the drug (usually referred to as the *Target*). In practice the databases contain the millions of structures of chemicals available for purchase from specialist manufacturers (*Diversity Suppliers*).[16, 17]

Where there is scope for invention in VHTS is in the way that the **similarity** of the target and the structures in the databases are measured. Many methods rely simply on molecular **shape**. Others take into account the **charge distribution** in the structures; this can lead to better replication of the hydrogen bonding character of the target.

One controversial aspect of VHTS is that it is often used by a second drug manufacturer to get a 'me-too' drug to rival a 'blockbluster' from another manufacturer, which has done all the hard work and incurred all the expense!

This description of **VHTS** trivializes the role of computers but it is the **High Throughput** aspects of the process that makes them indispensable. However, searching a database of a few million structures, where each individual comparison (target against database structure) is a trivial problem, makes the whole process, what is known as, *embarrassingly parallel*; if a cluster of computers is available (*i.e.*, a parallel computer) then it is relatively easy to assign each comparison to a different component machine in the cluster.

9.3.2 Docking

There is another tool that has a more rational basis, for determining whether a compound is likely to be a successful drug. The first requirement is that a structure be available for a protein that is involved in some way with the propagation of the disease, and is to be the target of the drug. It is best if the protein structure has a ligand at the active site. The aim is then to remove the ligand from the protein, and determine whether a proposed drug will fit in the cavity that is left. The task is known as *Docking*. Unfortunately there is no general consensus for the best algorithm for this task All of the available docking programs agree on the ultimate aim, but differ in how they go about achieving it. As might be expected, the performances of the programs also differ. The programs can also differ in the way that the results are scored.

In essence the docking process involves the following steps:

(1) Accessing the appropriate protein structure. Ideally this must be one that incorporates a ligand at the active site of the protein. The reason for this is that a structure of an *apo* protein (one without an incorporated ligand) is likely to differ from that of one with a ligand.

(2) Removing the ligand from the structure, so as to leave a cavity.

(3) Adding hydrogen atoms to the residual protein structure, since they are typically not present in a protein structure obtained from the Protein Data Bank.

(4) Defining the shape of the active site.

(5) Identifying the favourable interactions between protein and drug. These are likely to be hydrogen bonds, and possibly ionic interactions.

(6) Placing each of the potential drug structures that are to be tested in the active site. In practice multiple conformations of the drug structure would need to be generated, and tested separately.

(7) Adjusting the position of the ligand with respect to the cavity, so that the energy of the protein+drug complex is minimized. This implies six degrees of freedom; three positional and three rotational. Small variations in the drug conformation can also be allowed for, at this stage. The energy minimization is done by molecular mechanics calculations.

(8) Allowing changes in the conformation of the protein, to accommodate the binding of the drug.

Table 9.1 Part 1
Docking Programs

Name	Description, Features	Notes	Ref.	Rank
ADAM	An early program, identification of likely hydrogen bonds between ligand and protein.		34	15
Arguslab			35	Equal 17
AutoDock	Monte Carlo simulated annealing, traditional and Lamarkian genetic algorithms.	C	36	2
CDocker	Grid based Molecular Dynamics methods		37	Equal 1
Darwin	Flexible molecules, parallel version of CHARM molecular mechanics.		38	**
Divali			39	**
DOCK	Single graph matching algorithm. The first docking program that was published	C, F	40	1
DockVision	Two separate programs called Research and Gamma.		41	**
e-Hits	Fast, exhaustive, flexible ligand docking		42	**
EUDOC	User defined box enclosing the binding pocket is needed		43	**
FlexE	Permits variations of the receptor conformations		44	**
FlexX	Incremental construction of ligand in the binding pocket		45	4
FLOG	Atoms (except H) in the ligand are classed as anion, cation, donor, acceptor, polar etc.		46	Equal 17
FRED	Neither the source nor the algorithms in OpenEye's software are disclosed.	AR, AF	47	10
FTDOCK	Uses Fourier correlation algorithm for shape matching of ligand and cavity		48	**
GEMDOCK	Generic Evolutionary Method		49	Equal 11
GLIDE	Grid based LIgand Docking with Energetics		50	5
GOLD	Genetic Optimisation for Ligand Docking		51	3
Hammerhead	High affinity portion of ligand is the termed the Head		52	19
ICM	Internal Coordinate Modeling		53	7
Libdock	Site-Directed docking; ligand flexibility allowing conformational changes in ligand.		54	**
Ligandfit	Monte-Carlo modifications of ligand conformations to allow optimal fitting		55	9
LIGIN	Molecular docking using surface complementarity		56	**
LUDI	Incremental construction of ligand		57	8
MCDOCK	Monte Carlo Simulation technique		58	**
Molegro			59	16

Table 9.1 Part 2
Docking Programs (continued)

Name	Description, Features	Notes	Ref.	Rank
MOLSDOCK	Protein-ligand docking using Mutually Orthogonal Latin Squares		60	**
Ph4dock	Pharmacophore based protein-ligand docking		61	**
PLANTS	Use of an Ant Colony Model for Molecular Docking		62	***
Prodock	Use of internal coordinates for both protein and ligand		63	**
PRO_LEADS	Flexible docking using Tabu search and empirical estimate of binding affinity		64	**
QXP	Full conformational searching for flexible cyclic and acyclic structures		65	Equal 13
Rosetta	Protein-protein, rigid body, Monte-Carlo docking		66	Equal 13
SANDOCK	Guided matching algorithm to fit ligand atoms into protein binding pocket		67	**
SFDock	Genetic algorithm and 'tabu' search, protein-protein, small molecule-protein docking	C, PA	68	**
Soft docking	Soft docking by using attenuated form of the Lennard-Jones potential.		69	6
VINA	An improved scoring function for use with AutoDock.	A	70	**
YUCCA	Heuristic local search for rigid small molecule docking.		71	**

Notes for Table 9.3 (Parts 1 and 2)
The **Description, Features** column lists those features mentioned in a summary by the authors. Absence of a description usually means that the algorithms and other features are not disclosed.

Abbreviations in the **Notes** column:
Source language: C = C or C++;   F = Fortran;   Absence of details of source language usually means that the language is C
Availability of source code: A = Available;   PA = Partially available.
Licensing to Academic institutions: AF = Free;   AR = reduced charge
General licensing: OS = Open Source. In absence of any indication the source is assumed to be proprietary.

The **Rank** column refers to the 'popularity' of the program revealed by the number of references to the program *name* found in 'Web of Science'. The logical AND function with the string '*ligand docking*' was used in the search. Under these circumstances the program PLANTS (***) received a large number of hits, ranking third overall. We believe that this is spurious. The ranging of all the other programs were adjusted accordingly. The marking ** signifies that the number of hits was insignificant (<10

Of these individual steps, several (numbers 1, 2, 3 and to some extent 5 and 6) are straightforward. Number 4 is where there is most variation and scope for ingenuity, while numbers 7 and 8 imply iterative processes of the sort that are common in molecular mechanics structure optimizations.

There are a large number of different protein-ligand docking programs; to compare impartially the performance of the all of them is acknowledged to be difficult.[20] This is chiefly because it is difficult for any one person to become familiar (at the same time) with the operation of a range of, what by any standards are complex, programs. In the ambience of typical conferences and meetings, it may be that decisions are made on the basis of who makes the best presentation, or who has the glossiest brochures! We have listed the docking programs, which we know of, in Table 9.1. The comparisons that are made in the table are limited by the amount of detail of the methods that the authors disclose, and our ability to summarise this in one line! There *are* papers that make meaningful comparisons of these programs, particularly with regard to their performance with different classes of protein receptors.[21, 22, 23, 24, 25, 26] In the outline docking scheme (9.3.2) step number 6 is the one that offers most scope for new algorithms.[27, 28, 29, 30, 31, 32, 33] However, it is our belief that the last word has yet to be written on docking.

For those willing to try new docking algorithms, there are papers that offer lists of standard systems (protein-ligand pairs), lists of good ligands,[72] and list of detractor structures (decoys).[73] The assessment of the success of a docking process is best handled by a separate *scoring* method rather than the algorithm used by the process itself. To add to the confusion there are a large number of such scoring methods.[74]

If the steps of the docking process are carefully analysed, it is clear that crucial steps 4 and 5, are analogous to the processes of Virtual High Throughput Screening. Thus the shape of the protein cavity, (of a possibly unidentified protein when applying VHTS!), is being defined by the shape of the excised ligand. The process of docking is then equivalent to a shape matching process of the ligand (on the one hand) and the putative drug(s) on the other. Other authors have compared shape matching favourably with docking[26] particularly in the early stages of VHTS.

9.3.3 Additional Filtering Processes
There are a number of adjunct tools that are used as filters, to reduce the number of candidates that are submitted to the major design processes; VHTS, Docking, and also HTS.

For one of these tools we have (yet) another acronym *ADMET*; standing for **A**dsorption, **D**istribution, **M**etabolism, **E**xcretion, **T**oxicity. (Sometimes toxicity is treated separately; thus you may also encounter *ADME*). Treating these aspects of a *chemical* compounds *biochemical properties* together makes sense. Working by analogy from what is known about how the human (and/or animal) body handles ingested substances, it is possible to predict what will happen with new compounds. We have already met some of the parameters that are used here; ClogP and other Rule-of-Five quantities. To these we can add pK_a; (this can be calculated[75] by methods similar to those used for calculating ClogP), hydrogen-bond acceptor strengths,[76] and dipole moments, that are available from semi-empirical molecular orbital calculations (see Chapter 3). One way of dealing with toxicity is to look for signature toxic molecular fragments.[77]

Molecular Modelling and Medicinal Chemistry

In the forgoing discussion of molecular interactions, the underlying factor is *molecular shape*. Each of the docking programs uses it in some way, and there are newer algorithms that have yet too be incorporated into methods.[78] Whole books have been written about it, [79, 80] and more recent reviews,[81, 82, 83, 84] and papers are worth consulting.[85]

9.3.4 Isosteres - Variations on a theme!

The ideas of molecular shape are often taken further; if we have a promising lead compound - how can we modify it so that other leads can be found? One strategy is to replace a group of atoms by another group of similar number, shape, and size. Often the number + shape + size criteria are relaxed, or modified in other ways. The basic idea is not new, and underlies the thinking of different varieties of chemists. Such groupings of atoms are known as *Isosteres*. We illustrate the ideas in the Figures 9.1, 9.2, and 9.3.

The concept of isosterism is shown in these tables as a progression from the simple ideas of basic chemistry (the similar behaviour of halogen atoms in organic molecules) to what are termed ***bioisosteres***. A good account with many examples is given in the book by Silverman.[6] One classic example will suffice here. Because sulfanilamide (2) acts as an isostere of p-aminobenzoic acid (3), which is essential for the replication of bacteria, this results in sulfanilamide being bacteriostatic.

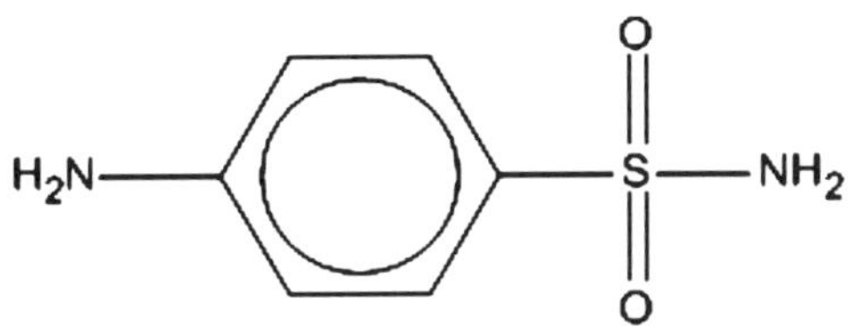

(2) 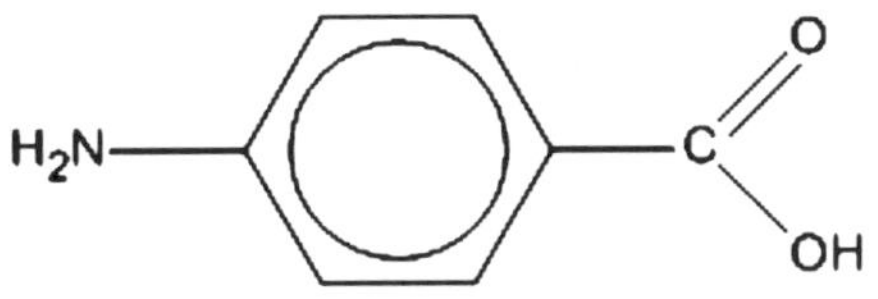   (3)

CO	N_2	Two atoms of similar size
CO_2	N_2O	Three atoms of similar size
CH_2N_2	CH_2O	Four or five atoms of similar size
Br^-	CN^-	Two anions of similar size
MnO_4^-	OsO_4	Two similar powerful oxidising agents

Figure 9.1
Classical Isosteres - The Basic Idea

Univalent Atoms or Groups

F	OH	NH_2	CH_3
Cl	SH	PH_2	
Br	iso-Propyl		
I	tert-Butyl		
CO_2R	CO.SR	$CO.CH_2R$	

Bivalent atoms or groups

O	S	Se	CH_2	NH

Tervalent atoms or groups

-N=	-P=	-As=	-CH=

Quadrivalent atoms

C	Si

Ring equivalents

-CH=CH-	-S-		*e.g.* Benzene ---> Thiophene
-CH=	-N=		*e.g.* Benzene ---> Pyridine
$-CH_2-$	-NH-	-O-	*e.g.* Cyclohexane ---> piperidine, dioxan

Figure 9.2
Classical Isosteres - More Ideas

Carbonyl group
\>CO
\>C=C(CN)$_2$ \>SO   \>SO$_2$
-SO$_2$NR-   -CO.NR-   -CH)CN)-

Carboxylic acid group
-CO$_2$H
-SO$_2$NHR-   -SO$_3$H   -PO(OH)NH$_2$
-PO(OH)OEt   -CO.NH.CN

Hydroxyl group
-OH
-NH.CO.R   -NH.SO$_2$R   -CH$_2$OH   -NH.CO.NH$_2$ -NH.CN
CH(CN)$_2$

Halogen
-F   -Cl   -Br   -I
-CN   -CF$_3$ -N(CN)$_2$ -C(CN)$_3$

Urea / thiourea
-NH.CO.NH$_2$ -NH.CS.NH$_2$
-NH.C(=NH).NH$_2$ -NH.C(=CH.NO$_2$).NH$_2$

Pyridine rings
Pyridine
p-Nitrophenyl

Spacer groups
-(CH$_2$)$_3$-
1,4 -Disubstituted benzene ring

Peptide linkage
-CO.NH-
-CH=CH-   -CH$_2$CH$_2$-   -CO.CH$_2$-   -SO$_2$NH-   -CO.NMe-   -CS.NH-
-NH.CO-   -CH$_2$S-
1,2-Disubstituted cyclopropane ring   1,2-Disustitued cyclopropenyl ring
Replacement of L-amino acids by D-amino acids (D-Ala in particular)

Figure 9.3
Non-classical Isosteres or Bioisosteres
It is to be understood that, within each italicised class of groups, a group or atom on the first line may be substituted by a group on the second or following lines, with the expectation that (roughly) similar space will be occupied.

References and Endnotes. Chapter 9

[1] Thalidomide was patented in 1957 by the firm Chemie Grünenthal (G.B.P. 768821). Thalidomide is chiral; that is to say it exists in two mirror image forms (1a and 1b in the main text). It was found that the toxicity was due to the *S-enantiomer* (1b). This discovery was largely responsible for the requirement in subsequent trials of chiral drugs, that both enantiomers should be tested separately, for both efficacy and toxicity. Although only one form of thalidomide is toxic, the fact that both forms are inter-convertible under physiological conditions would not have averted the tragic effects from the use of this drug, if only the *R*-enantiomer (1a) been used.

[2] R. Millin, *Chem. Eng. News,* 2011, 21 February, p.12. The total global expenditure on Research and Development is $140 billion (US).

[3] The abbreviation of ***pharmaceutical*** to ***pharma*** is common, and is used both as an adjective, and a noun. It is identical in the singular and plural forms (the plural is not *pharmae*!). It is also found as a qualifier in company names.

[4] F. D. King (ed.) *Medicinal Chemistry: Principles and Practice*, Royal Society of Chemistry, Cambridge, 1994.

[5] A. Gringauz, *Introduction to Medicinal Chemistry: How drugs Act and Why,* Wiley-VCH, New York, 1997.

[6] R. B. Silverman, *The Organic Chemistry of Drug Design and Drug Action, Second Edition,* Elsevier Academic Press, San Diego, 2004.

[7] M. J. Valler and D. Green, *Drug Discovery Today,* 2000, **5**, 286, define 'Hit' and 'Lead' in a more restricted fashion:

> ***Hit*** A Molecule with robust-response activity in a primary screen, and known, confirmed structure.

> ***Lead*** A representative of a compound series with sufficient potential (as measured by potency, selectivity, pharmokinetics, physiochemical properties, absence of toxicity, and novelty) to progress to a full development programme.

[8] C. A. Lipinski, F. Lombardo, B. W. Dominy, and P. J. Feeny, *Advanced Drug Reviews,* 1997, **23**, 3; 2001, **46**, 2.

[9] D. F. Veber, S. R. Johnson, H.-Y. Cheng, B. R. Smith, K. W. Ward, and K.D. Kopple., *J Med. Chem.*, 2002, **45**, 2615.

[10] A. Harvey, *Drug Discovery Today*, 2000, **5**, 294.

[11] *Dictionary of Natural Products on DVD v19.1*, Taylor and Francis, Boca Raton, 2010.

[12] J. W. Blunt, B. R. Copp, M. H. G. Munro, P. T. Northcote, and M. R. Prinsep, *Natural Product Reports,* 2011, **28**, 196.

[13] *MarinLit* database. http://www.chem.canterbury.ac.nz/marinlit/marinlit.shtml (accessed 25/7/2011).

[14] R. Verpoorte, *Drug Discovery Today*, 1998, **3**, 232.

[15] High Throughput Screening (HTS) is a collection of synthetic plus analytical techniques, run in a parallel fashion, so that large numbers of compounds may be studied. It is unlikely that bulk samples of all the compounds are retained. Compounds that show promise are then likely to be synthesised on a gram scale. The parallel synthesis operations are often referred to as ***Combinatorial Chemistry.***

[16] D. E. Clark and C. G. Newton, *Drug Discovery Today*, 2004, **9**, 492; D. E. Clark, *Drug Discovery Today*, 2007, **12**, 62; D. E. Clark, *Drug Discovery Today*, 2011, **16**, 147.

[17] H. Zhao, *Drug Discovery Today*, 2011, **16**, 158.

[18] W. L. Jorgensen, *Science,* 2004, **303**, 1813.

[19] *The Directory of Molecular Diversity Suppliers*, W. Warr, Dublin, Research and Markets.

It is likely that the first Diversity Supplier was the firm *Maybridge Chemical Company*, set up in 1962, in Tintagel, Cornwall, U.K., by Dr. R. J. Bridgewater and named after his wife, May. It moved to its present address in Trevillett, Cornwall in 1971, and is now a subsidiary of Thermo Fischer Scientific. Its original offerings were gleaned from the compounds stored in University Chemistry Departments made by Ph.D. students in the course of their work. It later evolved into making speciality chemicals, often based on particular organic structural themes.

However it is possible that the *Aldrich Library of Rare Chemicals* may predate this. This resulted from the forays made to Europe by the founder of the Aldrich Chemical Company, Dr. Alfred Bader, who gathered the material from various sources, including University research groups. The aim of this enterprise was somewhat different from the diversity suppliers as its name implies

A major boost to this micro-industry came about when the Soviet era ended in Russia. Many new Diversity Suppliers were then set up in Russia and Ukraine, with branches in the United States. These, as well as others throughout Europe and the United States, often followed the same evolutionary progress as Maybridge.

[20] J. C. Cole, C. W. Murray, J. W. M. Nissink, R. D. Taylor, and R. Taylor, *Proteins*, 2005, **60**, 325.

[21] D. Plewczynski, M. Laźniewski, R. Augustyniak, K. and K. Ginalski, *J. Comp. Chem.*, 2010, **32**, 742.

[22] M. D. Cummings, R. L. DesJarlias, A. C. Gibbs, V. Mohan, and E. P. Jaeger, *J. Med. Chem.* 2005, **48**, 962.

[23] S. F. Sousa, P. A. Fernandes, and M. J. Ramos, *Proteins*, 2006, **65**, 15.

[24] A. R. Leach, B. K. Shoichet, and C. E. Peishoff, *J. Med. Chem.*, 2006, **49**, 5851.

[25] J. A. Erikson, M. Jalaie, D. H. Robertson, R. A. Lewis, and M. Vieth, *J. Med. Chem.*, 2004, **47**, 45.

[26] P. C. D. Hawkins, A. G. Skillman, and A. Nicholls, *J. Med. Chem.*, 2007, **50**, 74.

[27] G. Hessler, M. Zimmermann, H. Matter, A. Evers, T. Naumann, T. Lengauer, and M. Rarey, *J. Med. Chem.*, 2005, 48, 6675.

[28] K. Lee, C. Czaplewski, S.-Y. Kim, and J. Lee, *J. Comp. Chem.*, 2005, **26**, 78.

[29] G. Bottegoni, I. Kufareva, M. Trotov, and R. Abagyan, *J. Med. Chem.*, 2009, **52**, 397.

[30] A. Volkamer, A. Griewel, T. Grombacher, and M. Rarey, *J. Chem. Inf. Model.*, 2010, **50**, 2041.

[31] R. G. Coleman and K. A. Sharp, *J. Chem. Inf. Model.*, 2010, **50,** 589.

[32] W. Shin, S. A. Hyun, C. H. Chae, and J. K. Chon, *J. Chem. Inf. Model.*, 2009, **49**, 1879.

[33] M. Rueda, G. Bottegoni, and R. Abagyan, *J. Chem. Inf. Model.*, 2010, **50**, 186.

References 34 to 71 are for Table 9.1

[34] M. Y. Mizutani, N. Tomioka, and A. Itai, *J. Mol. Biol.*, 1994, **243**, 310

[35] http://www.arguslab.com (accessed 5th April 2011).

[36] G. M. Morris, D. S. Goodsell, R. S. Halliday, R. Huey, W. E. Hart, R. K. Belew, and A. J. Olson, *J. Comp. Chem.*, 1998, **19**, 1639; G. M. Morris, D. S. Goodsell, R. Huey, and A. J. Olson, *J. Comp.-Aided Mol. Des.*, 1996, **10**, 293.

[37] G. Wu, D. L. Robertson, C. L. Brooks III, and M. Vieth, *J. Comp. Chem.*, 2003, **24**, 1549.

[38] J. S. Taylor and R. M. Burnett, *Proteins*, 2000, **41**, 173

[39] T. N. Hart and R. J. Read, *Proteins,* 1992, **13**, 206.

[40] T. J. A. Ewing and I. D. Kuntz, *J. Comp. Chem.*, 1997, **18**, 1175.

[41] http://dockvision.com/DV/dockvision.html (accessed 5th May 2011).

[42] Z. Zsoldos, D. Reid, A. Simon, S. B. Sadjad, and A. P. Johnson, *J. Mol. Graph. Mod.*, 2007, **26**, 198.

[43] Y.-P. Pang, E. Perola, K. Xu, and F. G. Prendergast, *J. Comp. Chem.*, 2001, **15**, 1750; E. Perola, K. Xu, T. M. Kollmeyer, S. H. Kaufmann, F. G. Prendergast, and Y.-P. Pang, *J. Med. Chem.*, 2000, **43**, 401.

[44] H. Claußen, C. Buning, M. Rarey, and T. Lengauer, *J. Mol. Biol.*, 2001, **308**, 377.

[45] M. Rarey, B. Kramer, T. Lengauer, and G. Klebe, *J. Mol. Biol.*, 1996, **261**, 470.

[46] M. D. Miller, S. K. Kearsley, D. J. Underwood, and R. P. Sheridan, *J. Comput.-Aided Mol. Des.*, 1994, **8**, 153.

[47] http://www.eyesopen.com (accessed 6th May2011); M. McGann, *J. Chem. Inf. Model.*, 2011, **51**, 578; G. B. McGaughey, R. P. Sheridan, C. J. Bayly, J. C. Culberson, C. Kreatsoulas, S. Lindsley, V. Maiorov, J.-F. Truchon, and W. D. Cornell, *J. Chem. Inf. Model.*, 2007, **47**, 1504; T. Tuccinardi, M. Botta, A. Giordano, and A. Martinelli, *J. Chem. Inf. Model.*, 2010, **50**, 1432.

[48] H. A. Gabb, R. M. Jackson, and M. J. E. Sternberg, *J. Mol. Biol.*, 1997, **272**, 106.

[49] J.-M. Yang and C.-C. Chen, *Proteins*, 2004, **55**, 288.

[50] R. A. Friesner, J. L. Banks, R. B. Murphy, T. A. Halgren, J. J. Klicic, D. T. Mainz. M. P. Repasky, E. H. Knoll, M. Shelley, J. K. Perry, D. E. Shaw, P. Francis, and P. S. Shenkin, *J. Med. Chem.*, 2004, **47**, 1739; R. A. Friesner, R. B. Murphy, M. P. Repasky, L. L. Frye, J. R. Greenwood, T. A. Halgren, P. C. Sanschagrin, and D. T. Mainz, *J. Med. Chem.*, 2006, **49**, 6177.

[51] G. Jones, P. Willett, and R. C. Glen, *J. Mol. Biol.*, 1995, **245**, 43; G. Jones, P. Willett, R. C. Glen, A. R. Leach, and R. Taylor, *J. Mol. Biol*, 1997, **267**, 727.

[52] W. Welch, J. Ruppert, and A. N. Jain, *Chemistry and Biology,* 1996, **3**, 449.

[53] R. Abagyan, M. Totrof, and D. Kuznetsov, *J. Comp. Chem.*, 1994, **5**, 488.

[54] D. J. Diller and K. M. Merz Jr., *Proteins*, 2001, **43**, 113; S. N. Rao, M. S. Head, A. Kulkarni, and J. M. LaLonde, *J. Chem. Inf. Mod.*, 2007, **47**, 2159.

[55] C. M. Venkatachalam, X. Jiang, T. Oldfield, and M. Waldman, *J. Mol. Graph. Modell.*, 2003, **21**, 289.

[56] V. Sobolev, R. C. Wade, G. Vriend, and M. Edelman, *Proteins*, 1996, **25**, 120..

[57] H.-J. Böhm, *J. Comp.-Aided Mol. Des.*, 1992, **6**, 61.

[58] M. Liu and S. Wang., *J. Comp.-Aided Mol. Des.*, 1999, **13**, 435.

[59] http://www.molegro.com (Accessed 5th May 2011)

[60] S. N. Viji, P. A. Prasad, and N. Gautham, *J. Chem. Inf. Mod.*, 2009, **49**, 2687.

[61] J. Goto, R. Kataoka, and N. Hirayama, *J. Med. Chem.*, 2004, **47**, 6804.

[62] O. Korb, T. Stützle, and T. Exner, *Lect Notes Com. Sci.*, 2006, **4150**, 247; O. Korb, T. Stützle, and T. Exner, *Swarm Intel.*, 2007, **1**, 115; O. Korb, T. Stützle, and T. Exner, *J. Chem. Inf. Model.*, 2009, **49**, 84.

[63] J.-Y., Trosset and H. Scheraga, *J. Comp. Chem.*, 1999, **20**, 412.

[64] C. A. Baxter, C. W. Murray, D. E. Clark, D. R. Westhead, and M. D. Eldridge, *Proteins*, 1998, **33**, 367; C. W. Murray, C. A. Baxter, and A. D. Frenkel, *J. Comp.-Aided Mol. Des.*, 1999, **13**, 547..

[65] C. McMartin and R. S. Bohacek, *J. Comp.-Aided Mol. Des.*, 1997, **11**, 333.

[66] J. J. Gray, S. Moughon, C. Wang, O. Schueler-Furman, B. Kuhlman, C. A. Rohl, and D. Baker, *J. Mol. Biol.*, 2003, **331**, 281.

[67] P. Burkhard, P. Taylor, and M. D. Walkinshaw, *J. Mol. Biol.*, 1998, **277**, 449.

[68] T. Hou, J. Wang. L. Chen, and X. Xu, *Protein Engineering*, 1999, **12**, 639.

[69] A. M. Ferrari, B. Q. Wei, L. Constantino, and B. K. Shoichet, *J. Med. Chem.*, 2004, **47**, 5076.

[70] O. Trott and A. J. Olson, *J. Comp. Chem.*, 2009, **31**, 455.

[71] V. Choi, *Chemistry and Biodiversity,* 2005, **2**, 1517.

References 72 onwards are for the main text

[72] M. J. Hartshorn, M. L. Verdonk, G. Chessari, S. C. Brewerton, and W. T. M. Mooij, *J. Med. Chem.*, 2007, **50**, 726.

[73] A. P. Graves, R. Brenk, and B.K. Shoichet, *J. Med. Chem.*, 2005, **48**, 3714.

[74] A. V. Ischenko and E. I. Shakhnovich, *J. Med. Chem.*, 2002, **45**, 2770; R. Wang, Y. Lu, and S. Wang, *J. Med. Chem.*, 2003, **46**, 2287; A. E. Klon, M. Glick, and J. W. Davies, *J. Med. Chem.*, 2004, **47**, 4356.

[75] B. J. Tehan, E. J. Lloyd, M. G. Wong, W. R. Pitt, J. G. Montana, D. T. Manallack, and E. Gancia, *Quant. Struct.-Act. Relat.*, 2002, **21**, 457; S. Jelfs, P. Ertl., and P. Selzer, *J. Chem. Inf. Model.*, 2007, **47**, 450.

[76] J. Schwöbel, R.-U. Ebert, R. Kühne, and G. Schüürmann, *J. Chem. Inf. Model.*, 2009, **49**, 956; M. Nocker, S. Handschuh, C. Tautermann, and K. R. Liedl, *J. Chem. Inf. Model.*, 2009, **49**, 2067.

[77] A. Chuprina, O. Lukin, R. Demoiseaux, A. Buzko, and A. Shivanyuk, *J. Chem. Inf. Model.*, 2010, **50**, 470.

[78] L. Mavridis, B. D. Hudson, and D. W. Ritchie, *J. Chem. Inf. Model.*, 2007, **47**, 1787.

[79] H. Kubinyi (ed.), *3D QSAR in Drug Design*, ESCOM, Leiden, 1993.

[80] P. M. Dean (ed), *Molecular Similarity in Drug Design*, Blackie, Bishopbriggs (Glasgow), 1995.

[81] S. W. Muchmore, J. J. Edmunds, K. D. Stewart, and P. J. Hajduk, *J. Med. Chem.*, 2010, **53**, 4830.

[82] R. J. Zauhar, G. Moyna, L. Tian, Z Li, and W. J. Welsh, *J. Med. Chem.*, 2003, **46**, 5674

[83] A. Nicholls, G. B. McGaughey, R. P. Sheridan, A. C. Good, G. Warren, M. Mathieu, S. W. Muchmore, S. P. Brown, J. A. Grant, J. A. Haigh, N. Nevins, A. N. Jain, and B. Kelley, *J. Med. Chem.*, 2010, **53**, 3862.

[84] C. Bissantz, B. Kuhn, and M. Stahl, *J. Med. Chem.*, 2010, **53**, 5061; D. K. Agrafiotis, D. Bandyopadhyay, J. K. Wegner, and H. van Vlijmen, *J. Chem. Inf. Model.*, 2007, **47**, 1279.

[85] M. J. Vainio, J. S. Puranen, and M. S. Johnson, *J. Chem. Inf. Model.*, 2009, **49**, 492.

[86] D. S. Sem, B. Bertolaet, B. Baker, E. Chang, A. D. Costache, S. Coutts, Q. Dong, M. Hansen, V. Hong, X. Huang, R. M. Jack, R. Kho, H. Lang, C.-T. Ma, D. Meininger, M. Pellacchia, F. Pierre, H. Villar and L. Yu, *Chemistry and Biology*, 2004, **11**, 185.

Chapter 10
Using Interprobe Software for Drug Discovery

10.1 Overview
First we show how the program QUICKSCAN can be used, together with INTERCHEM, to find likely analogues of known drugs, from a database of compounds in the catalogues of Diversity Suppliers. In a second series of experiments we investigate the use of the merge facilities provided in INTERCHEM to perform manual docking of drug candidates into protein cavities. These are not trivial tasks, and before you try the manual docking experiments it would be well to revise the techniques for splitting up and editing protein structure files (see Chapter 5).

10.2 Setting up the database using Cygwin.
One of the optical discs that are supplied with this book is a DVD. This contains 4.3 million 3D structures of small organic molecules in INTERCHEM 'D' format. These are derived from the 2D structures used in catalogues by Diversity Suppliers. In order to contain them on one disc it was necessary to compress them using the Linux *gzip* program. For the program QUICKSCAN to make use of this data it is necessary to uncompress the files. The program *gzip* is not available in the Windows operating systems; however the program *Cygwin* comes to the rescue, by allowing a version of Linux to run alongside Windows.

Refer to Chapter 2 for the philosophy behind *Cygwin* and details of how to install it. Henceforth we shall assume that it is installed and working correctly on your computer. There are several steps in transferring the database to your computer.

(1) Create a new folder on your system called `database43`. If you have partitioned the disc(s) on your system, place this new folder in a partition that you are using for these experiments. Thus the full *path* of the would be (say):
```
E:\database43
```

(2) On the other disc (a CD) that is supplied with this book, there is a folder (directory) called `scripts`. In this is a file called:
```
install_script_d
```
Copy this file to the folder that you have just created using standard Windows methods..

(3) Place the DVD containing the compressed database into the optical disc reader of your system. We assume that this reader has drive letter `D:`. [But see stage (6) below].

(4) Click on the Cygwin icon on your desktop. This will start the Cygwin program and provide a 'command line' window

(5) At the prompt ($ symbol) type the two commands (without the comments!)

Command	Comment
`cd ../../cygdrive`	Change the directory
`ls -l`	List the contents of this directory

This should give the following response or something similar

```
d---------+ 1 ????????  ????????  0 Mar 10 21:00  c
drwxr-xr-x 1 User_name Group name 0 Mar 16 20:09  d
d---------+ 1 ????????  ????????  0 Mar 10 21:00  e
```

The information in the final column of lines 1 and 3 shows that you have two partitions (or discs) on your hard drive system. These are the Windows partitions `C:` and `E:`. The entry in the final column of line 2 shows that the optical drive is present as Windows drive letter `D:` (this line will only be present if there is a disc in the drive).

(6) This step is *only* necessary if the optical drive has a label other than `D:`. In this case it will be necessary to edit the file `install_script_d` so that in the lines such as:

```
cp  /cygdrive/d/compressed/ambinter/*  .
```

are changed to (for a case where the drive letter is : `Q:`)

```
cp  /cygdrive/q/compressed/ambinter/*  .
```

There are eighteen of these lines; note that the full stop at the end of each of the lines *must* be present. Store the edited file as (say) `install_script_q`.

(7) Finally, set up the databases by executing the appropriate script by typing

```
cd /cygdrive/e/database43
./install_script_d  or ./install_script_q
```

The whole process takes about 30 minutes. The first (trivial) stage is to set up the eighteen folders that hold the separate file systems (one for each Diversity Supplier); the second stage is the copying of the compressed files to these folders. The most time consuming stage is the un-compression of these files using *gunzip*.

10.2.1 The importance of the full stop (and some other hints)

The full stop has been alluded to. Its presence means 'the current directory' in UNIX and LINUX systems; it is also used in the same way by Windows systems in command line work, but then Windows discourages this! Without that full stop the copy (`cp`) command in step (6) would not work.

The scripts mentioned in step (7) are one form of executable program. If the system recognises them as executable, *and the directory where they are, is recognised is also set up correctly,* typing the name of the script will cause it to run. But there can be a couple of problems; the phrase in italics in the last sentence implies that the so-called PATH is set up correctly, and that can cause problems itself. So preceding the name of the script by `./` means in step (7): 'in the current directory run the script'.

The second potential problem is this; the script itself needs to have permission to be treated as an executable. To ascertain if this is so, go to the appropriate directory, and list the contents:

```
cd /cygdrive/e/database43
ls -l
```

The script should appear in the list thus:

```
-rwxr-xr-x+ <user-name> <group-name> 2242 Mar 17 15.00 install_script_d
```

The 'permissions' for this file are shown in the first column of 11 characters. The significant characters are the first group of three `rwx` meaning that the user (you) has permission to read, write, and **execute** the file. If the 'x' is missing, the easiest way to make the correction is to type:

```
chmod 755 install_script_d
```

If the system responds by indicating that you do not have permission to make the change, then it is likely that the machine is not under your complete control, and you might have to ask an administrator for help!

10.2.2 How the databases are organised

To understand how the files are stored and used, and the rôle of Cygwin as an adjunct to Windows, look at the contents of the directory for the smallest of the databases (`peakdale`). Table 10.1 shows the listings obtained from both Windows and Cygwin. Since the database chosen is the smallest, you can see a directory listing showing all of the files. Note that the sizes of the individual files as recorded by the two operating systems are identical. The organisation of the eighteen individual databases is the same. Each database has a series of files containing the structures in concatenated form, all but the last contain structures for one thousand compounds. There are nine such files in the case of `peakdale`, named `peak000.DDD` to `peak008.DDD`. The last file of these is smaller because it contains less than 1000 structures. Note that the numbers embedded in the filenames also start at zero.

Table 10.1
The Files in the `peakdale` Database
(a) As listed in Cygwin using ls -l

```
-r--r--r--+ 1 cbas25 None 3331581 Mar 21 18:16 PEAK000.DDD
-r--r--r--+ 1 cbas25 None 3582909 Mar 21 18:16 PEAK001.DDD
-r--r--r--+ 1 cbas25 None 3540636 Mar 21 18:16 PEAK002.DDD
-r--r--r--+ 1 cbas25 None 3325967 Mar 21 18:16 PEAK003.DDD
-r--r--r--+ 1 cbas25 None 3769961 Mar 21 18:16 PEAK004.DDD
-r--r--r--+ 1 cbas25 None 3274887 Mar 21 18:16 PEAK005.DDD
-r--r--r--+ 1 cbas25 None 3758127 Mar 21 18:16 PEAK006.DDD
-r--r--r--+ 1 cbas25 None 3623264 Mar 21 18:16 PEAK007.DDD
-r--r--r--+ 1 cbas25 None 1878890 Mar 21 18:16 PEAK008.DDD
-r--r--r--+ 1 cbas25 None  760416 Mar 21 18:16 PEAKcatind.dat
-r--r--r--+ 1 cbas25 None 1196160 Mar 21 18:16 PEAKind.dat
```

(b) As listed in Windows

```
21/03/2011  18:16         3,331,581 PEAK000.DDD
21/03/2011  18:16         3,582,909 PEAK001.DDD
21/03/2011  18:16         3,540,636 PEAK002.DDD
21/03/2011  18:16         3,325,967 PEAK003.DDD
21/03/2011  18:16         3,769,961 PEAK004.DDD
21/03/2011  18:16         3,274,887 PEAK005.DDD
21/03/2011  18:16         3,758,127 PEAK006.DDD
21/03/2011  18:16         3,623,264 PEAK007.DDD
21/03/2011  18:16         1,878,890 PEAK008.DDD
21/03/2011  18:16           760,416 PEAKcatind.dat
21/03/2011  18:16         1,196,160 PEAKind.dat
```

Table 10.2
Details of data stored in `AMBIind.dat` file

1	2	3	4	5	6	7	8	9	10	11	12	13	14	15	16
0	6.9778	1.4363	2.3635	386.0000	−100.0000	−0.5400	302.3500	58.9200	2	4	5	0	3	2	0
1	9.3351	1.5927	1.9980	634.0000	2.0352	4.1940	492.5468	85.8900	2	7	7	1	11	7	0
2	7.0560	1.4434	2.7481	390.0000	3.4021	3.1710	259.3482	30.8200	0	3	3	0	4	2	0
3	10.1048	1.2794	2.0309	455.0000	3.2839	3.2660	339.3909	56.7900	1	5	5	0	7	6	0
4	9.3283	1.3425	2.6063	537.0000	3.6691	4.9020	379.4990	30.9300	0	4	4	0	7	5	0
.	.	.	.	.	.	.	.	.	.	.	.	.	.	.	.
951541	8.6431	0.9709	1.5839	437.0000	4.9389	7.2120	473.3251	92.3600	2	6	6	2	5	5	0
951542	12.3873	1.5312	4.0680	623.0000	5.0705	4.9370	514.6047	126.2600	1	9	8	0	9	6	0
951543	11.4405	2.1075	3.7310	506.0000	4.8036	3.8380	484.3623	136.0700	2	10	9	2	12	9	0
951544	10.1709	1.7403	3.1478	533.0000	4.6592	4.4590	423.5387	86.0800	0	7	7	0	10	6	0

The first five and last four entries in the file are shown

Column Number	Contents
1	Compound serial number
2	Length of major axis (Ångstrom Units)
3	Ratio of length of major axis to length of second axis to (Elipticity-1)
4	Ratio of length of major axis to length of third axis to (Elipticity-2
5	Volume ($Å^3$)
6	ClogP (Chapter 4, reference number 11)
7	XlogP (Chapter 4, reference number 12)
8	Formula weight
9	Polar surface area ($Å^2$)
10	Number of hydrogen bond donor atoms
11	Number of hydrogen bond acceptor atoms
12	Total number of Nitrogen and Oxygen Atoms
13	Total number of halogen atoms
14	Total number of rotatable bonds (includes CH_3, CF_3, tert-Butyl groups)
15	Total number of effective rotatable bonds (excludes CH_3, CF_3, tert-Butyl groups)
16	Total number of unusual atoms (Atoms other than C, H, N, O, Halogens, S, P)

In any one database there could be a maximum of 1000 files (numbered from 000 to 999). The individual files in a set of a thousand are also (conceptually) numbered from 000 to 999. Hence any one database could hold a maximum of 1000 ×1000 structures *i.e.* one million structures. Of the eighteen databases that are presented here, the one that holds the most structures is `ambinter`, with a total of 951,545 structures in 952 files.

In each of the databases there are two index files; the smaller of these, in this case `PEAKcatind.dat` contains cross-referencing to the catalogue numbers of the original Diversity Supplier. The larger file (`PEAKind.dat`) contains a one line entry for each compound in the database with These are the parameters that are used by the program QUICKSCAN. Table 10.2 shows excerpts from the file `AMBIind.dat` from the database `ambinter`.

Table 10.3
The file `databaselist.dat`

```
 952    951545 AmbInter                       C:\database43\ambinter                   AMBI
 233    232287 Asinex-Gold                    C:\database43\asinex_gold                ASAU
 114    113565 Asinex-Platinum                C:\database43\asinex_platinum            ASPT
 379    378158 ChemBridge                     C:\database43\chembridge                 CHBR
  10      9257 ChemDiv-New-Chemistry          C:\database43\chemdiv_newchemistry       CHDN
 555    554331 ChemDiv-Discovery              C:\database43\chemdiv_discovery          CHDD
 357    356009 Enamine-Stock                  C:\database43\enamine_stock              ENAS
 120    119337 IF-Labs                        C:\database43\iflabs                     IFLA
 368    367147 Inter Bio Screen               C:\database43\inter_bioscreen            INTB
  48     47368 Key Organics                   C:\database43\key_organics               KEYO
  60     59491 Maybridge-Screening            C:\database43\maybridge_screening        MAYS
 166    165159 Moscow-Med-Chem                C:\database43\moscow_med_chem            MOMC
  69     68897 Otava-In-House                 C:\database43\otava_inhouse              OTAI
  73     72149 Otava-Supplier                 C:\database43\otava_supplier             OTAS
   9      8544 Peakdale                       C:\database43\peakdale                   PEAK
 502    501543 TimTec-OVS                     C:\database43\timtec_ovs                 TIMO
 136    135038 TimTec-Stock                   C:\database43\timtec_stock               TIMS
 248    247599 National Cancer Institute      C:\database43\nci                        NCIA
####
#234x1234567x1234567890123456789012345678x12345678901234567890123456789012345678901x1234
#2345678901234567890123456789012345678901234567890123456789012345678901234567890123456789
#         1         2         3         4         5         6         7         8
```

Notes

The program detects the end of the file by reading the four hash characters (####). Any following line beginning with a hash character is ignored and treated as a comment. The three lines that are shown beginning with a hash character are rulers to aid in formatting. The data lines in the file **must** be formatted correctly in FORTRAN style (I4,1X,I7,1X,A30,1X,A40,1X,A4).

10.2.3 Setting up the databases for use by QUICKSCAN.

If you have set up the small molecule databases following the instructions in the preceding section, then you have done the first step to setting up QUICKSCAN.

Two files that reside in the same directory as the executable program (`quickscan.exe`) need to be edited. The file `databaselist.dat` must show, in the fourth column, the paths to the individual databases (including the Windows drive letter). The entries in this column must have exactly 40 characters (including spaces), and must agree with the formatting shown in the comment line that follows. You might have to rename the directory that holds the databases, if its name is too long. Table 10.3 shows the formatting of this file in detail.

The next step is to edit the file `STEER.dat`. This file has three lines holding the paths to files or directories that QUICKSCAN uses. For example:

```
.\databaselist.dat
C:\database43
C:\interprobe\interchem\version2010_12_26\mechanics.data
```

Line 1 should hold the location of the file you have just edited
Line 2 should show the directory that holds the database files.
Line 3 should point to the directory that holds the molecular mechanics data in the INTERCHEM program. (At the present time this is not used).

If you have followed these steps, and there is an icon for QUICKSCAN on your desktop, then clicking this icon will start the program. If the program fails to start, go back and check that you have made the alterations to `databaselist.dat` and `STEER.DAT` correctly.

10.2.4 Using QUICKSCAN

If QUICKSCAN loads correctly then there will appear on your computer a display similar to that in Figure 10.1. By typing *single letter or number, case insensitive, characters*, in response to this display and others, that are shown in Figures 10.2 and 10.3, you will be able to control the program. At appropriate places you will be invited to enter the ***Paths*** of files or directories; these should be entered as full paths including the disc drive letter. The three displays in these Figures are taken in turn.

In Figure 10.1

This requires a number response to one of seven options.

1 = Load a TARGET STRUCTURE and find and abstract the best MATCHES to it.
This option accesses the main function of the program. The TARGET structure could be an already known drug, or a ligand abstracted from a protein structure. In either case it should be a structure with correct bond orders and a full complement of hydrogen atoms. The structure must be in **INTERCHEM D** format.

2 = Set SEARCH PARAMETERS, find, and ABSTRACT ALL matching structures.
This option switches to the options of Figure 10.2 for the next stage .

3 = Abstract RANDOMLY selected STRUCTURE ENTRIES from a SPECIFIED CATALOGUE.
This option is provided so that control sets may be generated, for example to test the type and variety of structures offered by a supplier. There is a switch to the menu of Figure 10.3. You will be asked for the number of structures required, and the number for name of the Supplier. The number of structures requested should be reasonable in relation to the total number of compounds offered by the chosen supplier. This is because duplicates are eliminated; thus asking for 1000 structures from *peakdale* (total number = 8544) will leave you disappointed!

4 = Abstract SPECIFIED STRUCURE ENTRIES from a SPECIFIED CATALOGUE.
This option allows you to get the structure knowing the serial number in the Supplier Directory (*n.b.* this is not the Supplier's catalogue number).

In options 2, 3, and 4 , the program requires that the name of a new directory be supplied, This directory will hold the output structures. Option 6 allows a new directory to be created for other purposes. Option 5 is currently disabled.

7 = EXIT.
This option ends the program. You are required to manually erase the remains of the menus from the screen.

```
Output

Number/Letter                                   Structures  Files                                      Filenames
  1      A      AmbInter                           951545   952  C:\database43\ambinter                    AMBI
  2      B      Asinex-Gold                        232287   233  C:\database43\asinex_gold                 ASAU
  3      C      Asinex-Platinum                    113565   114  C:\database43\asinex_platinum             ASPT
  4      D      ChemBridge                         378158   379  C:\database43\chembridge                  CHBR
  5      E      ChemDiv-New-Chemistry                9257    10  C:\database43\chemdiv_newchemistry        CHDN
  6      F      ChemDiv-Discovery                  554331   555  C:\database43\chemdiv_discovery           CHDD
  7      G      Enamine-Stock                      356009   357  C:\database43\enamine_stock               ENAS
  8      H      IF-Labs                            119337   120  C:\database43\iflabs                      IFLA
  9      I      Inter Bio Screen                   367147   368  C:\database43\inter_bioscreen             INTB
 10      J      Key Organics                        47368    48  C:\database43\key_organics                KEYO
 11      K      Maybridge-Screening                 59491    60  C:\database43\maybridge_screening         MAYS
 12      L      Moscow-Med-Chem                    165159   166  C:\database43\moscow_med_chem             MOMC
 13      M      Otava-In-House                      68897    69  C:\database43\otava_inhouse               OTAI
 14      N      Otava-Supplier                      72149    73  C:\database43\otava_supplier              OTAS
 15      O      Peakdale                             8544     9  C:\database43\peakdale                    PEAK
 16      P      TimTec-OVS                         501543   502  C:\database43\timtec_ovs                  TIMO
 17      Q      TimTec-Stock                       135038   136  C:\database43\timtec_stock                TIMS
 18      R      National Cancer Institute          247599   248  C:\database43\nci                         NCIA
            Welcome to QUICKSCAN-PC  Version 2.0

This program allows you to abstract copies of SMALL MOLECULE Structures
from the catalogue DATABASES of various DIVERSITY PROVIDERS.
The different ways of using the program are selected from this MENU

1 = Load a TARGET STRUCTURE and find and abstract the best MATCHES to it
2 = Set SEARCH PARAMETERS, find, and ABSTRACT ALL matching structures
3 = Abstract RANDOMLY selected STRUCTURE ENTRIES from a SPECIFIED CATALOGUE
4 = Abstract SPECIFIED STRUCTURE ENTRIES from a SPECIFIED CATALOGUE
5 = Display the contact details of the DIVERSITY PROVIDERS
6 = Go to, and if necessary CREATE, SPECIFIED DIRECTORY
7 = EXIT
        Enter your choice NOW

1
Enter the name of the TARGET structure file
Press "Escape" and "Enter" to go to preceding menu

C:\Users\cbas25\quickscan_test\diazepamD.DAT
```

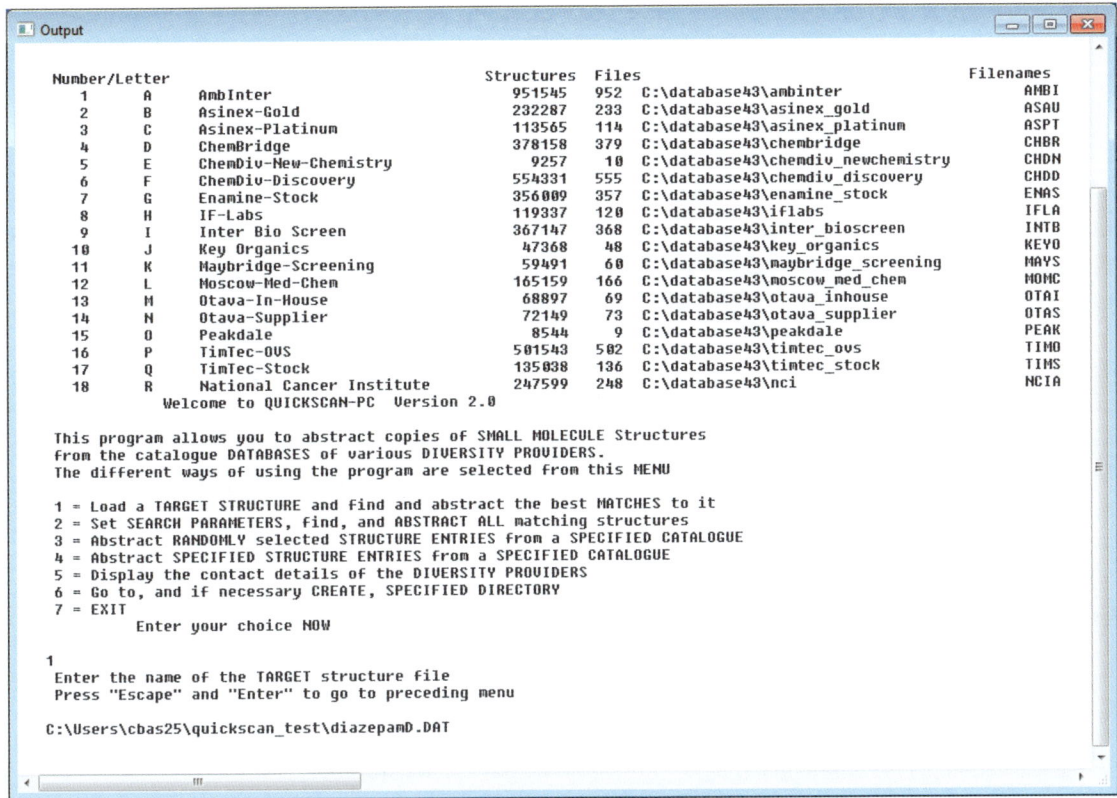

Figure 10.1
Start Menu in QUICKSCAN

```
Output

C:\Users\cbas25\quickscan_test\diazepamD.DAT
 The target structure had the following characteristics:

 |--------------Parameter---------------|  Current Minumum Current Maximum   Selected
Number          Name               Value    Search Value    Search Value
   1   Longest dimension (A)        6.23      95.0%          105.0%           YES
   2        Ellipticity-1           1.25      95.0%          105.0%           YES
   3        Ellipticity-2           2.09      95.0%          105.0%           YES
   4        Volume (A**3)         329.00      95.0%          105.0%           YES
   5   Polar surface area (A**2)   32.67      95.0%          105.0%           NO
   6       Formula weight         284.74     234.74          334.74           NO
   7          ClogP                 4.18       3.18            5.18           NO
   8          XlogP                 2.99       1.99            3.99           YES
   9     Hydrogen bond donors       0          0               1              NO
  10   Hydrogen bond acceptors      3          2               1              NO
  11   Oxygen & Nitrogen atoms      3          2               4              NO
  12      Halogen atoms             1          0               2              NO
  13    Total rotatable bonds       2          1               3              NO
  14  Effective rotatable bonds     1          0               2              NO
  15     Unusual atoms              0          0               1              NO

 To ACCEPT the SEARCH PARAMETERS shown and proceed to a database search type:  G  or  g
 To SWITCH a SEARCH PARAMETER ON or OFF                                 type:  S  or  s
 To SWITCH a SEARCH PARAMETER ON and CHANGE minimum and maximum values  type:  C  or  c
 To set ALL PARAMETERS to the DEFAULT values for this target            type;  D  or  d
 To set PARAMETERS for (almost) EXACT matching to target structure      type:  X  or  x
 To return to the MAIN menu                                             type:  M  or  m
 To go back to the preceding menu                                       type:  B  or  b
```

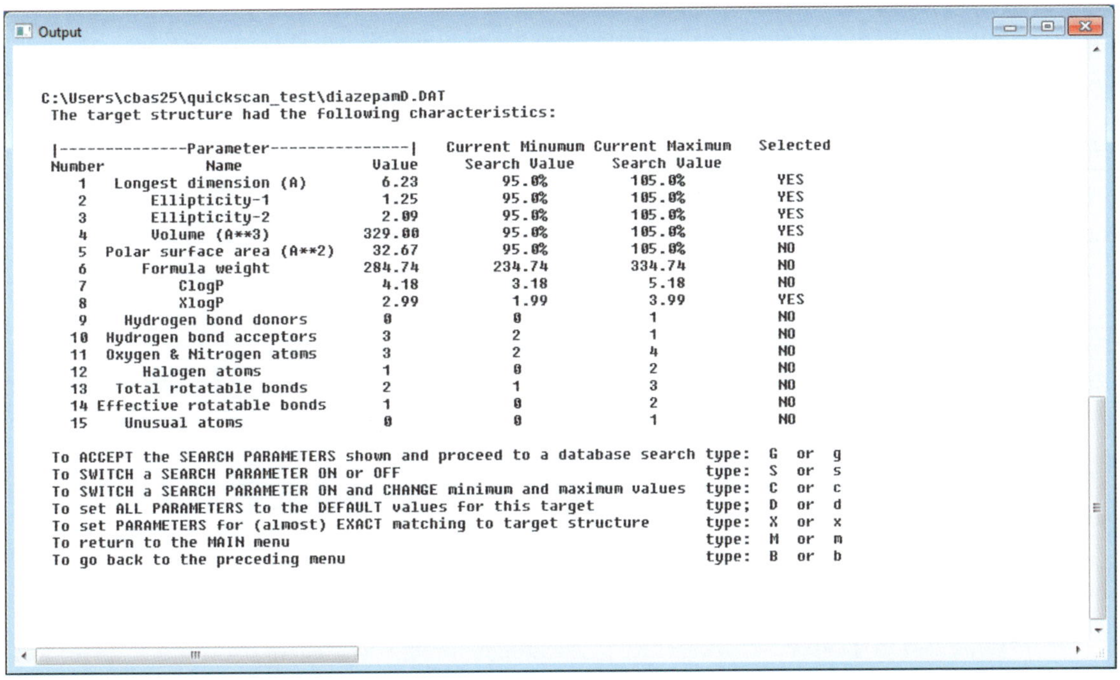

Figure 10.2
Second Menu in QUICKSCAN
Parameter Selection for Matching a Target Structure

In Figure 10.2

This provides the menu that follows on from option 2 of Figure 10.1 and is, in fact, at the heart of the program. This instance of the menu follows from the choice of `diazepamD.DAT` in the first menu.

The aim is to end up with a set of structures closely similar to the target. The ***number*** of these structure should be reasonable for the circumstances. Thus if the set of structures were to be used at the start of a VHTS process, 2000 might be appropriate. If no further pruning is envisaged, then a small number (<100) is more suitable.

There are numbered lines for each of the fifteen parameters that can be used to refine compound selection. In the third column are the values of the parameters ***calculated by the program*** for the ***target structure***. The fourth and fifth columns hold ***suggestions*** for ***minimal*** and ***maximal*** search values. The sixth column indicates whether the parameter is ***switched ON*** or ***OFF***.

The single letter (upper or lower case) menu options that follow allow:
- **G** To accept the current values in columns three and five, and the status in column six.
  This goes to the next step (Figure 10.3) requiring the selection of a particular database.
- **S** To alter the status of column six
- **C** To switch parameters ON and alter their minimal and maximal values.
  This requires the entry of three numbers followed by **RETURN**
  The first number is an **integer** (range 1 to 15) for the **parameter number**.
  The second and third are numbers that are interpreted as either
  integers (parameters **9 to 15**), or as
  floating point numbers (parameters **1 to 8**).
  The end of the list of changes is signalled by typing **three zeros**

- **D** This sets the parameter values to default (entry) values for the TARGET.
- **X** This sets the parameters to values that are designed to select structures that are
  very similar to the TARGET
- **B** Go back to the menu of Figure 10,2.

There are some rules that should be applied in selecting parameters:
Firstly the program will not allow you to proceed if no parameters are switched on. Logically this would give you the whole contents of any database selected subsequently. You are advised by an error message to have at least ***three parameters*** switched on.

However one application, that might warrant the use of only a single parameter, could be a search of a database for all structures containing unusual atoms. For this set only parameter `15 ON` and supply minimum and maximal values `1` and (say) `30`.

There are two alternative parameters for calculated logP values (parameters 7 and 8). It is advisable to have only one of these switched on.

To set the parameters for the rule-of-five tests, use the following :
- Parameter 6 (Formula weight)   Minimum = 0;   Maximum = 500.01
- Parameter 7 (CLogP)   Minimum = -2;   Maximum =   5.01
- Parameter 8 (XlogP)   Minimum = -2;   Maximum =   5.01
- Parameter 9 (H-Bond Donors)   Minimum = 0;   Maximum =   5
- Parameter 10 (H-Bond Acceptors)   Minimum = 0;   Maximum = 10

Using Interprobe Software for Drug Discovery 235

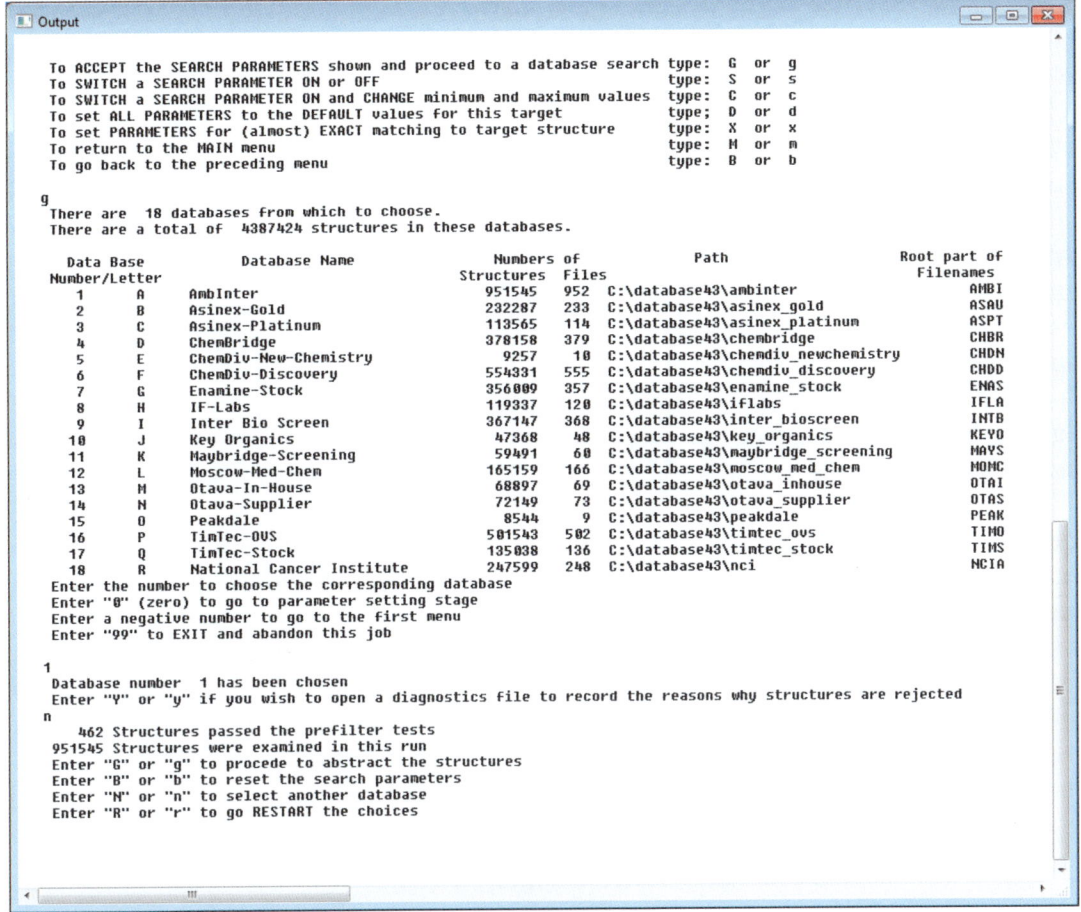

Figure 10.3
Results after an Initial Search for a Match to a Target Structure

In Figure10.3:
When you arrive at this menu from the preceding one, you are required to enter the number corresponding to one of the eighteen databases.

The program then searches the chosen database and after some time returns the total number of structures that pass the parameter test.

You can accept this result by typing **G** or **g**.
You will then be required to specify the full **PATH** of a directory to hold the copies of the files of the structures. This must be a new directory.

After this process finishes you are returned to this menu; you can then choose to extract structures from another database with the same parameters set

Alternatively, you may type **B** or **b** to go back to Menu of Figure10.2, and reset some of the parameter values, to get a more suitable number of structures for your purpose.

10.2.5 Using INTERCHEM to view the structures extracted using QUICKSCAN
After using QUICKSCAN to extract a set of structures to match a target, you can use INTERCHEM to view the results.

In DUAL Display choose *Series Display* (R10/C6). Set up the system from the Series Display menu. With *Set Working Folder* (R9/C1) choose the directory where you stored the structures. With *Load Target into A* (R9/C2) select the file that you used as TARGET. Next with *Set up B for Series* (R9/C3) choose the directory where the structures are stored. Having selected INTERCHEM D format, the system should return a message that shows the number of structures extracted. This should agree with the number returned by QUICKSCAN.

You can now scan through the series of structures, either one at a time (R9/C4) or automatically (R9/C7),

The structure of Diazepam (1) used as an example in the sections above, was chosen because it is known that structures of this and many similar Benzodiazepine drugs are present in several of the databases. Note that while Benzodiazepines are not chiral in the strict sense, the structures returned might be mirror images of the structure of Diazepam that has been used. Consequently, if you use *Align B with A* (R7/C7) to match two structures that should be identical, and do not get the expected result, try using *Select B* (R1/C2), then *Invert Structure* (R2/C6), and try again with *Align B with A*. It is recommended that you familiarize yourself with QUICKSCAN by using this example, before trying out other target structures.

(1)

10.3 Docking experiments using INTERCHEM.
It must be emphasised that INTERCHEM does not provide automatic docking tools, so that the restricted aim of the experiments that are described here, is to illustrate the steps needed to place a small molecular structure in the binding domain. The experiments are based on a real system,[1] the human oestrogen receptor site. The preliminary steps make use of techniques introduced in the preceding section and in Chapter5 (Section 5.3.5).

10.3.1 Getting the necessary files.
In this section and the following ones, we discuss in detail a specific example. The intention is that this shall serve by providing procedures that can be applied to other problems.

Using Interprobe Software for Drug Discovery 237

The experiments start with the Protein Data Bank structure 3ERT. The file that holds this structure 3ERT.pdb should be downloaded from the PDB, and stored in a suitably named folder on your system. Use the program PROTEINS to derive two INTERCHEM 'D' files. The first of these is PRO3ertD.DAT, that holds the structure of the receptor site with the bound-in ligand, 4-hydroxy-tamoxifen (2c), and is got by using PROTEINS option 7. The second file, PRH3ertD.DAT, contains the receptor site only (no ligand), but hydrogen atoms have been added to it (use PROTEINS option 12).

(2)

Tamoxifen and analogues

		R_1	R_2	R_3	R_4
(a)	Tamoxifen	H	H	CH_3	$N(CH_3)_2$
(b)	3-Hydroxytamoxifen (Droloxifene)	H	OH	CH_3	$N(CH_3)_2$
(c)	4-Hydroxytamoxifen	OH	H	CH_3	$N(CH_3)_2$
(d)	Idoxifen	I	H	CH_3	1-Pyrrolidine
(e)	Toremifene	H	H	CH_2Cl	$N(CH_3)_2$

The next step is to get a file that contains just the structure of the ligand. (The procedure has been used already; Refer to the Section 5.3.5. Editing Protein and Nucleic Acid Structures).

With the full structure of the protein plus ligand in *Area A* and in *Dual Display* mode, proceed as follows:

(1) Press *Highlighting* (R4/C9).

(2) Press *Set Highlights* (R4/C5).

(3) Choose from the submenu choose *Ligands* (R3/C10), then *ACCEPT/EXIT* (R6/C10)

The screen will now show the main protein chain in *white*, and the ligand coloured *cyan*. (If necessary, refresh the screen by pressing *Highlighting* R4/C9 again).

(4) Press *Set Highlights* (R4/C5).

(5) Choose from the submenu choose *Eliminate-->C* (R7/C6).

(6) Press *Set Highlights* (R4/C5).

(7) Choose from the submenu choose *Retain-->D* (R7/C7).

(8) Press *Quad Mode* (R1/C9).

You should then see:
In the *A* window: the original structure
In the *C* window: the protein structure *without* the ligand
In the *D* window: the ligand structure
It might be that only a part of the ligand structure is visible, in that case press (while in *Quad Mode*), *Select D* (R2/C2), *Centre Structure* (R15/C1), and then *Wire Frame* (R16/C1).

At this point it is important to *Store* (*Quad Mode,* R5/C1) the two structures in areas *C* and *D*, **before** you make any further changes to them. The choice of names for the files is open of course, but we suggest STR3ertD.DAT and LIG3ertD.DAT respectively.

If you attempt to display the ligand structure in a way that shows the type of bonding, you will be disappointed! The reason is that the original PDB holdings do not include this information. We can, however, construct the bonding information, and use this to add the missing hydrogens to the structure. The final steps in this will use INTERCHEM, but we can facilitate the early stage by using an editor. Proceed as follows:

(1) Suspend INTERCHEM via the *Restart, Diagnostics, Button, Exit* button. (Alternatively you could use a second screen if one is connected to your computer)

(2) Make a copy of the file LIG3ertD.DAT as LOR3ertD.DAT. (The reason is that we need the original *orientation* of the ligand for the subsequent trial docking experiments).

(3) Edit LIG3ertD.DAT using NOTEPAD, (alternatively, if you have installed CygWin, you could use the LINUX editor called *vim*). The file consists of a table of atom identifiers and coordinates, and a connection table.

In the atom table the second column is a list of integers that encodes atomic numbers, amino acid residue numbers, *etc.* in a modified INTERCHEM D format. For all of these eliminate all but the last digits of the integers and replace the last digit by the correct atomic number for the atom. For example:

replace: 314573062 by 6
replace: 314573063 by 7
replace: 314573064 by 8

You can tell what are the correct atomic numbers from the colour coded display of the ligand in *Ball and Cylinder* mode. (Most of the atoms will be carbons of course).

Then in the connection table edit the third column (the numbers are all 10) so that all the bonds are single (2 electrons) except for all the bonds in the three aromatic rings (3 electrons!), and the bond between atoms 3 and 10 which is a double bond (4 electrons).

See section 5.3.3 of this book and Appendix A of the INTERCHEM Manual for explanations of how bond types are coded in INTERCHEM 'D' files.

(4) Return to INTERCHEM and load the edited file `LIG3ertD.DAT` into *Area C*.

(5) Press *Build Structure* (R9/C1) and then (***Build Menu***) ***Copy C to Base*** (R1/C3).

(6) Check that the alterations that you have made using the editor are correct. If they are not use the facilities provided here to make corrections.

(7) Store the resulting structure as a file called `temporary1D.DAT` (in case you make mistakes later!).

(8) Press *Add Hydrogens* (R4/C4)

(9) Store the resulting structure as a file called `temporary2D.DAT`

(10) Press *Copy Base to D* (R2/C4)

(11) Exit the *Building* (R8/C10)

Note. It is permissible to rotate the structure while editing it using the *Build* facility.

(12) Using the usual steps while in *Dual Mode*, [*Select C* (R1/C3), *Load Structure* (R2/C7)], load the original pose of the ligand `LOR3ertD.DAT` into *Area C*.

(13) Press *Structure Merging* (R9/C4), and then (in *Merge Mode*), press *Merge D with C* (R1/C2).

(14) Move the *Guest Structure* (coloured *magenta*) until it overlays the *Host Structure* (coloured *cyan*). It will be necessary, in the final stages of this process, to make the **shift increment** very small. Do this by repeatedly pressing *Halve Shift* (R17/C2).

It is permissible to *move* the Guest Structure, (but not to arbitrarily rotate it), but if you are careful, you can use the *Orthogonal View* button (R14/C3) as a help, provided you go back to the original pose by a second use of the button.

(15) Press *Exit without Merging* (R30/C3).

(16) Store the contents of *Area D* (*Dual Mode, Store Structure* (R2/C6) as file `LRH3ertD.DAT`

It has to be acknowledged that the steps just described are rather complicated and involved. The take-home message is that simple facilities provided in the operating system, such as editors should not be despised. Knowledge of them and how to use them can often be the key to doing some jobs.

Your working directory should now contain the following files:

(1) 3ERT.pdb — The file down loaded from the PDB
(2) PRO3ertD.DAT — Whole structure including ligand from file (1) using PROTEINS
(3) PRH3ertD.DAT — Protein structure with added hydrogen atoms, no ligand, from file (1) with PROTEINS
(4) STR3ertD.DAT — Protein structure by editing using INTERCHEM
(5) LIG3ertD.DAT — Ligand structure by editing using INTERCHEM
(6) LRH3ertD.DAT — Ligand with correct bond order and hydrogen atoms added

If you now use the program PROTOCHECK (the shortcut should be available on your computer desktop), you can get information on the nature of the atoms, amino acid residues etc. contained in the expanded structure files. Thus:

From file (2) you would get (7) PRO3ertD_expanded.dat
From file (3) you would get (8) PRH3ertD_expanded.dat

If you use PROTEINS again, from file (1) you can derive further files relating to the whole structure:

(9) SEQ3ertI.DAT — Sequence (1-letter code) of amino acid residues of the protein
(10) PAL3ertD.DAT — Coordinates of alpha carbon atoms in the protein structure

10.3.2 Docking of the ligand into the protein.

The purpose of this first experiment is to verify that the ligand can be replaced in the cavity of the protein, without bad contacts between atoms of the two structures. Two experiments are possible, firstly using the structures 'as is' (i.e. without added hydrogen atoms) and secondly using structures of the ligand and protein, to both of which hydrogen atoms have been added.

Neither of these experiments should require extensive rotation of the ligand structure with respect to the protein structure; lateral motion along the three Cartesian axes should suffice. However, a preliminary requirement is knowledge of the serial numbers of atoms that are likely to be in close contact *via* hydrogen bonds. This information is available in the two expanded files numbered (7) and (8) in the preceding section, and the files containing the ligand only; files (5) and (6). Remember that when hydrogen atoms are added to a protein structure by PROTEINS, they are added at the *ends of the atoms for each residue*. In contrast, when hydrogen atoms are added to a ligand structure using INTERCHEM, they are added *at the end* of the file.

Useful information on the way a ligand interacts with the host protein structure can be obtained from the *Ligand Explorer* feature on the PDB site. Figure 10.4 shows the diagram for the present case, (to which some extra anotations have been added). The crucial interactions are between the OH group in the ligand and residues Arg394 and Glu353 of the protein. Using the 'expanded' files (7) and (8) we can identify the relevant atom serial numbers in the combined protein + ligand structure. Press the *Protein Analysis* button (*Dual Mode,* R8/C10) and then select *Distance between a pair of atoms in a Protein* (the penultimate 'radio' button). The important distances found, after some trial experiments to find the appropriate atoms, are:

Atom 356 (carboxyl O atom in Glu353) to atom 1947 (OH oxygen in ligand) 3.255Å
Atom 357 (carboxyl O atom in Glu353) to atom 1947 (OH oxygen in ligand) 2.421Å *
Atom 697 (terminal N atom in Arg394) to atom 1947 (OH oxygen in ligand) 4.075Å
Atom 699 (terminal N atom in Arg394) to atom 1947 (OH oxygen in ligand) 3.025Å *

The smaller of the distances is chosen (marked *) in each case.

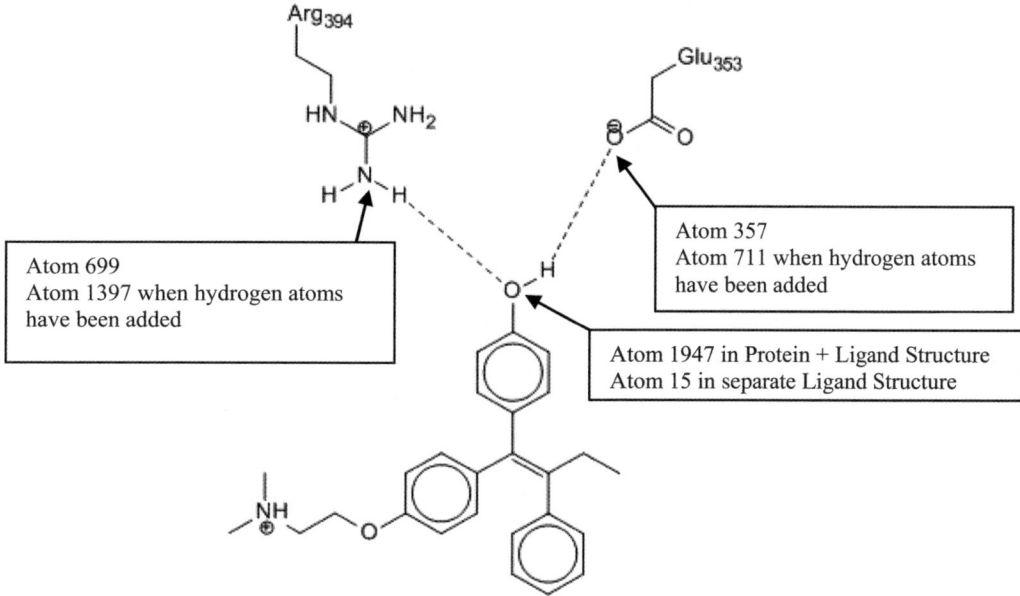

Figure 10.4
Information from the Protein Data Bank tool *Ligand Explorer*
Concerning PDB entry 3ERT.

To dock the (separated) ligand into the protein we need to use the serial number 15 for the OH oxygen in the ligand. Now set up INTERCHEM in the usual way with the protein structure contained in file STR3ertD.DAT loaded into *Area A*, and the ligand structure contained in file LIG3ertD.DAT in *Area B*. Select *Structure Merging* (*Dual Mode* R9/C4), and then *Merge B with A* (R1/C1). Select *Add Good Contacts* (R20/C1) and in the table set the first two lines (to start with) as follows:

```
Host SN   Guest SN   Low Limit   High Limit
  357        15         2.3         2.5
  699        15         2.9         3.2
```

Click *Accept*. Then click *Show Good Contacts* (C1/R21). There should appear two dotted lines that join atom 15 of the guest structure (magenta), with atoms 357 and 699 of the host protein structure (cyan). To start with these dotted lines will be coloured blue (indicating that the atom distances are greater than the set limits. The aim is to make the lines become green coloured so that the inter-atomic distances are within the set limits. To do this you can move the picture so that the interaction region is centred on the screen, and also increase the size of the picture (Button C2/R19). Then use the set of magenta coloured buttons in the second column (rows 2 to 7), to move the guest structure with respect to the host structure. For increasing precision press *Halve Shift* (C2/R18) as the process goes on. If the inter-atomic distances become too small the dotted lines will turn red. Because you are working in three dimensions, you need to view the interactions from a different direction occasionally, so press *Orthogonal View* (C3/R14); a second click will restore the original view.

As the alignment of the ligand with respect to the cavity in the protein progresses, you can set the limits for good contacts to narrower ranges (C1/R21). Finally you can *Show Bad Contacts* (C2/R21). The limits for *Bad Contacts* are preset to 3.0 Ångstroms, but you can reduce this number progressively to 1.2 Å using *Bad Contacts Limits* (C1/R21). Bad contacts show up as additional red dotted lines. A good result is achieved if you get the Good Contacts within a range of ±0.02 Å of the values measured in the PDB structure, and no Bad Contacts below 1.2 Å.

Next, click on *Complete Merging* (C1/R30) choose the option *Treat Guest as a Substrate*. This will place the combined structure in *Area A*. Store the structure in a suitably named file. The structure should be directly comparable with the original structure downloaded from the PDB, in that it should be possible to selectively highlight the ligand.

As an additional exercise, try docking the ligand into the protein where both structures have had hydrogen atoms added, (you will have to use the new serial numbers of the important atoms in the protein). To make the job easier it is best to temporarily hide the hydrogen atoms initially. The only difference in the criteria for acceptance is that there should be no Bad Contacts closer than 0.8 Å. The new atom serial number (after hydrogen atoms have been added) can be found by comparing the 'expanded' versions of the structure files. (The 'new' atom serial numbers are shown in Figure 10.4).

10.3.3 Docking of analogues of tamoxifen

4-Hydroxytamoxifen is the metabolite of tamoxifen that is responsible for its activity. The *Merck Index*[2] lists, besides tamoxifen, four analogues under the category *Antineoplastic, (Hormonal - Antioestrogen)*. These are listed under the general formula (2) on page 237. Try docking these structures into the protein of 3ERT.

We suggest that you get these structures of the analogues of tamoxifen by the use of SMILES strings. Choose the option to generate (say) 64 conformers of each structure, and select (say) four of the lowest energy structures by using the INTERCHEM *Series Display* facility.

10.3.4 Generating series of ligands for docking experiments

In a recent paper,[3] a scheme for generating a series of structures to serve as ligands for docking experiments is described. The concern is the design of antimalarial drugs. The target biological system is the dihydrofolate reductase (DHFR) enzyme of the malaria parasite *Plasmodium falciparum*, (*pF*). This section looks at the strategies proposed in the paper, seeking to make use of the AUTOBUILD tool that it is present in INTERCHEM. AUTOBUILD uses many of the facilities of the BUILD tool of INTERCHEM, and uses a modified version of the BUILD menu.

The ligand structures proposed in the paper are analogues of the established antimalarial drug Pyrimethamine (3, $R_1 = Cl$, $R_2 = CH_2 . CH_3$, $R_3 = R_4 = NH_2$), which is known to bind in the DHFR enzyme. They fall into several series, all of which can be generated by attaching fragments in a combinatorial fashion to templates. For this example a simplified series of compounds is suggested, rather than the more complex series described in the paper. The structures are generated by attaching to the template (5), appendages chosen from the four lists, A, B, C, and D. The more complex attachments in the lists are formulated here (as in the paper) as SMILES strings.

(c1ccccc1c2cncnc2)

(2)　　　　　　　　　　(3)　　　　　　　　　　(4)

(Before using AUTOBUILD, take a look at section 2.28 in the INTERCHEM manual).

Then for this example proceed as follows:

(1) Define a working directory for the project, and go to it (say `pyrimet_analogues`).

(2) Prepare the template structure (3) by using the SMILES string shown for the (unadorned) structure (4). To access the SMILES facility from *Dual Mode*, press *Build via Smiles* (R9/C2). Transfer the resulting structure to *Area A*.

(3) Press *Use AutoBuild* (R9/C5)

(4) Press *Set 2 Character Code* (*Autobuild Menu* R2/C7), and enter a distinct and memorable two letters! We will assume that the choice is 'PY'.

(4) Press *Store as TEMPLATE* (R2/C5). You will then be presented with up to eight pop up menus. If you have used the SMILES string *exactly* as written above, then to set up the attachment points (R_1 to R_4), the numbering will be as follows:

Attachment	Junction atom	Eliminate atom
R_1	3	(15)
R_2	8	(18)
R_3	10	(19)
R_4	12	(20)

In this particular case, because there is only one hydrogen atom attached to each of the junction atoms, you will not be prompted for the 'eliminate atoms', and so there will be only four pop up menus.

You will be asked for a name for the structure (say `pyrimettemplt` the limit is 13 characters).

At this point you may interrupt the set up process.

The next step is to set up the *Fragment Files*. One combinatorial scheme is set out in Table 10.4.

(5) There are four alternative ways of generating the fragment structures:

(5a) While in the AUTOBUILD menu page, load into one of the vacant areas the appropriate fragment from the catalogue (*e.g.* hydrogen fluoride for the fluorine atom, hydrogen molecule for the 'no substitution') case.
(5b) While in the AUTOBUILD menu page, use the BUILD facilities to make a structure in the normal fashion.
(5c) Before entering the AUTOBUILD menu page, use the SMILES facility to generate one or more structures (you are able to hold up to four structures in the *Areas A, B, C,* and *D*).
(5d) Load a structure that you have previously made into the Build *Base Area*.

Table 10.4
Specification of the Fragments for Generating Analogues of Pyrimethamine

R1	R2	R3	R4	Notes
H	CH_3	H	H	1, 2
F	CC	NH_2	NH_2	1, 2
Cl	CCC	CN	CN	1, 4
Br	C(C)C			1, 4
I	C(C)(C)(C)			1, 5
	CC1CCCC1			5
	CC1CCCCC1			5
	CC1CCOCC1			5

Notes
1 Whenever a substitution is **not** to be made for a particular R group, then AUTOBUILD handles this by performing a dummy substitution; the hydrogen molecule, from the fragment catalogue is used as the fragment!

2 The SMILES string for methane cannot be used as the parent for the methyl group (the SMILES parser rejects it as a error!), so use methane from the fragment catalogue.

3 Whenever a hydrogen atom in the template is to be replaced by another single atom, *e.g.* a halogen atom, or a small group (*e.g.* CN), the parent for the fragment is best obtained from the fragment catalogue. (HF, HI, NH_3, HCN, *etc.*).

4 Propane from the fragment catalogue can be used to generate both the fragments CCC (n-propyl) and C(C)C (isopropyl). Here they are shown as SMILES strings for convenience only).

5 These last three alternatives for R2 are best generated from SMILES strings.

(6) Load the appropriate structure into the BUILD *base* area (buttons R1/C1, R1/C2, R1/C3. R1/C4, R1/C5), and then store it as a fragment: (buttons R2/C1, R2/C2, R2/C3, R2/C4). Here the phase 'Next RA' means the next fragment for the next R_1 series *etc*. The same structure may be loaded into more than one series. (It does not matter in which order the fragments are stored). In each case you will first be asked for **one** *Junction atom*. You will only be asked for an *eliminate atom* if there is more than one hydrogen attached to the junction atom. If there is a choice, choose the hydrogen atom that will result in the least number of non-bonded interactions).

(7) Press button *Run AUTOBUILD* (R10/C4). This will start the combinatorial building process.

(8) When this finishes, press *Optimize Structures* (R10/C5). This will take all of the structures derived from the combinatorial process and optimize them using Molecular Mechanics.

(9) Exit AUTOBUILD (R8/C10), and view the results in INTERCHEM (*Dual Mode - Series Display* R10/C6), The optimized structures will be named (in this example) PYS0000D.DAT to PYS0359D.DAT. See the INTERCHEM manual for the naming of the other file series. The structures in a selection of these files can now be manually docked into the cavities of the relevant proteins (PDB entries 1J3L and 1J3K).

10.3.5 Other targets for tackling malaria.

That malaria has re-emerged as a problem disease in the tropics and sub-tropical areas of the world in well known, and is discussed in a recent paper from the World Health Organisation. (WHO).[4] One aspect of the problem is the difficulty of ensuring that patients complete a prescribed course of treatment. When such a course is not completed, there is a possibility that the malaria parasite, *Plasmodium falciparum*, becomes resistant to the drug being used. The recommended way of avoiding this is to use a cocktail of drugs (at least two), based on the principle that it is unlikely that resistance to more than one drug will result simultaneously. The WHO report[4] makes the point that the world is fast running out of reliable anti-malarials. In any case, design of new anti-malarials is well worth attention.

As an aid for this we show structures of representative drugs currently in use, and classify them according to the scheme used in that report.[4] The year that a drug was introduced is also mentioned; you will see that many of them can be traced back to the era of the Second World War and its immediate aftermath.

However the first three drugs (6), (7), and (8) are new and are classified as *sesquiterpene lactones*. Strictly, only the first of these (6) is a lactone. However their distinguishing feature is the presence of an *epidioxide (peroxide) linkage* that bridges either a seven- or an eight-membered ring. This appears to be the feature that makes them active. The first two are related to a natural product (artemisin), while the third (8) is synthetic. The anti-malarial activity appears to be due to the breakdown of the epidioxide group to yield free radicals that then interact with the iron atom of the *hemoglobin*. The *pF* parasite relies for its reproduction on the blood ingested by the Mosquito after it bites a malaria victim; thus the action of artemisinin and its analogues is unusual if not unique. This is in contrast to other anti-malarials that bind to proteins. The mode of the interaction between the epidioxide compounds and the heme moiety is not clear, and there are conflicting papers on the subject.[5,6] The *recognition* of artemisinin analogues as valuable drugs is relatively new (1972) although the Chinese herbal medicine ('qinghao') based on artemisin has been known for 2000 years.[7] There is also evidence that Chinese military authorities kept secret their use of artemisinin for some years!

Artemisinin (6)   Artesunate (7)   Arteflene (8)

The remaining classes of anti-malarials in the WHO review[4] are:

4-Aminoquinolines   Chloroquine (10), Amodiaquine (13), Piperaquine (19),
Amino-alcohols   Quinine (9), Mefloquine (17), Lumefantrine (16)
Sulphonamides   Sulphadoxine (15)
Biguanides   Chloroproguanil (12)
Diaminopyrimidine   Pyrimethamine (18)
8-Aminoquinolines   Primaquine (11)
Antibiotics   Tetracycline (14)

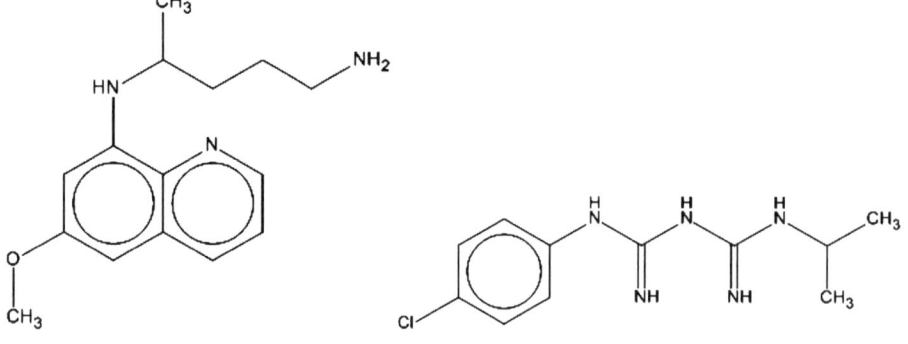

Quinine (9) (19th Century)   **Chloroquine (10) (1941)**

Primaquine (11) (1946)   **Chlorguanide (12) (1948)**

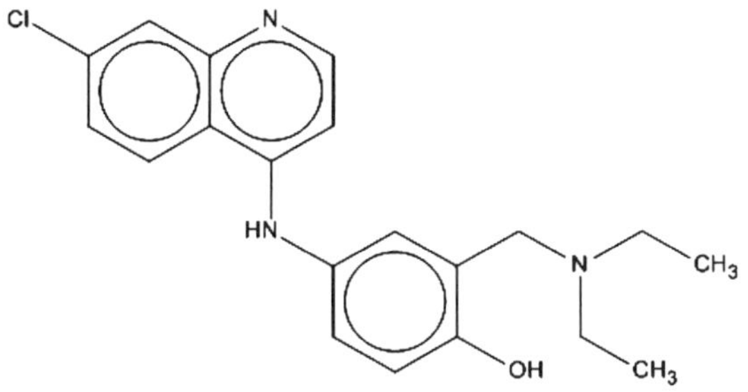

Amodiaquin (13) (1949)

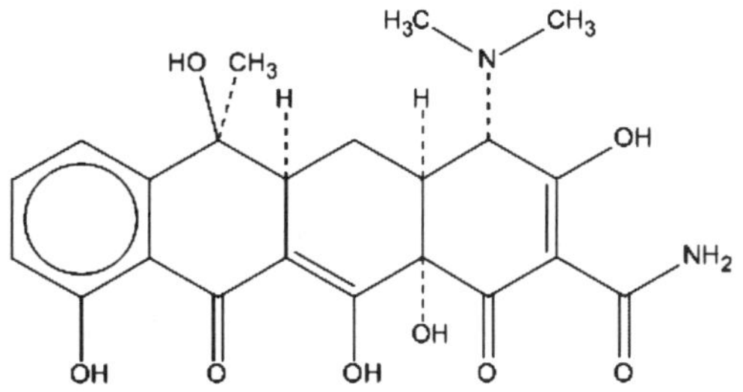

Tetracycline (14) (1953)

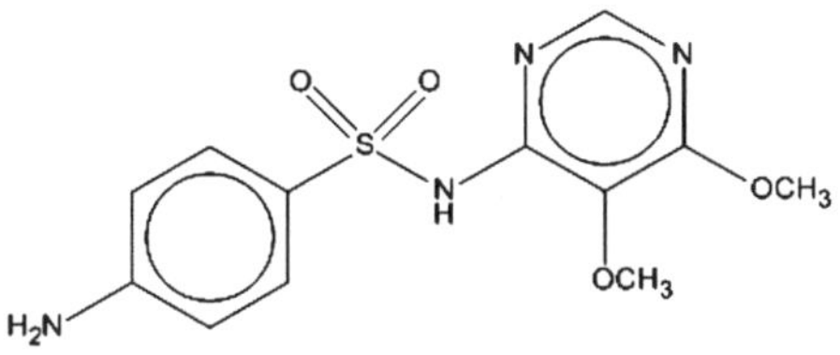

Sulfadioxine (15) (1962)

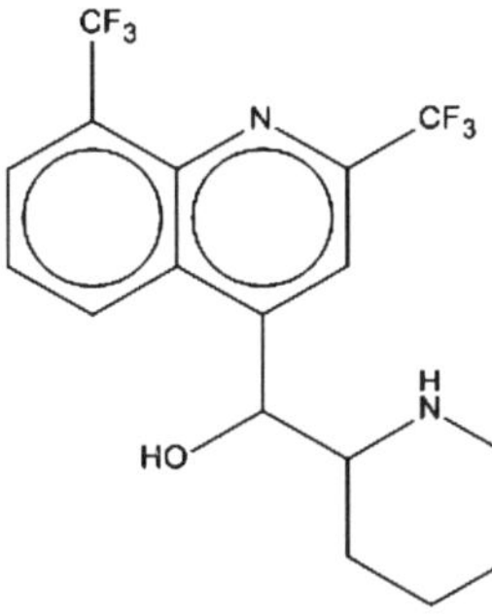

Mefloquine (16) (1978)

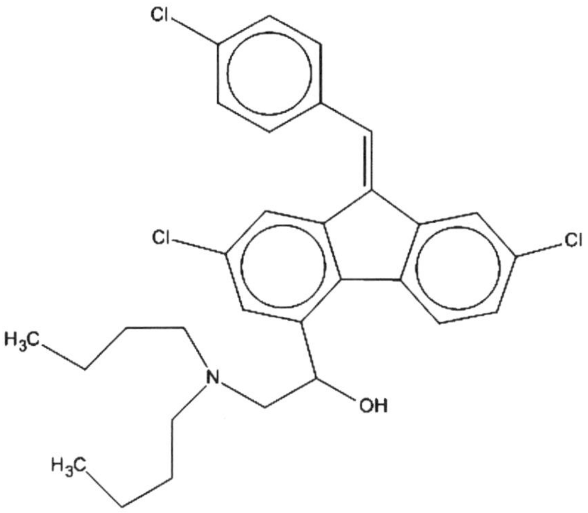

Lumefantrine (17) (1970)

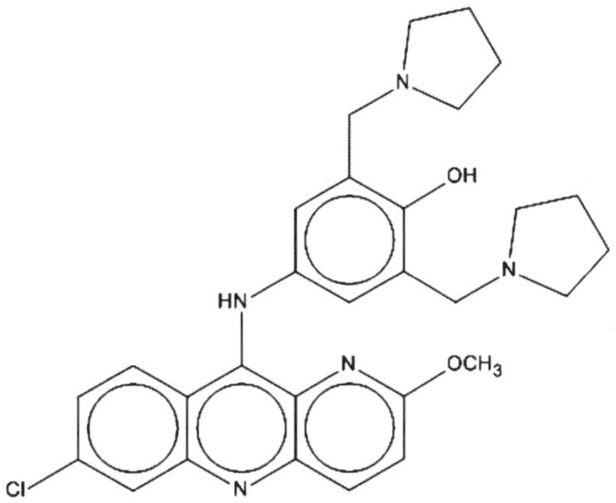

Pyronaridine (18) (1979)

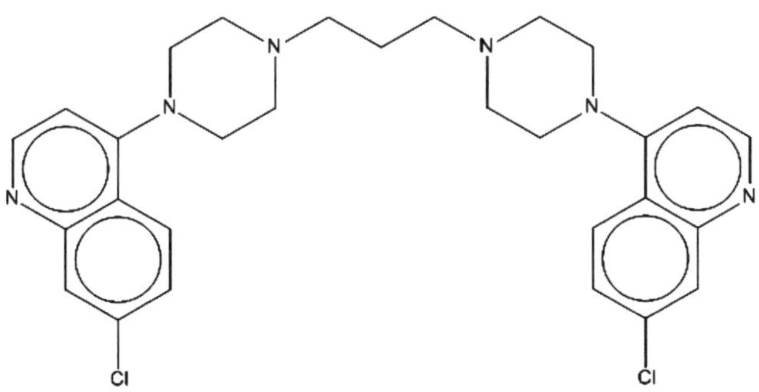

Piperaquine (19) (Reintroduced 2004)

If you look at the structures of all these compounds, you will see that, with the exceptions of the artemisinin and its analogues, quinine, and tetracycline, they have at most one chiral centre. What distinguishes the exceptions is that they are, or are related to, natural products. The dates show the year when they were reported, or when they were introduced as drugs, and these dates reveal that the non-natural products, the synthetic anti-malarials, were introduced at (or before) a time when the problems with thalidomide had only just been revealed. That many of these older drugs are still being used, as secondary treatments of malaria, is significant.

Tetracycline is primarily used as an antibiotic, and is typical in having several chiral centres. All of the structures are shown as planar structural formulae, which hides their true three dimensional shape.

10.4 Drugs from natural products.
Malaria (and AIDS) are two of the major concerns of tropical medicine, but in this section in this chapter we turn to what is another major concern for medicine - *cancer*.

Many of the cancer drugs introduced recently or now being developed are very large molecules. Many have been inspired by natural products that have anti-cancer activity, but for which there is no possibility that the demand could be supplied from natural resources. They certainly do not obey the rule-of-five. However cancer treatment usually takes place in a clinic, where the treatments need to be monitored and where drugs are rarely administered orally.

> **Vinblastine** and **Vincristine** are examples of bis-indole alkaloids. They were isolated from *Vinca rosea* Linn.(*Catharansus roseus* Don).(The Madagascan periwinkle).
>
> Vinblastine is now made by synthesis from separate mono indole alkaloid components, (catharanthine and vindoline) that are more readily available from other *Vinca* species.[8]

Vinblastine (20, R$_1$ = CH$_3$)
Vincristine (20, R$_1$ = CHO)

Paclitaxel (Taxol) (21)

Docetaxel (Taxotere) (22)

Paclitaxel is the natural alkaloid isolated from the stem bark of the Western Yew (*Taxus brevifolia*),[9] and was found to have antileukemic and tumor inhibitory properties. The low yield from the source plant (a protected species) required that other sources be found. In 1989 Pierre Potier's group in France published a method of synthesis from 10-deacetylbaccatin, a compound isolated from the European Yew.[10] A useful review on the chemistry of paclitaxel has been provided by David Kingston.[11]

Docetaxel does not occur naturally, but is a synthetic drug that is an intermediate in the conversion of 10-deacetylbaccatin into paclitaxel. It has anti-cancer activity, and is sold under the name **Taxotere.**

The name Taxol formerly used for paclitaxel was abandoned, partly because of confusion with another, completely unrelated, drug. The name still persists but its use is discouraged.

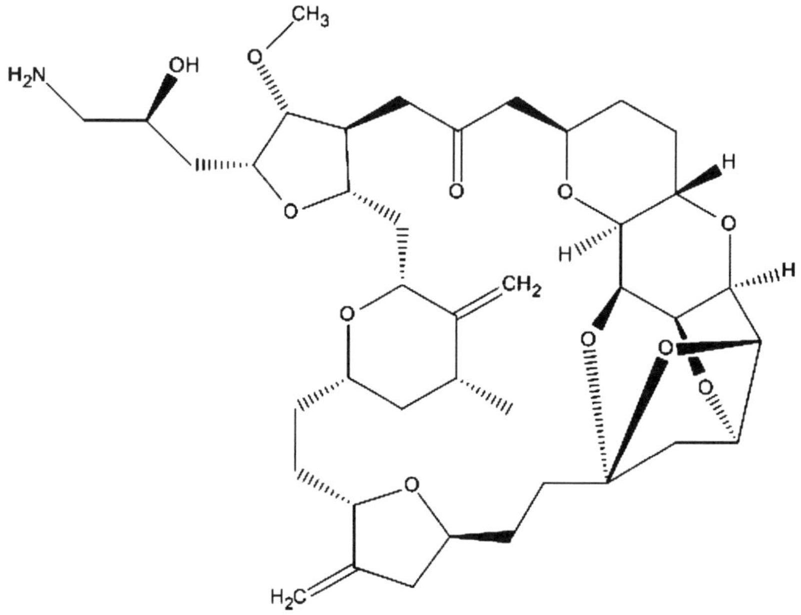

Eribulin (23)

Eribulin is a synthetic analogue[12] of a marine natural product Halichondrin B,[13] that showed promise as an anti-cancer drug. The activity was found in this smaller compound as well, and so this has become the drug being used.

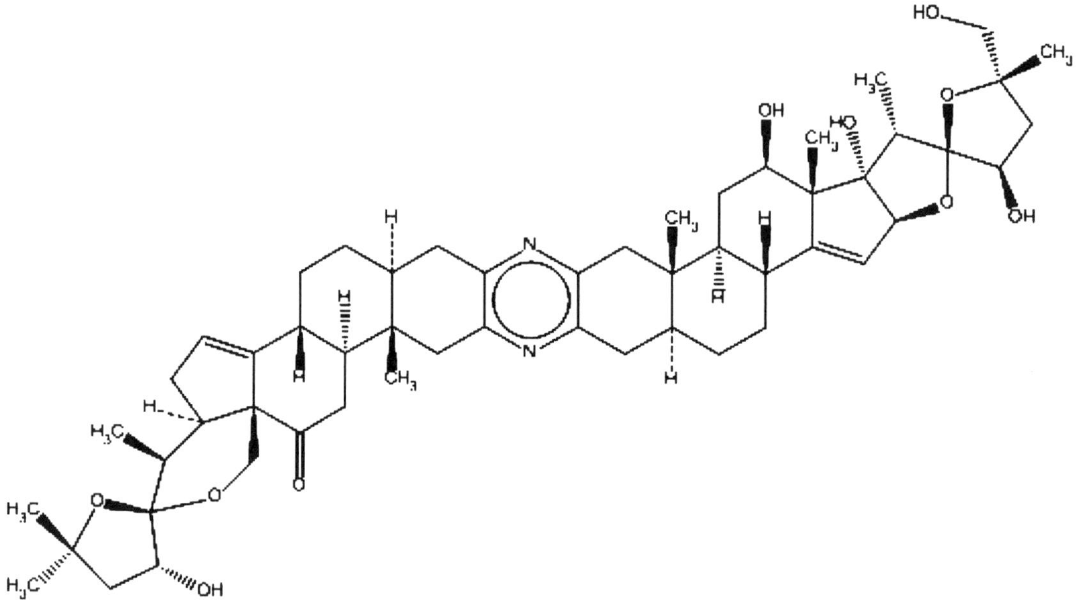

Cephalostatin-1 (24)

Cephalostatin-1 is another marine natural product. It is remarkable as being composed of two steroid components, joined at their A-rings. It has been synthesised[14] by joining appropriate precursor compounds (*western half* and *eastern half*) to form the central pyrazine ring.

Realization that natural products still have a role to play in drug discovery, even if only as sources of lead compounds, has stimulated interest in this area again. Marine natural products are particularly attractive for providing lead compounds, since there is the added incentive of collecting the specimens by snorkelling! However there are environmental considerations that limit the use of the marine sources for actual drug *production*.

It is clear from the structures of the examples that there are subtle differences between these compounds and what, for want of a better word, we have to call conventional drugs. Some differences are fairly obvious; the large number of oxygen atoms, but relatively small numbers of nitrogen atoms is one such. The large number of chiral centres is another obvious difference. To better understand all of these factors there are programs that seek to design potential drugs of this type.[15]

10.5 Problems for you to solve

For all of the structures in this Chapter that are shown as planar structural formulae, you should generate 3D structures. You should choose a method based on SMILES strings, building from fragments, or use of the SKETCH facility, as seems most appropriate in each case.

Structures derived by building from fragments should be optimised using molecular mechanics methods first, and then using semi-empirical molecular orbital methods (MOPAC). Try to find crystal structures for each example for comparison. In many cases crystal structures of a derivative of the drug may be all that is to be found.

Finally, compare the chirality of the structures that you generate with the chirality of structures reported in the literature, or in the answers provided at the end of the book.

10.6 Envoi

The design of a new drug, and bringing it to the market place, is a complex and expensive process. This fact is generally acknowledged by all concerned. Is there a niche for small research groups, independent of the large pharmaceutical companies, to contribute to the discovery process? These large companies, for the most part still profitable, and still viewed as acceptable by ethical investment standards, are often open to collaboration with small organisations, either affiliated with a university, or independent. That such independent groups can be less encumbered with administrative concerns is an advantage, and the receptiveness of pharma to collaboration is attested to by recent reviews.[16, 17, 18] Ideally initial stage development should progress beyond the modelling stage to small scale synthesis of potential drugs, but there could be openings for participation involving only the molecular modelling stage.

The progress of research in drug discovery is reviewed in *Annual Reports in Medicinal Chemistry*.[19] The bulk of each of these books is devoted to reviews on specific drug-type topics, but each year there is a review (called *To Market, To Market*) of the New Chemical Entities finally gaining approval to be used as drugs. The decline in the numbers of these NMEs over the recent years is shown by comparisons of the annual volumes.

References and Endnotes. Chapter 10

[1] A. K. Shiau, D. Barstad, P. M. Loria, L. Cheng, P. J. Kushner, D. A. Agard, and G. L. Greene, *Cell,* 1998, **95**, 927.

[2] M. J. O'Neil (ed.), *The Merck Index, An Encyclopedia of Chemicals, Drugs, and Biologicals, Fourteenth Edition*, Merck & Co., Whitehouse Station, 2006.

[3] D. Hecht and G. B. Fogel, *J. Chem. Inf. Model.*, 2009, **49**, 1105.

[4] E. Heseltine (ed.), *Global Report on Antimalarial Drug Efficacy and Drug Resistance: 2000-2010*. World Health Organization, Geneva, 2010.

[5] A. Robert and B. Meunier, *J. Am. Chem. Soc.*, 1997, **119**, 5968.

[6] K. L. Shukla, T. M. Grund, and S. R. Meshnick, *J. Mol. Graph.*, 1995, **13**, 215

[7] S. R. Meshnick, *Int. J. Parasit.*, 2002, **32**, 1655.

[8] P. Mangeney, R. Z. Andriamialisoa, N. Langlois, Y. Langois, and P. Potier, *J. Am. Chem. Soc.*, 1979, **101**, 2243

[9] M. C. Wani, H.L.Taylor, M. E. Wall, P.Coggon, and A. T. McPhail, *J. Am. Chem. Soc.*, 1971, **93**, 2325.

[10] L. Mangatal, M.-T. Adeline, D. Guénard, F. Guéritte-Voegelein, and P. Potier, *Tetrahedron*, 1989, **45**, 4177.

[11] D. G. I. Kingston, *Pharmacology and Therapeutics*, 1991, **52**, 1

[12] D.-S. Kim, C.-G. Dong, J. T. Kim, H. Guo, J. Huang, P. S. Tiseni, and Y. Kishi, *J. Am. Chem. Soc.*, 2009, **131**, 15636.

[13] Y. Hirata and D. Uemura, *Pure and Applied Chemistry*, 1986, **5**, 701.

[14] K. C. Fortner, D. Kato, Y. Tanaka, and M. D. Shair, *J. Am. Chem. Soc.*, 2010, **132**, 275.

[15] M. J. Yu, *J. Chem. Inf. Model.*, 2011, **51**, 541.

[16] H. J. Federsel, *Expert Opinion - Drug Discovery*, 2010, **5**, 813.

[17] F. L. Douglas, V. K. Narayanan, L. Mitchell, and R. E. Litan, *Nature Reviews - Drug Discovery*, 2010, **9**, 683.

[18] R. Kneller, *Nature Reviews - Drug Discovery*, 2010, **9**, 867.

[19] The latest issue is: *Annual Reports in Medicinal Chemistry,* Volume 45, Ed. J. E. Macor. Elsevier, Oxford, 2010.

Appendices

Appendix A1
Basic Mathematics of Molecular Modelling

A1.1 Introduction
There are many books that deal in depth with the mathematics involved in molecular modelling; here we describe those parts that have particular relevance to the subjects we have covered. There are three such areas:

(1) Calculations on molecular objects
(2) Aspects of the display of pictures of such objects
(3) The special requirements of molecular objects derived from crystallographic data

The first two of these areas have much in common; being dealt with in the familiar surroundings of orthogonal Cartesian space. The third area requires that we extend our thinking into the realms of non-orthogonal coordinates systems.

A1.2 Vectors, coordinates and distances
To perform molecular modelling we must be able to calculate the distances between atoms in a structure. The rectilinear displacement and direction from the centre of an atom, i, to the centre of an atom, j, can be described by a vector denoted $\mathbf{r}_{ij}$ where $|\mathbf{r}_{ij}|$ is the distance between the atoms. Vectors provide a complete description of the positions of the atoms with respect to each other however, we must introduce the idea of a *basis* for the vector $\mathbf{r}_{ij}$ so that a set of coordinates can be supplied to describe the vector.

We are so familiar, intuitively, with the idea of a Cartesian basis that we tend to forget its significance. Firstly a single point must be selected in space to serve as an origin for a set of coordinates. Secondly in a three-dimensional space we introduce three reference vectors to which all other vectors in that space can be referred. A Cartesian basis is an example of an orthonormal basis, one in which the reference vectors are all mutually perpendicular further, it is customary to chose the reference vectors to be unit vectors. Hence the Cartesian basis vectors of a three-dimensional space can be denoted, $\hat{\mathbf{i}}, \hat{\mathbf{j}},$ and $\hat{\mathbf{k}},$ respectively (the basis is sometimes denoted $(\mathbf{X}, \mathbf{Y}, \mathbf{Z})$ as in section A1.10).

Now that a Cartesian basis has been established the vector $\mathbf{r}_{ij}$ can be fully described by a coordinate triple (x, y, z) often denoted, as here, as a row vector hence Equation A1.1. Further, the magnitude of the vector and hence the separation distance between atoms i and j is given by Equation A1.2. A unit vector, $\hat{\mathbf{r}}_{ij}$, is obtained by dividing each of the components by the magnitude of the vector as in Equation (A1.3).

$$\mathbf{r}_{ij} = x\hat{\mathbf{i}} + y\hat{\mathbf{j}} + z\hat{\mathbf{k}} \tag{A1.1}$$

$$|\mathbf{r}_{ij}| = \sqrt{x^2 + y^2 + z^2} \tag{A1.2}$$

$$\hat{\mathbf{r}}_{ij} = \frac{x}{|\mathbf{r}_{ij}|}\hat{\mathbf{i}} + \frac{y}{|\mathbf{r}_{ij}|}\hat{\mathbf{j}} + \frac{z}{|\mathbf{r}_{ij}|}\hat{\mathbf{k}} \tag{A1.3}$$

A1.3 Calculation of bond angles

To calculate the angle, θ, between two bonds to a common atom, the Cartesian coordinates of the common atom are subtracted from the coordinates of the other two atoms to form two vectors directed along the covalent bonds joining the atoms. Thus for atoms 1 and 3 bonded to atom 2, with position vectors $\mathbf{r}_1$, $\mathbf{r}_3$ and $\mathbf{r}_2$ respectively, the cosine of the angle θ subtended by the two vectors is:

$$\cos\theta = \frac{(\mathbf{r}_1 - \mathbf{r}_2)\cdot(\mathbf{r}_3 - \mathbf{r}_2)}{|(\mathbf{r}_1 - \mathbf{r}_2)||(\mathbf{r}_3 - \mathbf{r}_2)|} \quad (A1.4)$$

Note on the right hand side of Equation A1.4 the vector dot product is divided by the magnitudes of the vectors. Hence, the cosine of the bond angle is given by the vector dot product of the two unit vectors. If the cosine is positive then the bond angle, θ, should be returned in the range $0° \leq \theta < 90°$ otherwise if the cosine is negative the angle should be returned in the range $90° < \theta \leq 180°$ and if zero the angle is exactly $90°$.

A1.4 Calculation of torsion angles

To calculate a torsion angle, ϕ, the Cartesian coordinates of four atoms are required. Two of these atoms, atom 2 and atom 3, define the central bond which can be thought of as a line of intersection of two planes. The first plane contains the atoms defining the central bond and atom 1 (bonded to atom 2) similarly, the second plane contains the atoms defining the central bond and atom 4 (bonded to atom 3). If the position vectors of the four atoms are, $\mathbf{r}_1$, $\mathbf{r}_2$, $\mathbf{r}_3$ and $\mathbf{r}_4$ then the unit normal-vectors to the first plane, $\hat{\mathbf{n}}_1$, and the second plane, $\hat{\mathbf{n}}_2$, are given in Equation (A1.5). The cosine of the torsion angle is found from the vector dot product of the plane unit-normal vectors. Given this definition of the cosine of the torsion angle, the torsion angle is positive if the vector cross product $\hat{\mathbf{n}}_1 \wedge \hat{\mathbf{n}}_2$ is a vector in the same direction as the bond vector $\mathbf{r}_{23} = \mathbf{r}_3 - \mathbf{r}_2$ and otherwise is negative.

$$\hat{\mathbf{n}}_1 = \frac{(\mathbf{r}_2 - \mathbf{r}_1)\wedge(\mathbf{r}_3 - \mathbf{r}_2)}{|(\mathbf{r}_2 - \mathbf{r}_1)||(\mathbf{r}_3 - \mathbf{r}_2)|} \quad \hat{\mathbf{n}}_2 = \frac{(\mathbf{r}_3 - \mathbf{r}_2)\wedge(\mathbf{r}_4 - \mathbf{r}_3)}{|(\mathbf{r}_3 - \mathbf{r}_2)||(\mathbf{r}_4 - \mathbf{r}_3)|} \quad (A1.5)$$

$$\cos\phi = \hat{\mathbf{n}}_1 \cdot \hat{\mathbf{n}}_2$$

Conventionally the valid range for a torsion angle is $-180° < \phi \leq 180°$.

A1.5 Calculation of angle between two planes (up to 6 atoms defined)

The calculation of the angle between two arbitrary planes, χ, is very similar to that of a torsion angle. However, now up to six independent atoms may be specified, three for the first plane (position vectors $\mathbf{r}_1$, $\mathbf{r}_2$ and $\mathbf{r}_3$) and three for the second plane (position vectors $\mathbf{r}_4$, $\mathbf{r}_5$ and $\mathbf{r}_6$). Equation A1.6 specifies how the plane unit-normal vectors are calculated.

$$\hat{\mathbf{n}}_1 = \frac{(\mathbf{r}_2 - \mathbf{r}_1)\wedge(\mathbf{r}_3 - \mathbf{r}_2)}{|(\mathbf{r}_2 - \mathbf{r}_1)||(\mathbf{r}_3 - \mathbf{r}_2)|} \quad \hat{\mathbf{n}}_2 = \frac{(\mathbf{r}_5 - \mathbf{r}_4)\wedge(\mathbf{r}_6 - \mathbf{r}_5)}{|(\mathbf{r}_5 - \mathbf{r}_4)||(\mathbf{r}_6 - \mathbf{r}_5)|} \quad (A1.6)$$

$$\cos\chi = \hat{\mathbf{n}}_1 \cdot \hat{\mathbf{n}}_2$$

A1.6 Implementation of rigid-body rotations

To apply an arbitrary rotation to a group of atoms, treated as a rigid body, the orientation of the rotation axis must be defined and the centre of rotation for the group of atoms, (x_0, y_0, z_0) in Cartesian coordinates, must be specified. The coordinates of the ith atom in the group, $(x_i, y_i, z_i)^T$, are transformed to the point, $(x'_i, y'_i, z'_i)^T$, by rotation through an angle, θ, about a rotation-axis parallel to the unit vector (l_x, l_y, l_z) according to the rotation matrix given by Equation A1.7. Note that the coordinates of the centre of rotation are subtracted before the rotation matrix is applied and then added on subsequently.

$$\begin{pmatrix} x'_i \\ y'_i \\ z'_i \end{pmatrix} = \begin{pmatrix} \cos\theta + l_x^2(1-\cos\theta) & l_x l_y(1-\cos\theta) + l_z\sin\theta & l_x l_z(1-\cos\theta) - l_y\sin\theta \\ l_x l_y(1-\cos\theta) - l_z\sin\theta & \cos\theta + l_y^2(1-\cos\theta) & l_y l_z(1-\cos\theta) + l_x\sin\theta \\ l_x l_z(1-\cos\theta) + l_y\sin\theta & l_y l_z(1-\cos\theta) - l_x\sin\theta & \cos\theta + l_z^2(1-\cos\theta) \end{pmatrix} \begin{pmatrix} x_i - x_0 \\ y_i - y_0 \\ z_i - z_0 \end{pmatrix} + \begin{pmatrix} x_0 \\ y_0 \\ z_0 \end{pmatrix} \quad (A1.7)$$

If the rotation is applied about a Cartesian axis then the rotation matrix simplifies substantially. For example the matrix for rotation about the x axis $l_x=1$, $l_y=0$ and $l_z=0$, is given in A1.8.

$$\begin{pmatrix} 1 & 0 & 0 \\ 0 & \cos\theta & \sin\theta \\ 0 & -\sin\theta & \cos\theta \end{pmatrix} \quad (A1.8)$$

A1.7 Special considerations applying to crystallographic systems

When describing the structures of crystals, the presence of periodicity and symmetry and the concept of the crystal lattice have dictated the use of a non-Cartesian and frequently a non-orthogonal basis to describe the vector $\mathbf{r_{ij}}$. In terms of the simplifications that can thereby be afforded and the elegance of the mathematical treatment of symmetry, the use of non-Cartesian coordinates is perfectly understandable. Practically speaking, molecular modelling is best facilitated by the use of a Cartesian basis however, we must consider these other, non-Cartesian coordinate systems as crystal structures derived from the X-ray analysis of crystals are specified using them.

A1.8 Fractional coordinates

There are two systems of fractional coordinates employed to describe the positions of atoms in crystals. The first system relates to direct space and the second to reciprocal space. In both cases a vector basis is selected, for direct space this is usually denoted as (**a**, **b**, **c**) and for reciprocal space as (**a***, **b***, **c***). The important aspect here is that the three basis vectors **a**, **b** and **c** are not necessarily mutually orthogonal (and similarly the three basis vectors **a***, **b*** and **c***). If a direct-space basis and a reciprocal-space basis are being used to describe the same crystal lattice, then the two sets of basis vectors are interrelated by the identity given in Equation A1.9 where **I** is the identity matrix.

$$\begin{pmatrix} \mathbf{a}^* \\ \mathbf{b}^* \\ \mathbf{c}^* \end{pmatrix} \begin{pmatrix} \mathbf{a} & \mathbf{b} & \mathbf{c} \end{pmatrix} = \begin{pmatrix} \mathbf{a}^* \cdot \mathbf{a} & \mathbf{a}^* \cdot \mathbf{b} & \mathbf{a}^* \cdot \mathbf{c} \\ \mathbf{b}^* \cdot \mathbf{a} & \mathbf{b}^* \cdot \mathbf{b} & \mathbf{b}^* \cdot \mathbf{c} \\ \mathbf{c}^* \cdot \mathbf{a} & \mathbf{c}^* \cdot \mathbf{b} & \mathbf{c}^* \cdot \mathbf{c} \end{pmatrix} = \mathbf{I} \quad (A1.9)$$

The reciprocal lattice is of particular significance in the elucidation and description of crystal structures. When X-rays are used to probe crystal structures the electron density is mapped on planes that are defined on a reciprocal-space lattice and it is useful to consider directions both perpendicular and parallel to these planes. The individual planes are designated by sets of indices, the Miller Indices, denoted (h k l). Given the basis (**a**, **b**, **c**) for fractional coordinates in direct space, a Miller plane (h k l) intersects the a, b and c axes at the points $|\mathbf{a}|/h$, $|\mathbf{b}|/k$, $|\mathbf{c}|/l$. This notation is also important as it is used do designate

Appendices 257

the external surfaces on individual crystals so that these surfaces can then be correlated directly with the underlying molecular structure and packing arrangement.

A1.9 Use of the asymmetric unit to generate fractional coordinates for all atoms
The fractional coordinates of all the atoms in the unit cell of a crystal can be calculated from the fractional coordinates of the asymmetric unit (supplied in the crystallographic structure file). The calculation requires sets of symmetry operations associated with the 230 crystallographic space groups and these are defined with respect to a fractional basis in direct space. For the triclinic, monoclinic, hexagonal, trigonal and rhombohedral crystal systems the basis vectors are non-orthogonal. For the other crystal systems the basis vectors are mutually orthogonal but are not generally unit vectors.

Every symmetry operator can be represented by the combination of a (3 x 3) matrix, **W**, and (3 x 1) column vector, **w**. The matrix represents the point-symmetry part (rotation, reflection or inversion) of the coordinate transformation and the column vector the translational part. A general symmetry operation that transforms the fractional coordinates $(x, y, z)^T$ into new fractional coordinates $(x', y', z')^T$ but still referring to the same basis vectors is given by Equation A1.10 (note a superscript T denotes the transpose of a vector or matrix).

$$\begin{pmatrix} x' \\ y' \\ z' \end{pmatrix} = \begin{pmatrix} W_{11} & W_{12} & W_{13} \\ W_{21} & W_{22} & W_{23} \\ W_{31} & W_{32} & W_{33} \end{pmatrix} \begin{pmatrix} x \\ y \\ z \end{pmatrix} + \begin{pmatrix} w_1 \\ w_2 \\ w_3 \end{pmatrix} \qquad \text{A1.10}$$

Taking the example of the tetragonal space group $P4_12_12$ there are eight symmetry operations required to generate the fractional coordinates of all the atoms in a unit cell from the asymmetric unit. Table A1.1 shows the symmetry operations, as denoted in the International Tables for Crystallography Volume A, and the corresponding matrix **W** and vector **w**.

A1.10 Transformations between fractional and Cartesian coordinates
There are many possible choices for selecting the mutual orientation of fractional and Cartesian axis systems in computational crystallography. The transformation matrices for converting from fractional to Cartesian coordinates for three such choices are given below where the origins of the fractional and Cartesian coordinate systems are coincident. The first matrix is for the choice of **a** parallel to **X**, **b*** parallel to **Y** and **a^b*** parallel to **Z** (caret [^] denotes the vector cross product). The second matrix is for the choice of **b** parallel to **Y**, **c*** parallel to **Z** and **b^c*** parallel to **X**. The third matrix is for the choice of **c** parallel to **Z**, **a*** parallel to **X** and **c^a*** parallel to **Y**. The direct-space fractional coordinates have a non-orthogonal basis hence the angle between **a** and **b** is denoted γ, between **a** and **c** β, and between **b** and **c**, α. Similarly for the basis vectors in reciprocal space the angle between **a*** and **b*** is denoted γ^*, between **a*** and **c***, β^* and between **b*** and **c***, α^*.

To convert from a coordinate triple which refers to a fractional, direct-space basis, expressed as a column vector, to the corresponding coordinate triple that refers to the Cartesian basis, the column vector is pre-multiplied by the appropriate three-by-three matrix, **Q**, (see A1.11) according to the orientation required. Similarly to convert coordinates which refer to the Cartesian basis back to coordinates which refer to the fractional basis, the column vector recording the Cartesian coordinate-triple is pre-multiplied by the inverse of the same matrix, $\mathbf{Q}^{-1}$. The forms of the inverse matrices $\mathbf{Q}^{-1}$ are not reproduced here as it is a simple manner to generate the inverse matrix from $\mathbf{Q}$ numerically using a spreadsheet facility such as Origin © OriginLab or Microsoft Excel.

Table A1.1 Symmetry operations for Space Group 92, $P4_12_12$, represented as a Matrix, W, and Vector, w (see Equation A1.10)

Symmetry Operation	Matrix W	Vector w
x, y, z	$\begin{pmatrix} 1 & 0 & 0 \\ 0 & 1 & 0 \\ 0 & 0 & 1 \end{pmatrix}$	$\begin{pmatrix} 0 \\ 0 \\ 0 \end{pmatrix}$
$\bar{x}, \bar{y}, z+\tfrac{1}{2}$	$\begin{pmatrix} -1 & 0 & 0 \\ 0 & -1 & 0 \\ 0 & 0 & 1 \end{pmatrix}$	$\begin{pmatrix} 0 \\ 0 \\ \tfrac{1}{2} \end{pmatrix}$
$\bar{y}+\tfrac{1}{2}, x+\tfrac{1}{2}, z+\tfrac{1}{4}$	$\begin{pmatrix} 0 & -1 & 0 \\ 1 & 0 & 0 \\ 0 & 0 & 1 \end{pmatrix}$	$\begin{pmatrix} \tfrac{1}{2} \\ \tfrac{1}{2} \\ \tfrac{1}{4} \end{pmatrix}$
$y+\tfrac{1}{2}, \bar{x}+\tfrac{1}{2}, z+\tfrac{3}{4}$	$\begin{pmatrix} 0 & 1 & 0 \\ -1 & 0 & 0 \\ 0 & 0 & 1 \end{pmatrix}$	$\begin{pmatrix} \tfrac{1}{2} \\ \tfrac{1}{2} \\ \tfrac{3}{4} \end{pmatrix}$
$\bar{x}+\tfrac{1}{2}, y+\tfrac{1}{2}, \bar{z}+\tfrac{1}{4}$	$\begin{pmatrix} -1 & 0 & 0 \\ 0 & 1 & 0 \\ 0 & 0 & -1 \end{pmatrix}$	$\begin{pmatrix} \tfrac{1}{2} \\ \tfrac{1}{2} \\ \tfrac{1}{4} \end{pmatrix}$
$x+\tfrac{1}{2}, \bar{y}+\tfrac{1}{2}, \bar{z}+\tfrac{3}{4}$	$\begin{pmatrix} 1 & 0 & 0 \\ 0 & -1 & 0 \\ 0 & 0 & -1 \end{pmatrix}$	$\begin{pmatrix} \tfrac{1}{2} \\ \tfrac{1}{2} \\ \tfrac{3}{4} \end{pmatrix}$
$y, x, \bar{z}$	$\begin{pmatrix} 0 & 1 & 0 \\ 1 & 0 & 0 \\ 0 & 0 & -1 \end{pmatrix}$	$\begin{pmatrix} 0 \\ 0 \\ 0 \end{pmatrix}$
$\bar{y}, \bar{x}, \bar{z}+\tfrac{1}{2}$	$\begin{pmatrix} 0 & -1 & 0 \\ -1 & 0 & 0 \\ 0 & 0 & -1 \end{pmatrix}$	$\begin{pmatrix} 0 \\ 0 \\ \tfrac{1}{2} \end{pmatrix}$

$$\begin{pmatrix} a & b\cos\gamma & c\cos\beta \\ 0 & b\sin\gamma\sin\alpha^* & 0 \\ 0 & -b\sin\gamma\cos\alpha^* & c\sin\beta \end{pmatrix} \text{(choice 1 } a||X \text{ and } b^*||Y \text{ and } a^\wedge b^* || Z)$$

$$\begin{pmatrix} a\sin\gamma & 0 & -c\sin\alpha\cos\beta^* \\ a\cos\gamma & b & c\cos\alpha \\ 0 & 0 & c\sin\alpha\sin\beta^* \end{pmatrix} \text{(choice 2 } b^\wedge c^*||X \text{ and } b||Y \text{ and } c^* || Z) \quad (A1.11)$$

$$\begin{pmatrix} a\sin\beta\sin\gamma^* & 0 & 0 \\ -a\sin\beta\cos\gamma^* & b\sin\alpha & 0 \\ a\cos\beta & b\cos\alpha & c \end{pmatrix} \text{(choice 3 } a^* || X \text{ and } c^\wedge a^*||Y \text{ and } c||Z)$$

For practical purposes, equations are needed that inter-relate the direct and reciprocal lattice parameters so that the elements of the matrices in A1.11 can be evaluated from the unit cell parameters. Equations A1.12 give the definitions of the reciprocal-lattice basis vectors in terms of the direct-lattice basis vectors.

$$\mathbf{a}^* = \frac{\mathbf{b}^\wedge \mathbf{c}}{\mathbf{a}\cdot\mathbf{b}^\wedge \mathbf{c}}, \mathbf{b}^* = \frac{\mathbf{c}^\wedge \mathbf{a}}{\mathbf{b}\cdot\mathbf{c}^\wedge \mathbf{a}}, \mathbf{c}^* = \frac{\mathbf{a}^\wedge \mathbf{b}}{\mathbf{c}\cdot\mathbf{a}^\wedge \mathbf{b}} \quad (A1.12)$$

The unit cell volume, V, can be calculated from Equations A1.13 and this is used to calculate the reciprocal lattice parameters a*, b*, c* and the sine and cosine of the reciprocal lattice parameters α^*, β^* and γ^* using Equations A1.14).

$$V = \mathbf{a}\cdot\mathbf{b}^\wedge\mathbf{c} = \mathbf{b}\cdot\mathbf{c}^\wedge\mathbf{a} = \mathbf{c}\cdot\mathbf{a}^\wedge\mathbf{b}$$
$$V = abc\left(1 - \cos^2\alpha - \cos^2\beta - \cos^2\gamma + 2\cos\alpha\cos\beta\cos\gamma\right)^{\frac{1}{2}} \quad (A1.13)$$

$$a^* = \frac{bc\sin\alpha}{V} \quad \sin\alpha^* = \frac{V}{abc\sin\beta\sin\gamma} \quad \cos\alpha^* = \frac{\cos\beta\cos\gamma - \cos\alpha}{\sin\beta\sin\gamma}$$
$$b^* = \frac{ca\sin\beta}{V} \quad \sin\beta^* = \frac{V}{abc\sin\gamma\sin\alpha} \quad \cos\beta^* = \frac{\cos\gamma\cos\alpha - \cos\beta}{\sin\gamma\sin\alpha} \quad (A1.14)$$
$$c^* = \frac{ab\sin\gamma}{V} \quad \sin\gamma^* = \frac{V}{abc\sin\alpha\sin\beta} \quad \cos\gamma^* = \frac{\cos\alpha\cos\beta - \cos\gamma}{\sin\alpha\sin\beta}$$

A1.11 Calculation of the spacing between crystal lattice planes

The most general form of the equation (for triclinic crystals) relating the inter-planar spacing between reciprocal lattice planes, d_{hkl}, to the Miller index, (h, k, l) and the reciprocal lattice parameters is given by Equation A1.15. Compare this to Equation 7.15 for a cubic crystal. Equation A1.15 is required to implement the BFDH approach for predicting a crystal habit in the case of a crystal in the triclinic system.

$$\frac{1}{d_{hkl}^2} = h^2 a^{*2} + k^2 b^{*2} + l^2 c^{*2} + 2hka^*b^*\cos\gamma^* + 2klb^*c^*\cos\alpha^* + 2hla^*c^*\cos\beta^* \quad (A1.15)$$

A1.12 Symmetry operations in Cartesian coordinates

The same symmetry operator matrix, **W**, as described in Section A1.9 can be applied to manipulate Cartesian coordinates in the way shown in Equation A1.16 where the coordinates $(X, Y, Z)^T$ in the Cartesian basis (note the use of upper case for Cartesian coordinates) are equivalent to the coordinates $(x, y, z)^T$ in the fractional basis and are transformed to the new Cartesian coordinates $(X', Y', Z')^T$. Note that since all symmetry transformations can be centred on the origin of the coordinate system, the Cartesian coordinates of the centre of the asymmetric unit, $(X_0, Y_0, Z_0)^T$, are subtracted from the Cartesian coordinates of a general atom before the transformation matrix, **W**, is applied. In Equation A1.16, **Q** is the matrix for transforming fractional coordinates to Cartesian coordinates.

$$\begin{pmatrix} X' \\ Y' \\ Z' \end{pmatrix} = \mathbf{QWQ}^{-1} \begin{pmatrix} X - X_0 \\ Y - Y_0 \\ Z - Z_0 \end{pmatrix} + \mathbf{QWQ}^{-1} \begin{pmatrix} X_0 \\ Y_0 \\ Z_0 \end{pmatrix} + \mathbf{Q} \begin{pmatrix} w_1 \\ w_2 \\ w_3 \end{pmatrix} \qquad \text{A1.16}$$

Appendix A2 Data Tables

Table A2.1
Standard Bond Lengths User in INTERCHEM

Atom 1	Atom 2	Type	Å	σ
C sp³	H	single	1.066	
C sp²	H	single	1.077	
C sp	H	single	1.058	
C ar	H	single	1.083	
C sp³	C sp³	single	1.530	0.015
C sp³	C sp2	single	1.507	0.015
C sp³	C sp	single	1.434	0.006
C sp³	C ar	single	1.513	0.014
C sp²	C sp²	single	1.460	0.015
C sp²	C sp	single	1.431	0.014
C sp²	C ar	single	1.487	0.007
C ar	C ar	single	1.487	0.007
C sp²	Csp²	double	1.316	0.015
C sp	C sp	triple	1.181	0.014
C sp³	N sp³ (+)	single	1.499	0.018
C sp³	N sp³	single	1.469	0.014
C sp³	N sp²	single	1.465	0.011
C sp²	N sp²	single	1.335	
C sp²	N sp²	double	1.280	
C ar	N ar	aromatic	1.335	
C sp	N sp	triple	1.136	
C sp³	O sp³	single	1.428	
C sp²	O sp²	double	1.210	
C sp³	S sp³	single	1.820	
C sp²	S sp²	double	1.660	
C sp³	P	single	1.885	

Atom 1	Atom 2	Type	Å	σ
C sp³	F	single	1.399	0.017
C sp²	F	single	1.340	0.009
C ar	F	single	1.363	0.008
C sp³	Cl	single	1.790	0.007
C sp²	Cl	single	1.713	0.011
C ar	Cl	single	1.739	0.010
C sp³	Br	single	1.966	0.029
C sp²	Br	single	1.883	0.015
C ar	Br	single	1.899	0.012
C sp³	I	single	2.162	0.015
C ar	I	single	2.095	0.015
N sp³ (+)	H	single	1.033	
N sp³	H	single	1.009	
N	O	oximes	1.394	0.018
N	O	N-oxides	1.304	0.015
N	O	nitro group	1.212	
O sp³	H	single	0.967	0.011
S sp³	H	single	1.336	
S	O	single	1.577	0.015
S	O	double	1.497	0.013
S	S	Single	2.048	0.015
P	H	single	1.437	
P	O	single	1.573	0.011
P	O	double	1.467	0.007

Sources of the data:
International Tables for Crystallography, Volume C, Mathematical, Physical and Chemical Tables. Ed. E. Prince, Kluwer, Dordrecht, 2004

Tables of Interatomic Distances and Configurations of Molecules and Ions. Special Publication No. 11, Ed. H. J. M. Bowen, J. Donohue, D. G. Jenkin, O. Kennard, P. J. Wheatley, and D. H. Whiffen. The Chemical Society, London, 1958.

Tables of Interatomic Distances and Configurations of Molecules and Ions. Supplement 1956-1959. Special Publication No. 18, Ed. L. E. Sutton. The Chemical Society, London, 1965.

Uses of the Data
The bond lengths are used to decide on the bond order when crystallographic lacks information

Table A2.2
Crystallographic Space Groups

#	Symbol	Type	No.	#	Symbol	Type	No.	#	Symbol	Type	No.	#	Symbol	Type	No.
1	P1	TC*	10	33	Pna2$_1$	OR	7	65	Cmmm	OR	105	97	I422	TE*	155
2	P-1	TC	2	34	Pnn2	OR	73	66	Cccm	OR	134	98	I4$_1$22	TE*	137
3	P2	MC*	96	35	Cmm2	OR	224	67	Cmma	OR	146	99	P4mm	TE	230
4	P2$_1$	MC*	5	36	Cmc2$_1$	OR	26	68	Ccca	OR	60	100	P4bm	TE	227
5	C2	MC*	12	37	Ccc2	OR	107	69	Fmmm	OR	150	101	P42cm	TE	223
6	Pm	MC	182	38	Amm2	OR	179	70	Fddd	OR	42	102	P42nm	TE	189
7	Pc	MC	19	39	Abm2	OR	156	71	Immm	OR	118	103	P4cc	TE	202
8	Cm	MC	65	40	Ama2	OR	95	72	Ibam	OR	64	104	P4nc	TE	117
9	Cc	MC	9	41	Aba2	OR	38	73	Ibca	OR	77	105	P4$_2$mc	TE	229
10	P2/m	MC	103	42	Fmm2	OR	119	74	Imma	OR	106	106	P4$_2$bc	TE	130
11	P2$_1$/m	MC	16	43	Fdd2	OR	43	75	P4	TE*	160	107	I4mm	TE	212
12	C2/m	MC	17	44	Imm2	OR	124	76	P4$_1$	TE*	39	108	I4cm	TE	186
13	P2/c	MC	15	45	Iba2	OR	57	77	P42	TE*	120	109	I4$_1$md	TE	168
14	P2$_1$/c	MC	1	46	Ima2	OR	114	78	P4$_3$	TE*	47	110	I4$_1$cd	TE	68
15	C2/c	MC	3	47	Pmmm	OR	192	79	I4	TE*	76	111	P-42m	TE	218
16	P222	OR*	171	48	Pnnn	OR	148	80	I4$_1$	TE*	80	112	P-42c	TE	185
17	P222$_1$	OR*	121	49	Pccm	OR	198	81	P-4	TE*	84	113	P-42$_1$m	TE	79
18	P2$_1$2$_1$2	OR*	18	50	Pban	OR	142	82	I-4	TE	28	114	P-42$_1$c	TE	30
19	P2$_1$2$_1$2$_1$	OR*	4	51	Pmma	OR	147	83	P4/m	TE	161	115	P-4m2	TE	228
20	C222$_1$	OR*	24	52	Pnna	OR	40	84	P4$_2$/m	TE	111	116	P-4c2	TE	188
21	C222	OR*	145	53	Pmna	OR	100	85	P4/n	TE	41	117	P-4b2	TE	144
22	F222	OR*	195	54	Pcca	OR	59	86	P4$_2$/n	TE	29	118	P-4n2	TE	90
23	I222	OR*	83	55	Pbam	OR	74	87	I4/m	TE	56	119	I-4m2	TE	173
24	I2$_1$2$_1$2$_1$	OR*	152	56	Pccn	OR	20	88	I4$_1$/a	TE	21	120	I-4c2	TE	136
25	Pmm2	OR	207	57	Pbcm	OR	35	89	P422	TE*	217	121	I-42m	TE	89
26	Pmc2$_1$	OR	97	58	Pnnm	OR	46	90	P42$_1$2	TE*	158	122	I-42d	TE	54
27	Pcc2	OR	200	59	Pmmn	OR	70	91	P4$_1$22	TE*	149	123	P4/mmm	TE	109
28	Pma2	OR	203	60	Pbcn	OR	11	92	P4$_1$2$_1$2	TE*	23	124	P4/mcc	TE	126
29	Pca2$_1$	OR	13	61	Pbca	OR	6	93	P4$_2$22	TE*	219	125	P4/nbm	TE	194
30	Pnc2	OR	108	62	Pnma	OR	8	94	P4$_2$2$_1$2	TE*	92	126	P4/nnc	TE	86
31	Pmn2$_1$	OR	50	63	Cmcm	OR	36	95	P4$_3$22	TE*	164	127	P4/mbm	TE	125
32	Pba2	OR	93	64	Cmca	OR	31	96	P4$_3$2$_1$2	TE*	25	128	P4/mnc	TE	132

Appendices

129	P4/nmm	TE	85
130	P4/ncc	TE	62
131	P4$_2$/mmc	TE	197
132	P4$_2$/mcm	TE	201
133	P4$_2$/nbc	TE	175
134	P4$_2$/nnm	TE	163
135	P4$_2$/mbc	TE	141
136	P4$_2$/mnm	TE	94
137	P4$_2$/nmc	TE	113
138	P4$_2$/ncm	TE	123
139	I4/mmm	TE	98
140	I4/mcm	TE	157
141	I4$_1$/amd	TE	91
142	I4$_1$/acd	TE	61
143	P3	TR*	82
144	P3$_1$	TR*	45
145	P3$_2$	TR*	49
146	R3	TR*	32
147	P-3	TR	34
148	R-3	TR	14
149	P312	TR*	214
150	P321	TR*	140
151	P3$_1$12	TR*	196
152	P3$_1$21	TR*	44
153	P3$_2$12	TR*	206
154	P3$_2$21	TR*	53

155	R32	TR*	67
156	P3m1	TR	222
157	P31m	TR	205
158	P3c1	TR	139
159	P31c	TR	71
160	R3m	TR	75
161	R3c	TR	37
162	P-31m	TR	213
163	P-31c	TR	69
164	P-3m1	TR	138
165	P-3c1	TR	51
166	R-3m	TR	66
167	R-3c	TR	27
168	P6	HE*	191
169	P6$_1$	HE*	52
170	P6$_5$	HE*	58
171	P6$_2$	HE*	143
172	P6$_4$	HE*	167
173	P6$_3$	HE*	48
174	P-6	HE	174
175	P6/m	HE	193
176	P6$_3$/m	HE	33
177	P622	HE*	226
178	P6$_1$22	HE*	81
179	P6$_5$22	HE*	88
180	P6$_2$22	HE*	162

181	P6$_4$22	HE*	181
182	P6$_3$22	HE*	135
183	P6mm	HE	225
184	P6cc	HE	216
185	P6$_3$cm	HE	187
186	P6$_3$mc	HE	110
187	P-6m2	HE	211
188	P-6c2	HE	215
189	P-62m	HE	199
190	P-62c	HE	104
191	P6/mmm	HE	166
192	P6/mcc	HE	131
193	P6$_3$/mcm	HE	178
194	P6$_3$/mmc	HE	87
195	P23	CU*	210
196	F23	CU*	159
197	I23	CU*	112
198	P2$_1$3	CU*	55
199	I2$_1$3	CU*	172
200	Pm3	CU	209
201	Pn3	CU	184
202	Fm3	CU	180
203	Fd-3	CU	116
204	Im-3	CU	129
205	Pa3	CU	43
206	Ia-3	CU	133

207	P432	CU*	221
208	P4232	CU*	220
209	F432	CU*	190
210	F4132	CU*	204
211	I432	CU*	208
212	P4332	CU*	169
213	P4$_1$32	CU*	165
214	I4$_1$32	CU*	183
215	P-43m	CU	154
216	F-43m	CU	151
217	I-43m	CU	78
218	P-43n	CU	102
219	F-43c	CU	128
220	I-43d	CU	72
221	Pm-3m	CU	127
222	Pn-3n	CU	115
223	Pm-3n	CU	177
224	Pn-m	CU	176
225	Fm-3m	CU	63
226	Fm-3c	CU	170
227	Fd-3m	CU	101
228	Fd-3c	CC	122
229	Im-3m	CU	99
230	Ia-3d	CU	153

Notes

This table shows the 230 three-dimensional Space Groups. It is split over two pages into four major columns per page. The four columns in each major column show:

(1) The space group number; (2) The abbreviated Hermann-Mauguin Symbol; (3) The crystallographic system to which the group belongs:

TC = Triclinic; MC = Monoclinic; OR = Orthorhombic; TE = Tetragonal; TR = Trigonal; HE = Hexagonal; CU = Cubic.

(4) The ranking for population of the space group from 587,899 structures in the Cambridge Structural Database on 1 Jan 2011.

An asterisk (*) in the third column means that the group is a *chiral space group*, i.e. it is one of the 65 space groups in which a chiral molecule can crystallise. This does not imply that all structures that crystallise in such a group are chiral.

Appendices

Appendix A3
Numbering of Steroid Skeletons

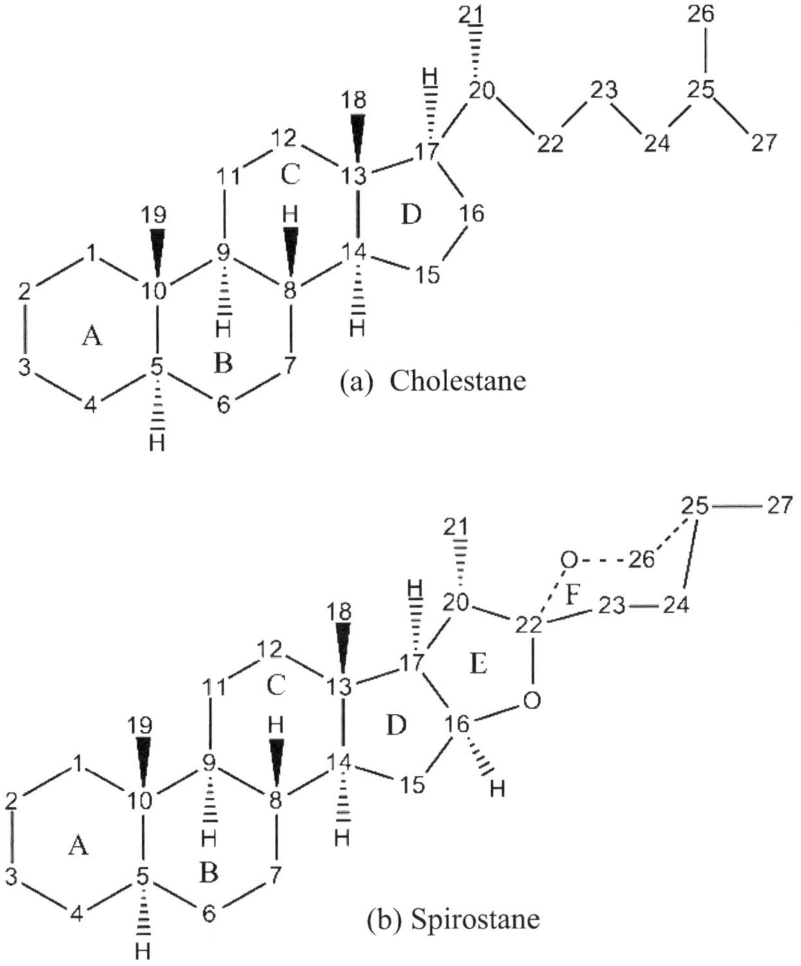

Figure A3.1
Numbering of the carbon atoms in typical steroid structures

Figure A3.1 shows the (IUPAC approved) numbering schemes for the carbon atoms in (a) cholestane, and (b) spirostane. Usually only the carbon atoms are assigned numbers, and the atoms that are shown as their symbols (oxygen and hydrogen), when they need to be identified, are referred to by the carbon atoms to which they are attached. The structures also indicate the way the rings are identified (A to F). The importance of the steroid numbering schemes arises from the presence of steroid motifs in many natural products and drugs. The schemes are shown here because the answers to some of the set problems in our book make implicit reference to them.

Another reason for including this topic is the (almost universal) convention of displaying steroid structural diagrams in one particular way: that is with the rings (also) conventionally labelled: A, B, C, D (and in the spirostane system), E, and F going from bottom left to top right, and the numbering scheme (as shown) also going in the same overall direction.

The other convention that has arisen is that of indicating the configuration of substituent atoms or groups as either *alpha* (α) or *beta* (β) according to whether they are below or above the plane of the paper. Thus in (a) the methyl group (shown as carbon number 19) that is attached to carbon 10 is said to be a 10β-methyl group.

What has all of this to do with modelling? The answer is that the modeller (human or machine) has the job of faithfully transforming the *name* of a chemical compound into a 3D *structure*.

A3.1 Stereochemistry in steroid ring systems
In (a) there are eight chiral centres; and for the structure shown, these are respectively;
5*R*, 10*S*, 9*S*, 8*S*, 14*S*, 13*R*, 17*R*, and 20*S*.

In (b) there are eleven chiral centres, and these are respectively;
5*R*, 10*S*, 9*S*, 8*R*, 14*S*, 13*S*, 17*R*, 16*S*, 20*S*, 22*R*, and 25*R*.

Note that the stereochemistry *apparently* changes significantly between the two related structures. This is due to the way the CIP scheme works; in this case the configuration at carbon 13 is influenced by the extra oxygen atom that joins carbon 16 and carbon 22 in spirostane.

The numbers of stereoisomers that are possible for structure (a) is 2^8 (= 256), whereas for (b) it is 2^{11} (= 2048). However in this latter case, it is likely that a *cis* arrangement of hydrogen atoms at the junction between rings D and E (as shown) is the only one that would be stable. In that case only 1024 isomers would be possible..

Appendices

Appendix A4
Essential Information on Mounting Interprobe Software

A4.1 Software that is included on the compact disc
There are four types of software included on the compact disc:

(1) The program INTERCHEM and PRESTO that rely on graphical user interfaces for the operation.

(2) The programs CONVERT, PROTEINS, and QUICKSCAN that use text based interfaces.

(3) Programs that are in the public domain that were developed by other organisations.

(4) Scripts that are used in mounting software and databases.

In addition a DVD is provided that holds structure files for 4.3 million small molecules

A4.2 Copying the contents of the CD onto the hard-drive of your computer
We assume that your computer in running a recent version of the Microsoft ®Windows operating system (2000, XP, Vista, or Windows 7), and that it has a single hard disc that is not partitioned and that has the drive letter C. We assume that there is a single optical disc drive and that this has the drive letter D. If you have partitioned your disc, or have multiple discs or optical drives, we will assume that you are capable of modifying these instructions accordingly.

Step (1) Mount the Interprobe software disc by inserting it into drive D.

Step (2) Search the disc to find the folder **interprobe_software**.

Step (3) Copy the entire contents of this folder into a folder on your hard drive.
There are two options at this stage:
    (a) You may copy the software into a folder in your personal user area.
    (b) You may copy the software into an area accessible to all users.

A4.3 Setting up the program INTERCHEM
Go to the folder on the hard disc where the Interprobe software is stored, and search for the folder **interchem,** go to this folder and search for the folder that holds the latest version of INTERCHEM, and open this folder to reveal its contents. The file listing will probably look like that in Figure A4.1.

If there are any short-cuts listed you may safely delete them, as they probably refer to a system on another machine.

There are four executable programs in this folder labelled **interchem**, **interchem_stereo, autobuild,** and **protocheck. Right-click** on **interchem** and choose **Creat Shortcut**. Copy the short-cut and paste it onto your desktop. If you have facilities for running stereo viewing you should also do the same for **interchem_stereo**. You can usefully create and paste short cuts for **protocheck** and **autobuild.**

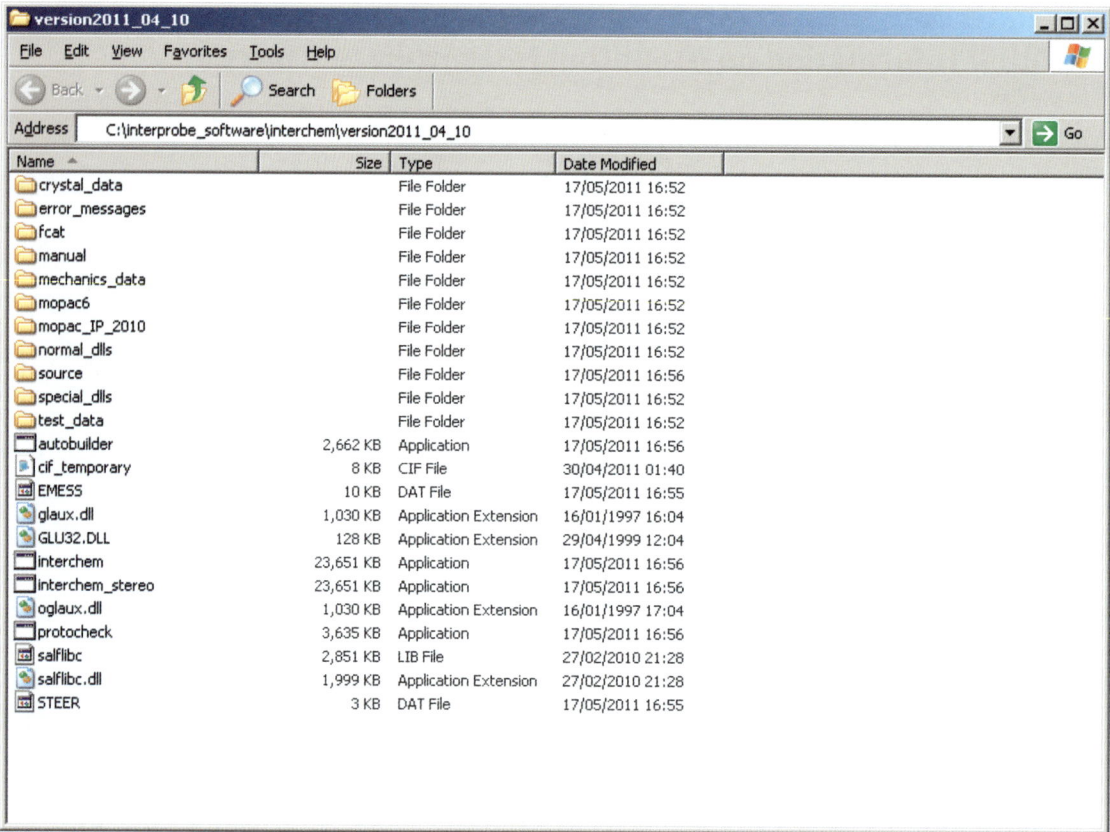

Figure A4.1
Listing of the Contents of Folder
C:\interprobe_software\interchem\version2011_04_10

A4.3.1 Testing the program INTERCHEM

Click on the shortcut for INTERCHEM on your desktop. The program should start with a display of the Welcome Message (in its entirety) and then after you press the Escape key you should get the DUAL display (See Figure 5.1 in Chapter 5). You should be able to do the initial experiments suggested in Chapter 1. If any of these conditions does not prevail, you have got a problem. Broadly speaking problems at this stage, that prevent the program working properly, can be of three kinds. We provide you with remedies for these faults and moreover explain why they occur.

A4.3.2 Problems due to mismatched dynamic-linked libraries.

The symptoms of problems of this type are the appearance of a blanked-out window, and failure to load a structure in the simple experiments of Chapter 1. The cause is the inadequacy of the dynamic-linked library (DLL) called **OpenGL32.DLL** supplied by Microsoft. The cure is simple; go to the folder that holds the executable program (Figure A4.1) and then to the folder `normal_dlls`. Copy the file **OpenGL32.DLL** to the folder that holds the executable program INTERCHEM. Try running INTERCHEM, and the tests again. This will usually clear the fault. The explanation of this fault, and the cure is that when the program is run, links to dynamic-linked libraries are first attempted to any that exist in the same folder as the program. Only if the required library is not present locally, is the corresponding system library used. Note that there are three other DLLs that are used by INTERCHEM; `glaux.dll`,

Appendices

GLU32.DLL, and salfibc.dll. The first two of these are needed by extensions to OpenGL, while the third is used by the utility Clearwin+. The fact that some of these libraries are labelled '32', to indicate that they are 32 bit programs, does not matter; they work on 64 bit systems as well.[1]

A4.3.3 Problems due to the file STEER.DAT

The file **STEER.DAT** is crucial to the operation of INTERCHEM; it allows INTERCHEM to access other programs and the important data files. Typical contents are shown in Figure A4.2. Each line in this file shows the address of a file or folder *relative* to the address of the executable program. An extreme error condition could say that the file is missing; in that case it is relatively easy to construct one based on this figure.

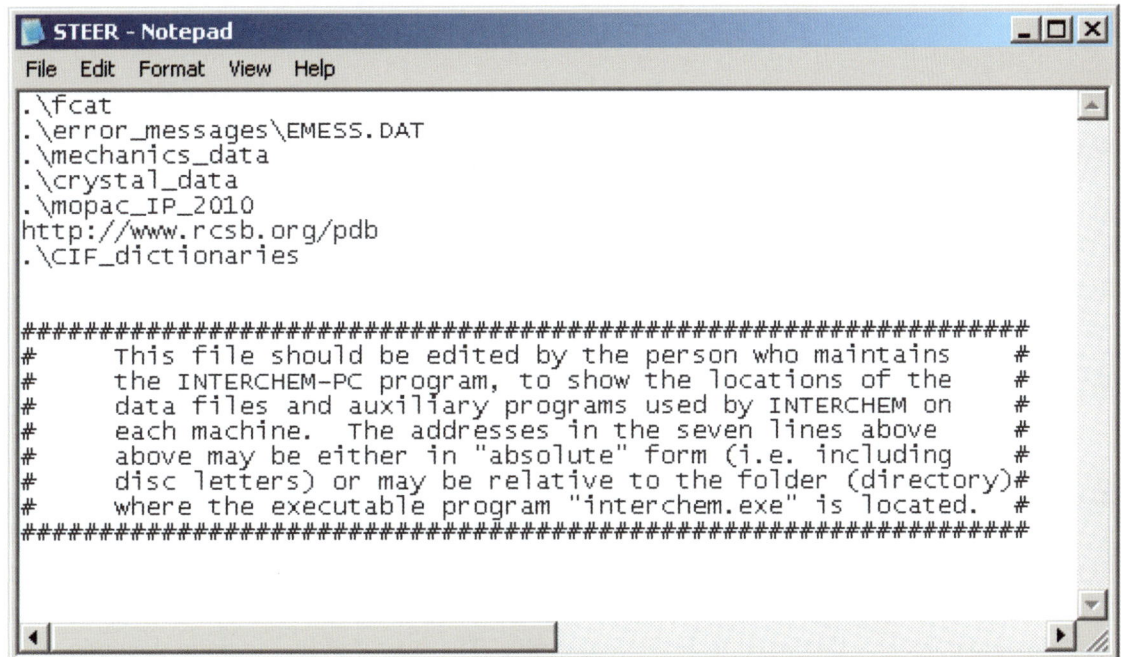

Figure A4.2
The file STEER.DAT used by INTERCHEM

The Figure A4.2 shows the STEER.DAT file used by INTERCHEM for the particular setup of the program shown in Figure A4.1. All the paths in the first five lines point to sub-folders in the folder where the executable program is held. (The way the path is constructed using ' .\ ' to preface the name of a file [or folder] is interpreted by the computer to mean ' look for a file [or folder] in the current folder'). Doing things this way means that there is a high probability that the system will work on most computer set-ups.

Errors of this type usually occur when a version of the program has been copied from a system where the sub-folders have been moved. This latter type of arrangement can be beneficial in some circumstances, and is permissible.

The sixth line offers a link to the Protein Data Bank, and may be changed if your site has a local version, or if there is a national mirror site that nearer. The seventh line of the file is not used at present.

A4.3.4 Errors due to the settings on your display screen

The program INTERCHEM is designed to work on screens with well defined resolutions.

When the program is started the screen resolution is measured and the appropriate parameters are set to allow the program to use the whole of the screen area or an appropriate part of it. Table A4.1 shows the screen resolutions that are acceptable.

A series of tests are applied (in the order implied by the rows, and using the conditions specified by the first three columns of the table). The screen area is then allocated according to columns four to six. This arrangement allows all screen resolutions equal to or better than 1280 x 1024 to be accommodated, plus three others that are common in Lap Top machines. If the screen resolution is not satisfactory an error message will appear. You must then exit the program and set the resolution to meet the conditions.

Table A4.1
Acceptable Screen Resolutions for INTERCHEM

Measured Screen Resolution			Screen Area Used			Notes
X	Y	Acronym	X	Y	Shifted	
≥ 1920	≥ 1080	VESA TV	1280	1024	To top left	
≥ 1920	≥ 1200	WUXGA+	1280	1024	To top left	
≥ 1680	≥ 1048	WSXGA	1280	1024	To top left	
≥ 1400	≥ 1050	SXGA+	1280	1024	Centred	
$=1280$	$=1024$	SXGA	1280	1024	Full Screen	
$=1024$	$=768$	XGA	1024	768	Full Screen	Lap top
$=1024$	$=800$	WXGA	1024	768	To top	Lap top
$=1024$	$=900$	WXGA+	1024	768	To top	Lap top

A4.3.5 Errors in other display parameters

In addition to fixing the correct resolution of the screen, there are some Microsoft Windows parameters that need to be set correctly. The way that this is done depends on the version of the operating system that you are using. These can be classified as either *Before Vista (BV)* or *After (and including) Vista (AV)*.

The errors that can occur and necessary parameter adjustments are:

(a) The size of the characters. This is evident when the whole of the *Welcome Message* does not appear. The last line concerning 'Hardware Stereo Viewing' must be present. The correction is to use the smallest size characters (which may be the default) using:

*(BV) Control Panel >> Display >> Settings >> Advanced >> **Set to Small Fonts***
(AV) Control Panel>>Display>>Smaller 100% (default)

Appendices 271

(b) Failure of the last line of the DUAL display menu to appear. This can only appear when one of the Full Screen settings (Table A4.1) is being used. The correction is achieved by:

(BV) Control Panel>>Display>>Settings>>Task Bar and Start Menu>>Uncheck 'Always on Top'
(AV) Control Panel>>Task Bar and Start Menu>>Taskbar>>Check Auto-hide the task bar

(c) Flickering of the borders of the display windows when the focus (active window) is changed. This is corrected by selecting:

(BV) Control Panel>>Display>>Appearance>>Windows Classic Style+Windows Standard Color Scheme
(AV) Control Panel>>Personalization
There is a profusion of options here, some which will completely change the INTERCHEM colour scheme; experiment at your own risk!! The best advice is: **Avoid any High Contrast Themes!!!**

With the older versions of Windows *(BV)* you may have to restart your computer (or logout) before the changes [(for (a), (b), and (c)] take place. With Vista and Windows 7, the changes you make are acted upon more or less immediately.

A4.4 Setting up the program PRESTO
Go to the folder on the hard disc where the Interprobe software is stored, and search for the folder **presto,** go to this folder and search for the folder that holds the latest version of PRESTO and open this folder to reveal its contents. The file listing will probably look like that in Figure A4.3. If there are any short-cuts listed you may safely delete them, as they probably refer to a system on another machine.

There is one executable program in this folder labelled **presto. Right-click** on **presto** and choose **Create Shortcut.** Copy the short-cut and paste it onto your desktop.

A4.4.1 Testing the program PRESTO
Check that the introductory exercise outlined in Chapter 5 (5.51) can be carried out. This will test a selection of the functions. Some features need the presence of sequence databases, and these require that the correct form of the STEER.DAT file is present. It may be convenient for some of these relatively large files to be located on other discs or partitions other than the ubiquitous C: partition. Figure A4.4 shows a typical version of this file.

A4.4.2 Problems in running PRESTO
Problems similar to those found with INTERCHEM are sometimes encountered, for example those when DLL files are mismatched. However, PRESTO is supplied with the correct DLLs in the folder when the executable program is held.

A4.4.3 Effects of screen resolution on PRESTO
When a wide screen is available PRESTO makes use of it to display the longest possible stretch of protein sequences. The system is interrogated to determine the screen resolution, and the maximum sequence length allocated accordingly.

A4.4.4. Effect of other screen settings when using PRESTO
The recommendations made in section A4.3.5 for setting character sizes and screen styles also apply to the running of PRESTO.

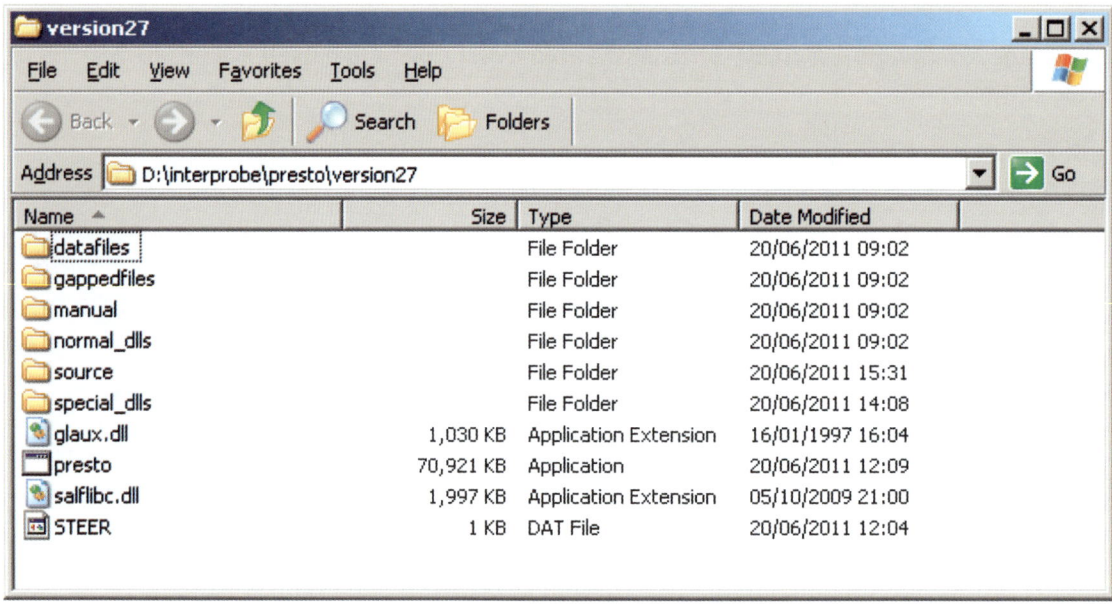

Figure A4.3
Listing of the Contents of Folder
C:\interprobe\presto\version27

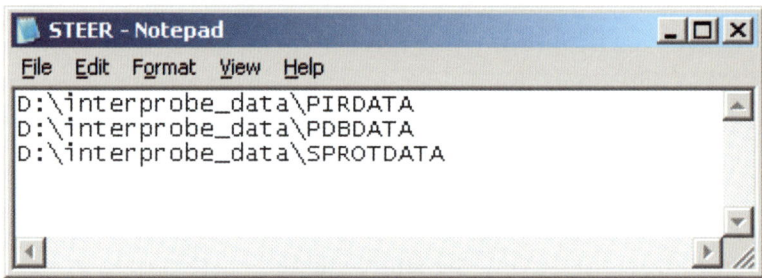

Figure A4.4
Listing of the Contents of the File
STEER.DAT for use by PRESTO

Table A4.2
Acceptable Screen Resolutions for PRESTO

Measured Screen Resolution			Screen allocation	
X	Y	Acronym	Residues	Sequences
≥1920	≥1080	VESA TV	140	60
≥1680	≥1050	WSXGA	110	60
≥1400	≥1050	SXGA+	90	60
≥1280	≥1024	SXGA	80	60
≥1024	≥768	XGA	60	40

Appendices

A4.5 Setting up the other Interprobe programs.
The programs CONVERT, PROTEINS, and QUICKSCAN differ from INTERCHEM and PRESTO in that they do not use a **G**raphical **U**ser **I**nterface (GUI). This makes the setting up much simpler. Each program is present as a folder that is a subfolder of **interprobe_software**. Navigate to the appropriate subfolder, and create a (new) shortcut to the executable program, and place the shortcut on your desktop. Click on this shortcut to start the program. A new text window will appear with instructions for continuing. In most cases, the first step should be navigation to the folder (directory) where the subsequent action will take place. This will either be a directory that exists already, or if not, one that the program will create for you.

Figure A4.5 shows the rolling menu used by CONVERT. In this context 'rolling' means that there is no refreshing of the screen, which has the advantage that you can scroll back the output if necessary. The chief use of CONVERT is in the splitting up of multi-structure files of various sorts, and for bulk conversion between formats. To convert a single structure file INTERCHEM has facilities that are easier to use. For the file formats that are not provided either by INTERCHEM or CONVERT, the public-domain program Open Babel is recommended.[2]

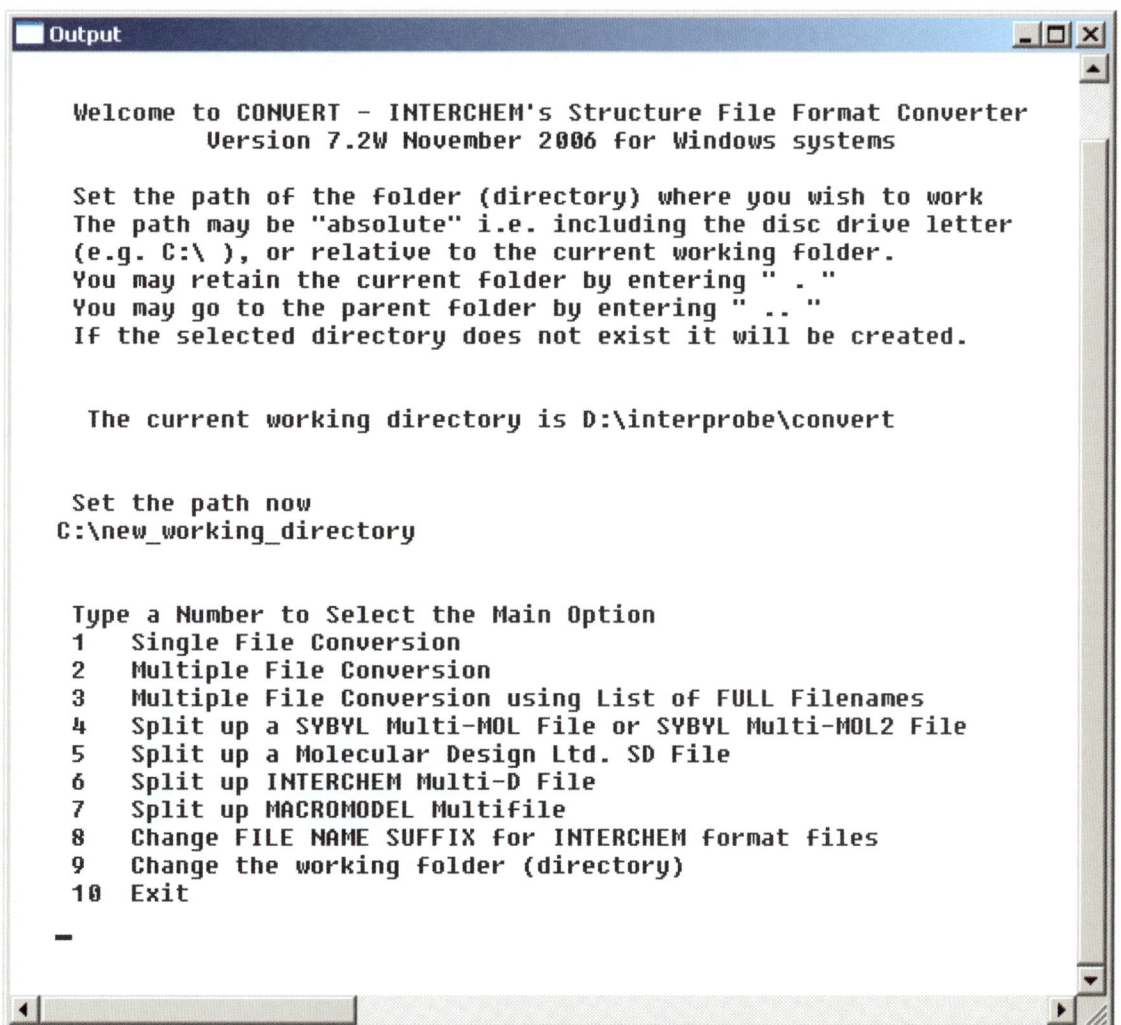

Table A4.5
The Rolling Menu of CONVERT

The use of the program PROTEINS is detailed in Chapter 5. Figure A4.6 shows the main text menu of the program.

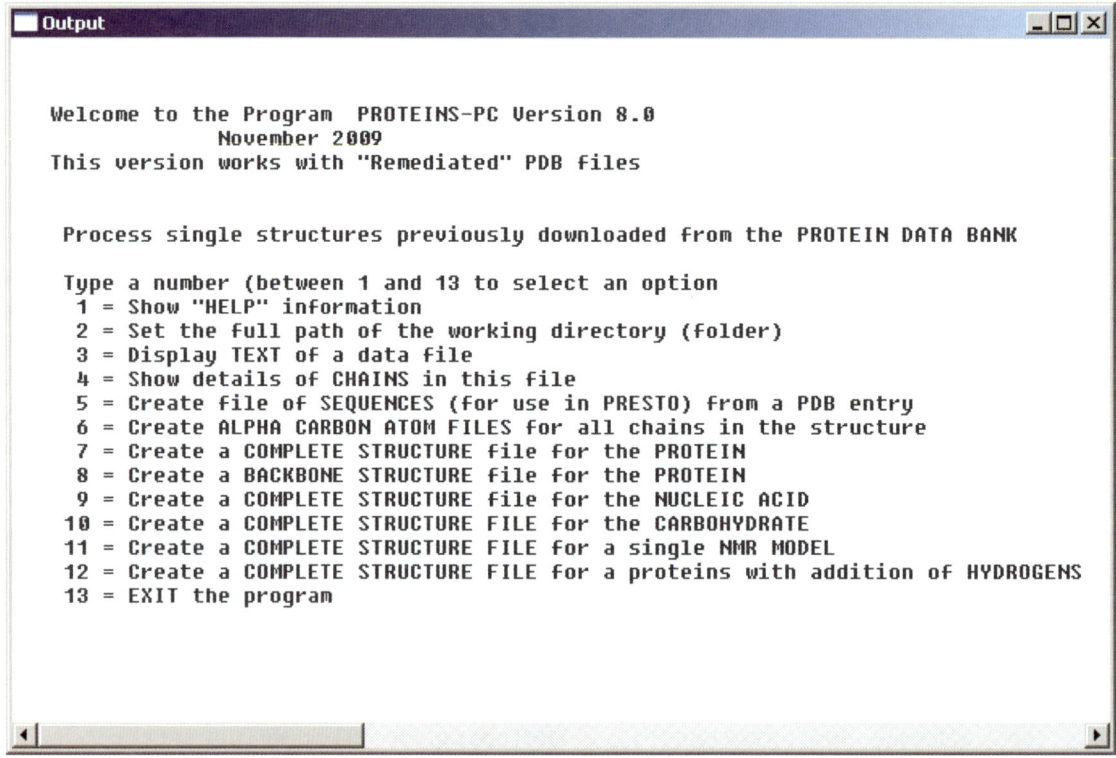

Table A4.6
The Main menu of PROTEINS

The setting up of QUICKSCAN is detailed in Chapter 9 (Paragraph 9.4 ff.).

A4.6 Setting up other programs
The compact disc that accompanies the book also has folders that feature software from sources independent of Interprobe. These are either programs that are issued under the GNU Licenses, or are in the public domain for other reasons. MOPAC is a program in the latter category, for reasons that are explained in Chapter 3 (Paragraph 3.3.1).

A4.6.1 Setting up MOPAC.
There are two separate folders on the disc that have implementations of MOPAC. The first of these is a subfolder of the INTERCHEM folder (see Figure 4.1) and is called MOPAC_IP_2010. If you open this subfolder you will find four versions of the executable program called MOPAC_40_40, MOPAC_80_80, MOPAC_120_120, and MOPAC_160_160. The two numbers incorporated in the names refer to the maximum number of light atoms (essentially hydrogen atoms) and heavy atoms (atoms of all other nuclei) that the version can handle. These executable programs are stored in a folder that is referred to by the INTERCHEM STEER.DAT file, so that they can be accessed by INTERCHEM. When MOPAC is started from within INTERCHEM (see Chapter 4, paragraph 4.3.1), the version appropriate for the structure size is

Appendices

invoked, and the results are accessible from within INTERCHEM immediately after the job has finished.

The same implementation of MOPAC, in all four size options, can be made available for use independently of INTERCHEM, if shortcuts are made and deposited on the desktop of your computer. One use of this facility is to rerun a job using modifications of the KEYWORDS on the first line of the input file (which should of course be renamed). Then dragging and dropping the new input file onto the appropriate shortcut of MOPAC will start the rerun.

A4.6.2 Miscellaneous programs and scripts
These additional files are to be found in the folder called **additional_items.** They include:

(1) A script provided for setting up the database of small molecular structures. The use of this is described in Chapter 9 (Paragraph 9.4.1)

(2) A script for running MOPAC on large numbers of structures with identical sets of KEYWORDS.

(3) Fortran programs for batch conversion of 2D structural data for small structures as SDF files into 3D data in INTERCHEM format. These include versions for running on a cluster of computers. These are designed to run under the Linux operating (or Cygwin), and are provided in source code only. Instructions for compiling the programs are provided as README files.

A4.7 Other problems - errors in the programs
It was stated in paragraph A4.3.1, that there were three types of problems that might occurring setting up INTERCHEM. There is a fourth type of problem that comes about when using a program. No computer program is completely free of errors, and to claim that the Interprobe Software is error-free would be ridiculous. The programs as supplied to you still have links to the source code. These are there so, that in development, errors that are present reveal quickly where the faults really are. The programs supplied to you are fully developed, they are **not** so-called Beta versions. However, the source code links have not been removed so that you, the user, can let the author of the program know when faults do arise. If this happens, please send an email message to:

cbas25@strath.ac.uk

Indicate in the subject line of your email the name of the program, and include a copy of any message that appears on the screen, that includes the reference to the line numbers in the source code. Include other information as well that you think might help or might be relevant.

Be aware that if what appears to be a catastrophic error happens, all is not necessarily lost. If an error of the type that refers to the source code does occur, **do not panic!** In the particular cases of INTERCHEM and PRESTO, if you can reveal the row of buttons that are hidden behind the main menus, (pressing 'Escape', and return with the mouse over a blank part of the screen can help) and then restart the program, nothing will be lost. If the restart button itself is not visible, clicking on *Exit the Program*, and then *DO NOT EXIT* will cause a restart.

A4.8 Removing and updating Interprobe software on your computer

The way that Interprobe software is designed and distributed means that to remove any single item, or indeed the whole package, from your system requires that you go to the appropriate directory and click on *Delete*. This will remove the folder and all subfolders.

When it is necessary to upgrade a component part of the software, the whole component program should be replaced. Each new distribution is self contained, and while it is permissible to have more than one version of a program on your system, the components parts should not be considered to be interchangeable (with the possible exception of DLLs, for the reasons mentioned in paragraph A4.3.2).

References and Endnotes. Appendix A4

[1] Problems associated with Dynamic-Linked Libraries (DLLs) are well known (see DLL-Hell in Wikipedia). The solution, adopted in the Interprobe Software, of having the correct DLL along side the executable program, is acknowledged to be the best. The whole aim of having general-purpose libraries came about when Digital Equipment Corporation (DEC), in the 1980s invented ways of extending the (then limited) core computer memories by using disc memory. Both core memory and disc memory were limited, and it made sense to have (for example) only one implementation of a square-root routine available to all programs. Nowadays, the amounts of memory and disc memory on any computer will be so large, that there is no need for shared libraries in any software; all programs could be self contained. That this is indeed possible is seen by the ability to run programs in virtual machine environments. The only reason for Microsoft to hold to its present designs of operating systems is to ensure that users are locked in to its present ideo-technology.

[2] Open Babel. See openbabel.org/wiki/Open_Babel:About (Accessed 20th June 2011)

Appendix A5
File Compression and Transfer of Files between Computers

A5.1 File compression

When large libraries of structures are to be handled, compressed files are often encountered. UNIX operating systems have three different facilities for compressing files: *compress*, *pack*, and *gzip*. The first two of these have fallen out of favour, in part because the software had been patented. *gzip* is available under the free software foundation license, and as such is part of the Linux operating system. If required on a Windows system, the use of *Cygwin* is recommended. The three facilities each have three commands (see Table A5.1).

Table A5.1
Commands for Lossless File Compression in UNIX and Linux

Facility Name	Compression	Expansion	One-time Expansion	File suffix
`compress`	compress	uncompress	zcat	.Z
`Pack`	pack	unpack	pcat	.z
`Gzip`	gzip	gunzip	gzcat	.gz

The compression and expansion commands work by replacing the existing file by the compressed or expanded file. The 'one-time expansion' commands allow the uncompressed version of a compressed file to be displayed on the computer screen *without* altering the existing compressed file. If you require a copy of this output in a file then *redirection*, such as:

$$\texttt{gzcat xxxxxx.gz >temporary} \qquad (A5.1)$$

will do that. (The 'greater than' symbol is used in Unix and Linux to redirect the output to a file, in this case with the name `temporary`, rather than to the screen). Constructs such as the command A5.1 can be incorporated in scripts and in compiled programs.

Note that compression or expansion of files requires that enough free memory is available during the process to hold both versions (compressed and expanded) of the file, and that the time taken is proportional to the size of the file. Hence, if you are working with large data sets of structures, but only need access to a few of them, and wish to use compression, it is better to have a number of relatively small files. For example, for 1 million structures it would be better to have a thousand compressed files each containing a thousand structures, rather than one large file. This consideration is at the heart of the design of the Interchem database of small structures

In UNIX and Linux operating systems sets of files are often collected together in '`tar`' files. The command:

$$\texttt{tar -c -f <filename>.tar <directory\_name>} \qquad (A5.2)$$

will collect all of the files in the named directory ***and dependent subdirectories*** and place them in the named ***file***.

After transferring `filename.tar` to another system, the command

```
tar -x -f <filename>.tar
```
(A5.3)

will reverse the process, re-establishing an appropriate directory in the process.

The `tar` command is often combined with the compression facility, so that you will often see files named (*e.g.*): `<filename>.gz.tar`.

The implication in this case is that 'untaring' of the file should be performed before uncompression (`gunzip`). However, on some systems the need for the two commands is recognised, and both are performed automatically.

The `tar` facility is unfortunately not available on Windows systems.

In the foregoing description we use the 'less than' and 'greater than' symbols as 'chevron brackets'. The meaning is that the name enclosed between these brackets can be replaced by any valid file name or directory name. The brackets are not to be included of course!

Printable files (*i.e.* those using ASCII characters) such as those used for all of the file formats so far discussed, differ in one important way when they are created and used on UXIX or Linux systems on the one hand, and on Microsoft Windows systems (DOS) systems on the other. In UNIX and Linux files each line of text is terminated by a 'line feed' character (`LF`, ASCII code 10, control+J on a keyboard) while in DOS systems two characters are used 'carriage return' (`CR`, ASCII code 13, control+M on a keyboard) and 'line feed'... In addition DOS files are terminated by an 'end of file' character (`SUB`, ASCII code 26, control+Z on a keyboard).

UNIX has utilities called *to_dos* and *to_unix*, that perform the conversions; Linux has similar utilities called *unix2dos* and *dos2unix*. These utilities are invoked with either one or two operands (file names) for example:

```
to_dos <unix source file> [<dos destination file>]
```
(A5.4)

Here the second operand is optional (implied by enclosing it in square brackets). When only one operand is present the source file is replaced by the destination file; when both operands are present the source file is retained.

A5.2 Transfer of files between computers and different operating systems

The file transfer utility *ftp* can be used to copy files between computers. When the source and destination computers are of different types (UNIX, Windows) then the addition or deletion of the extra characters (mentioned above) is taken care of, if the transfer is done in 'ASCII' mode. The tool *ftp* is now regarded with suspicion as being insecure, particularly when used between computers at different physical sites. A more secure replacement is *scopy* because this requires that encrypted passwords be used.

Answers to 'Problems for you to solve and questions for you to answer'

The nature of the questions asked and problems set is such that, in many cases, there is no single answer or solution. What we provide here are hints on how to approach the tasks, and references to published papers where answers might be found.

Chapter 4

In many of the problems, the best approach is by using SMILES strings, and going for the option of creating a series of 512 structures. With smaller molecules these will be generated in a few seconds, but in some cases the run will take perhaps one hour. (Time for a break and a cup of coffee. Also a reminder that your computer is not just a source of instant responses, and can be used for real computing!). When dealing with reasonably rigid structures, the results, after having been sorted according to energy, should fall into clearly defined sets. The members of a set that have the same energy represent one conformer (sometimes a pair of enantiomers). You can safely ignore the outliers that have very large energies.

[1] The lowest energy set for cyclododecane corresponds to the structure B shown as a rather cryptic diagram in U. Burkert and N. L. Allinger's book.[1] This structure is chiral (hence the adjacent picture B' for the enantiomer).

[2] Experiments aimed at finding the structures of *trans*-1,2-dichlorocyclohexane and *trans*-1,2-difluorocyclohexane by electron-diffraction studies have been reported.[2] If you tackle this problem by not specifying stereochemistry in the SMILES strings, you will get the results (for example) for both *cis* and *trans*. isomers of 1,2-difluorocyclohexane in one experiment. After sorting the results on the energies, the next step is to determine which sets of results correspond to the various isomers, by displaying the results on the computer screen. Results of some calculations are shown in Table ANS.1.

[3] The suggested method of answering this question involves using INTERCHEM as a substitute for hardware mechanical models! This is in itself instructive. The *full* answer is given in Chapter 6 (section 6.3.1).

[4] If a series of 512 structures is generated from the SMILES string for hexahelicene, you should get roughly half of the structures with excess energies of -27.7 kjoules/mol, and another group of structure with excess energies of +20.14 kjoules/mol. Both of these groups of structures comprise roughly equal numbers of enantiomers. The lower energy structures are the expected stable helical structures, while the higher energy structures are likely intermediates on the racemization path. A paper that deals with racemization of helicenes is available.[3]

[5] One of many possible SMILES strings for buckminsterfullerene C_{60} is:
```
C%11%21(=C%22C(=C%12%13)C(=C%14%15)C%23=C%24C%26=C%15C%16=C%17%27)C%21%3
1(=C%32C(=C%22%23)C(=C%24%25)C%33=C%34C%36=C%25C%26=C%27%37)C%31%41(=C%4
2C(=C%32%33)C(=C%34%35)C%43=C%44C%46=C%35C%36=C%37%47)C%41%51(=C%52C(=C%
42%43)C(=C%44%45)C%53=C%54C%56=C%45C%46=C%47%57)C%51%11(=C%12C(=C%52%53)
C(=C%54%55)C%13=C%14C%16=C%55C%56=C%57%17)
```

This is a single string of characters without spaces that, because it is so long has to extend over several lines. If you are using INTERCHEM, to enter this string do not attempt to type it, but copy it from the appropriate location on your computer (*e.g.* the folder that holds the test data for INTERCHEM):

`~\versionyyyy_mm_dd\test_data\SMILES_strings\fullern4.smi`

using WORDPAD and 'copy and paste'. (Do *not* use NOTEPAD!
The string was constructed by using a *Schlegel diagram* as starting point. If you wish to make models of the higher fullerenes, the book written by Roger Taylor[4] shows Schlegel diagrams for them, useful for building using INTERCHEM sketchpad.
It is not possible to construct an *unbranched* SMILES string for the C_{60} molecule. This is due to the need to specify single and double bonds. If you require a structure with only single bonds, it is possible to do this in a single string. In contrast building structures using SMILES strings is relatively easy for the hydrocarbons that have the symmetry of the *platonic polyhedra*. These include:

tetrahedrane (C_4H_4):	`C12C3C2C13`
cubane (C_8H_8):	`C12C3C4C2C5C4C3C15`
dodecahedrane ($C_{20}H_{20}$):	`C12C3C4C5C2C6C7C5C8C4C9C3C3C4C9C8C7C4C6C13`

Note also that the string for the last compound is made with ring break labels ≤9, by reusing the labels `'3'` and `'4'`. It thus avoids the use of two-digit labels like `'%11'`.

You will now realize that use of SMILES strings is perhaps not the best way to construct fullerenes!.

Table ANS.1
Results of Energy Calculations on 1,2-Dihalogenated Cyclohexanes

Compound	Conformer	Molecular Mechanics	MOPAC-AM1
1,2-*trans*-Difluorocyclohexane	aa	-3.03	-532.10
	ee	-2.47	-538.30
Difference (aa - ee)		-0.56	+6.20
1,2-*cis*-Difluorocyclohexane	ae	-2.75	-532.66
1,2-*trans*-Dichlorocyclohexane	aa	-1.25	-210.05
	ee	-3.18	-210.22
Difference (aa - ee)		+1.93	+0.17
1,2-*cis*-Dichlorocyclohexane	ae	-2.14	-208.24

All energies are in kjoule/mole.
Energies from Molecular Mechanics are 'excess energies'.
Energies from MOPAC are 'Heats of formation'.
Note that in the case of *trans*-1,2-Difluorocyclohexane, MOPAC predicts that the di-equatorial conformer will be more stable, whereas with Molecular Mechanics the di-axial conformer is predicted to be more stable.
The results show that there is little difference in energies between the *trans* and *cis* isomers of both series of compounds.

Chapter 5

The type of problems set in connection with this chapter are not capable of simple answers. If you have followed through even a few of the suggestions made, you will have gained insight into the complexities of the structures and origins of proteins. You will also have realized the wealth of information that is freely available from the Protein Data Bank.

Chapter 10

It was suggested that you find or generate the 3D structures for some of the anticancer drugs. Two of particular interest are *Eribulin* and *Cephalostatin-1*.

There does not appear to be an X-Ray crystal structure of Eribulin or a derivative of it. The *Sketch* facility in INTERCHEM, followed by optimization of the structure with MOLAC (AM1) has yielded the 3D structure shown in Figures ANS.1 and ANS.2 The numbering of the atoms in Figure ANS.1 has been arrived at by use of the *Morgan* algorithm,[5] (accessible in the *Fragment Building* menu in INTERCHEM). If you have built the structure in a different way, renumbering it using this algorithm should give an identical numbering, and so facilitate the comparison of the results. The seventeen chiral carbon atoms have the following configurations: 1*S*, 2*S*, 8*S*, 9*S*, 10*R*, 11*R*, 13*S*, 14*S*, 17*S*, 23*R*, 30*R*, 35*S*, 36*R*, 38*S*, 39*R*, 44*R*, and 49*R*.

There is an X-Ray crystal structure of Cephalostatin 1.[6] The structure shown in Figures ANS.3 and ANS.4 is derived from the crystal structure but has been further refined using MOPAC (AM1). The carbon atoms of the bi-steroid skeleton have been renumbered using the standard steroid scheme (See Appendix A3); the carbons in the first steroid system are numbered 1 to 27, those in the second system 28 to 54. There are a total of twenty one chiral centres: 5*S*, 8*R*, 9*S*, 10*S*, 12*R*, 13*R*, 16*S*, 17*S*, 20*S*, 22*S*, 23*R*, 25*S*, 32*S*, 35*S*, 36*S*, 37*S*, 40*R*, 44*R*, 47*S*, 49*R*, and 50*R*.

282 · Molecular Modelling: Computational Chemistry Demystified

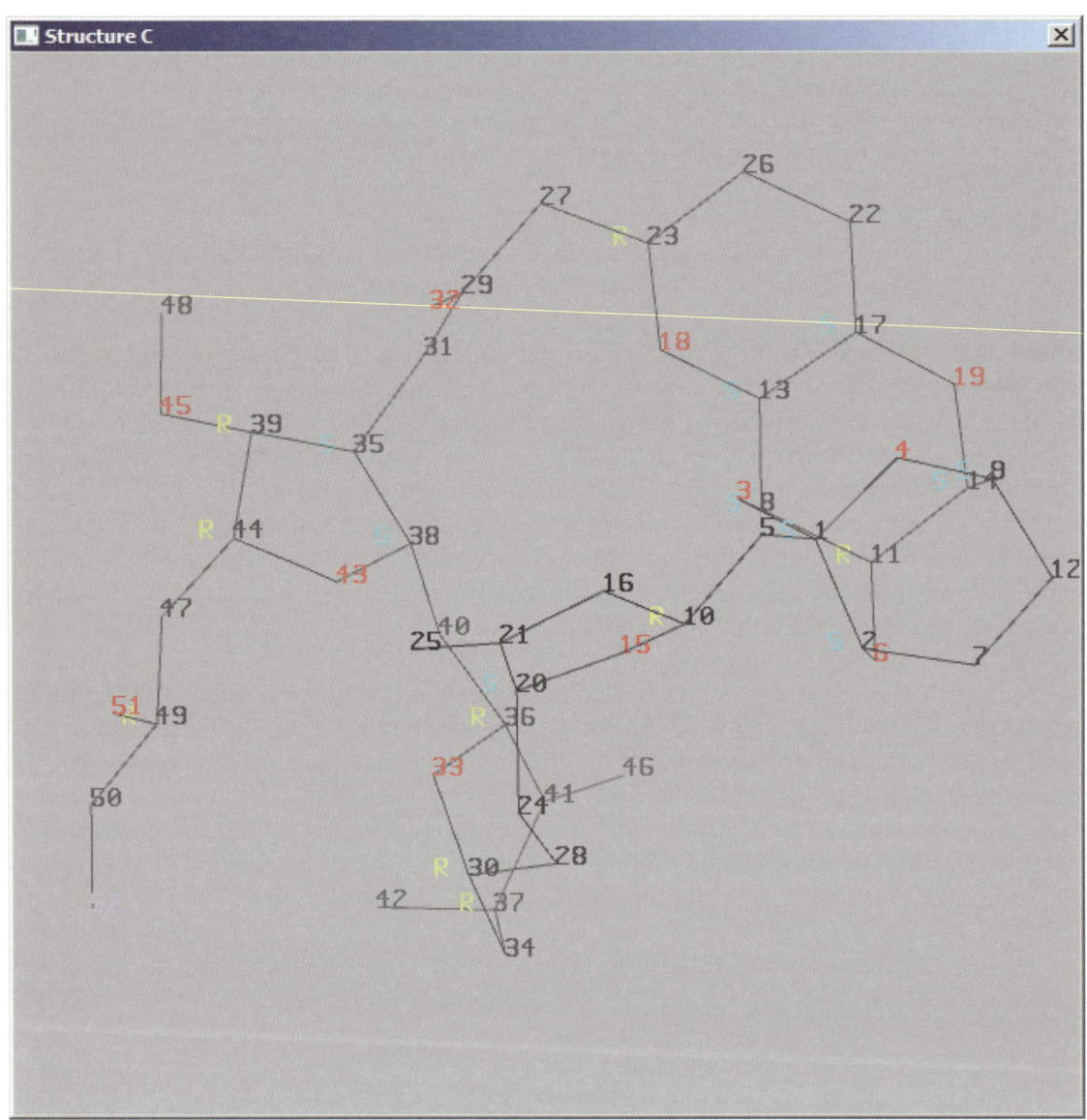

Figure ANS.1
The structure of Eribulin
This 3D structure was optimized using MOPAC (AM1).
Display of hydrogen atoms is suppressed to give a clearer picture

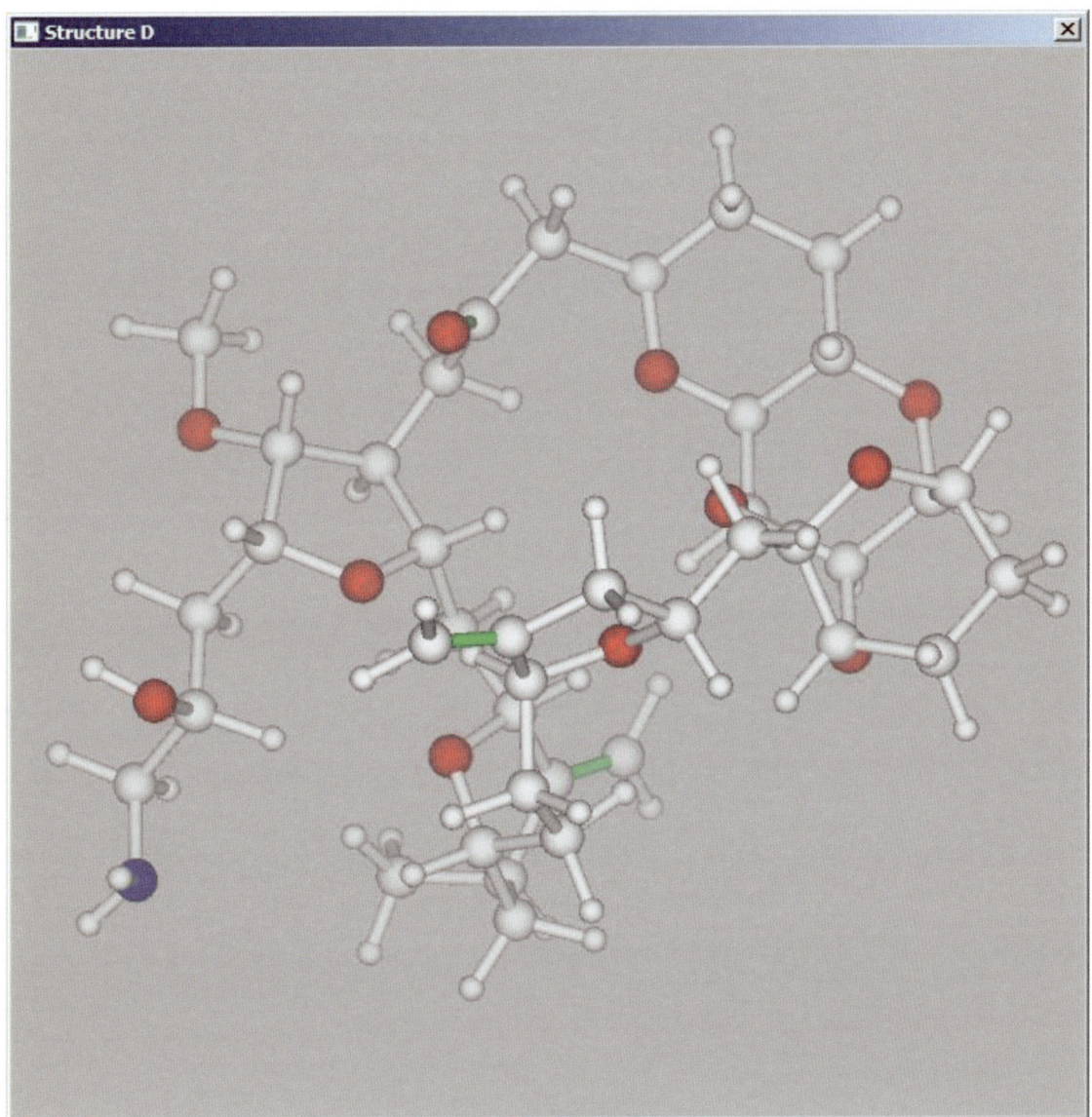

Figure ANS.2
The structure of Eribulin
This 3D structure was optimized using MOPAC (AM1).
The pose of this picture is similar to that in Figure ANS.1

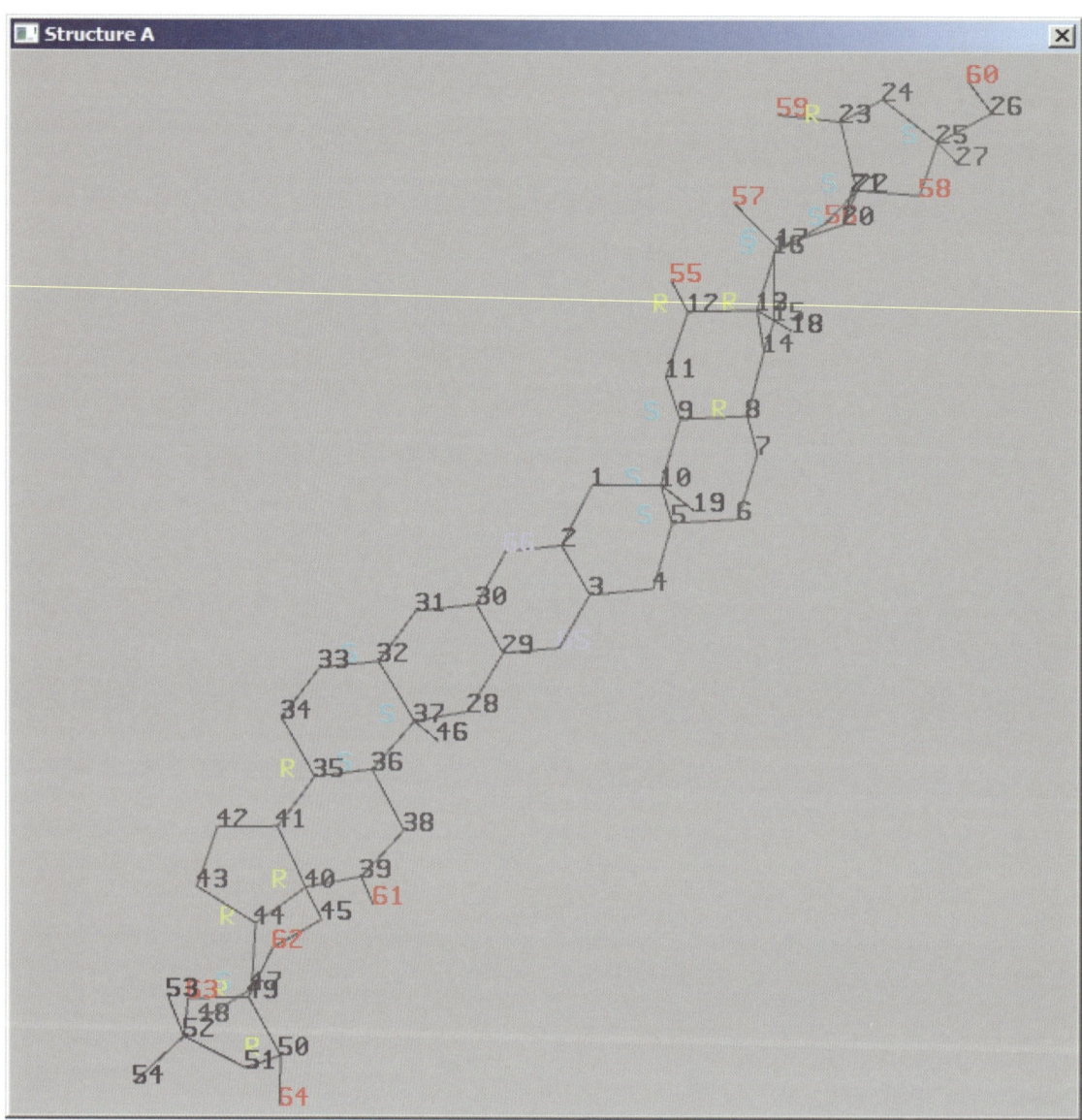

Figure ANS.3
The structure of Cephalostatin-1
This 3D structure was optimized using MOPAC (AM1).
Display of hydrogen atoms is suppressed to give a clearer picture
The carbon atoms in the right-hand steroid skeleton are numbered 1 to 27; those in the bottom left hand skeleton are numbered 28 to 54

Answers to 'Problems for you to solve and questions for you to answer'   285

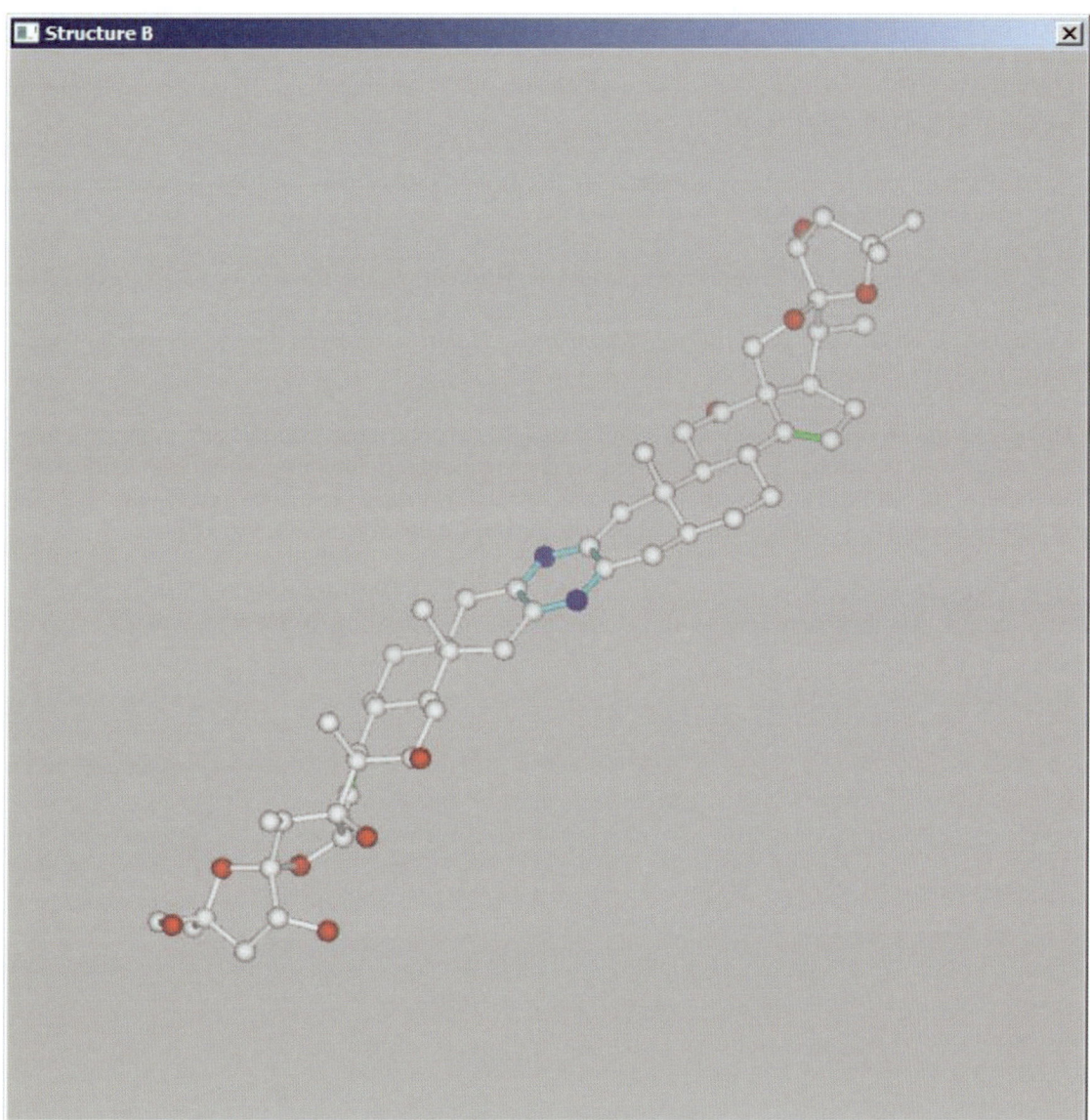

Figure ANS.4
The structure of Cephalostatin-1
This 3D structure was optimized using MOPAC (AM1).
Display of hydrogen atoms is suppressed to give a clearer picture
The pose in this picture is similar to that in Figure ANS.3
Note that the pose shown in the formula of cepalostatin -1 on page 250 differs from that shown here by being rotated through 360° on an axis perpendicular to the page.

The files from which the Figures are generated are available on the software disc. If you load these in INTERCHEM you can more easily examine the 3D structures.

References

[1] U. Burkert and N. L. Allinger, *Molecular Mechanics*, ACS Monograph 177. American Chemical Society, Washington, 1982, p. 107ff.

[2] A. D. Richardson, K. Hedberg, K. Utzat, R. K. Bohn, J.-X. Duan, and W. R. Dolbier, Jr., *J. Phys. Chem. A*. 1006, **110**, 2053.

[3] R. H. Janke, G. Haufe, E.-U. Würthwein, and J. H. Borkent, *J. Am. Chem. Soc.*, 1996, **118**, 6031.

[4] R. Taylor, *Lecture Notes on Fullerene Chemistry, A Handbook for Chemists*, Imperial College Press, London, 1999.

[5] This algorithm ignores hydrogen atoms and then seeks out the most 'central' atom in a structure and assigns it atom number 1. The remaining atoms are assigned numbers, working towards the periphery of the structure, using strict rules. INTERCHEM uses an extended version of the algorithm, due to W. T. Wipke, and T. M. Dyott, *J. Am. Chem. Soc.*, 1974, **96**, 4834.

[6] G. R. Pettit, M. Inoue. Y. Kamano, D. L. Herald, C. Arm, C. Duffresne, N. D. Christie, J. M. Schmidt, D. L. Doubek, and T. S. Kruba, *J. Am. Chem. Soc.*, 1988, **110**, 2006.

Subject Index

Page numbers in *italics* refer to illustrations. The suffix (T) indicates a table.

ab initio calculations 38
Accelyrys Inc. 49
acetophenone (methyl phenyl ketone) *54*, 55
acetophenone oxime *54*, 55
active stereo viewing 45, 49
ADMET 218–219
alicyclic structures 55–56
alignment matrix 117
alignments (protein sequences) 117–124
aliphatic structures 55–56, 57–58
allenes 139
AM1 (Austin Model One) 38
AMD 27
amino acids 89–93, 107, 111
   D/L system 128
p-aminobenzamide 63
AMPAC program 37–38
anaglyph stereo viewing 45
analyzing structures 79–83, 109–116
androstane *54*, 56, 58
angles *see* bond angles; inter planar angles; torsion angles
animation 44
anti-cancer drugs 248–250
anti-malarial drugs 245–248
anti-parallel beta sheets 94, *98*
anti-virus protection (computers) 24
Apple computers 27
Apple systems 23
archive data sources 74–78, 104–107, 189–191, 227–232
aromatic structures 53, 58–59, 60(T)
artemisin 245, *246*, 248
asymmetric units 150, 156, 257
ATI 27
atomic coordinates 39–41, 204–205, 254, 256–259, 260
atomic units 42
attachment energy (crystal morphology) 181–182
AUTOBUILD 242–245
axes of symmetry 145–146, 151–154
axial bonds 135

Balasubramanian plots 113, *116*
Barton, D. H. R. 135, 140
basis sets 39
benzene 40–41, 58–59
benzoic acid 72–73
benzophenone 162–164, 175–177
Beowulf clusters 24
BFDH (Bravais Friedel Donnay Harker) crystal morphology 178–181, 259
'Big-Oh' notation 42
bioisosteres 219–221
biphenyls 139
boat conformations 135, *136*
bond angles 79–80, 255
bond lengths 79, 261(T)
bonds
   basic types 6
   sketching 53, 55
   SMILES 57, 58, 60(T)
   stereochemistry 84, 135, *136*, 138–139
   strain energy 32–33
   see also disulfide bonds; hydrogen bonds; peptide bonds
Born-Oppenheimer principle 31
bowsprit bonds *136*
branching structures 58, 59
Bravais Friedel Donnay Harker crystal morphology *see* BFDH
Bravais lattices 153, 155(T)
Brookhaven database *see* PDB (Protein Data Bank)
Buckingham potential 166
buckminsterfullerene 67–68, *69*
building structures
   fragment assembly 62–65
   merging 65–68
   sketching 53–56
   SMILES 8–9, 57–62

caffeine 75, 76–77
Cahn-Ingold-Prelog (CIP) rules 131–132, 138–139
calcite ($CaCO_3$) *183*

Cambridge Crystallographic Data Centre (CCDC) 74, 189–190
Cambridge Structural Database (CSD) 189–190
cancer drugs 248–250
Cartesian coordinates 204–205, 254, 257–259, 260
   and internal coordinates 40
cases, computer 15
Cathode Ray Tube (CRT) devices 18
CCDC (Cambridge Crystallographic Data Centre) 74, 189–190
CDS (Chemical Database Service) 190
central processing units (CPU) 16
centred plane lattices 146–148
centrosymmetric crystals 155(T), 162–164
cephalostatin-1 250
chain structures 53
   branching 58, 59
chair conformations 135, *136*
changing atoms or bonds 55
Chemical Database Service (CDS) 190
chirality 83–84, 131–135, 139, 209–210
   computer data 205
   crystal structures 162–164
chloride ion
   as hydrogen bond acceptor 185
cholestane 265–266
cholic acid 75, 77–78
circular dichroism CD 140
CIF files 74, *200–201, 202*
CIP (Cahn-Ingold-Prelog) rules 131–132, 138–139
cis-trans isomers 84, 138–139
cluster analysis 111–112, *112,* 113
clusters (of computers) 24
cocrystals 161–162
cohesive forces in solids 164–166
colour 43, 44
comparing structures 70–72
computer display 18–20, 43, 270–271, 272(T)
computers, choosing 13–25
configuration 8
conformation 8, 59, 62, 135–137
CONVERT program 273
coordinates, atomic 39–41, 204–205, 254, 256–259, 260
coordinates, text 205
CPUs (central processing unit) 16

crystal structures 142–155
   archive data sources 74–78, 189–190
   chirality 162–164
   coordinate systems 204–205, 256–257
   examined with INTERCHEM 156–162
   hydrogen bonds 159–161
   lattice concept 143–144
   shapes (morphology) 177–182, 259
   three dimensional lattices 151–155
   two dimensional lattices 144–151
   see also lattice energy
crystallographic data 74–78, 189–190
crystallographic information files (CIF) 74, *200–201,* 202
CSD (Cambridge Structural Database) 189–190
cubic crystals 152–153, 155(T), 178–180
CUDA 28
cumulenes 139
cyclic structures
   alicyclic 55–56, 58
   aromatic 53, 58–59, 60(T)
   see also individual compounds
cyclododecane 62
cyclohexane 55–56, 58, 135, *136*
cyclopentane *136*
Cygwin 21–23, 25(T), 227–230
cytochrome-C sequences *120, 121,* 123–124

data, structural
   archive sources 74–78, 104–107, 189–191, 227–232
   interpretation by computer 205
data tables 261–263(T)
decalin *54,* 56, 58, 135, *136*
deleting atoms or bonds 55
deleting software 276
density functional theory (DFT) 38–39
depth cueing 44
desktop computers, choosing 14–18
diastereomers 132–135
dielectric constant 33, 35
dihedral angles *see* torsion angles
Dirac, P. A. M. 1
disc drives, optical 17
display, computer 18–20, 43, 270–271, 272(T)
distances, inter atomic 79
distances, inter residue 113, *114*

Subject Index

distances, intra-sequence *see* SID analysis
disulfide bonds 94
Diversity supplier 223
DNA 100, *101, 102, 103*
docetaxel (Taxotere) 249
docking (drug design) 214, 215–218
   INTERCHEM experiments 236–245
double bonds 55, 57, 84, 138–139
double buffering 44
double helix structure 100, *101, 102, 103*
Download sites (T) 25
drawing 53–56
DREIDING force field 36
drugs (medicines)
   database setup using Cygwin 227–232
   designing 210–213
   economics 210
   molecular modelling 214–221
   from natural products 248–251
   safety requirements 209–210
   virtual screening 190
DVD, small molecules database 227–232

E/Z isomers 84, 138–139
electron density 38
electrons 31
enantiomers 132–135, 209–210
enantiomorphous crystals 155(T), 162–164
enantiotropes 169
energies of structures 32–33, 72–73, 81
   see also attachment energy; lattice energy
energy minimization (optimization) 32–33, 34, 36–37, 39–41
   failures in 81
equatorial bonds 135
equipartition energy 185
eribulin 250
error messages *see* troubleshooting
ethylbenzene 53–54, *54*

FDAT files 74
Fermi 28
ferrocene 34–35, 65–67
   history 87
file compression 277–278
file formats 74, 191–204
   introduction 191
   CIF 74, 200–*201,* 202

FDAT 74
INTERCHEM D 202–204
MDL 198–199
multi-structure files 204
PDB 195–198
STAR 202
TRIPOS MOL2 192, *194*
XR format 74, 192, *193*
files
   8.3 86
   drugs database 229–231
   names 86
   STEER.DAT file 269, 272
   transferring 278
filtering (drug design) 214, 218–219
Fischer projections 132–133
flagpole bonds *136*
force field method *see* molecular mechanics
forms (crystal faces) 178–182
fractional coordinates 205, 256–259
fragment structures 62–65, 244
frame-sequential stereo 49
full stop, in programming 228
fullerene C_{60} (buckminsterfullerene) 67–68, *69,* 87
fumaric acid 138
functional groups 59

GAMESS program 38
GAUSSIAN program 38
GeForce 28
genetic uniqueness 111
glide lines 148–149
global alignment 117, *120,* 122–123
GLUT (Open GL Utility Toolkit) 44
glycine 90, 169–170
graph-set analysis 185
graphical display software 42–45
graphics cards 16
gzip 227, 277

habits, crystal 178–182, 259
hard discs 16–17
hardware, choosing 14–20
helix structures 94, *95, 96, 97*
   DNA 100, *101, 102, 103*
hexagonal crystals 152, 155(T)
hexagonal plane lattice 145, 146
hexamine 156–157, *158*

high energy structures 81
high throughput screening (HTS) 213
highlighting tools 108–109
hints, sketching 56
hits (drug design) 211–212
homologous sequences 117
hydrogen atoms 54, 59, 240
hydrogen bonds, in crystals 159–161
hydrophobicity 83
hypofluorous acid (HOF) 149–151

InChi system 8
INDO (Intermediate Neglect of Differential Overlap) 37
inorganic crystals *see* crystal structures; lattice energy
Intel 27
inter atomic distances 79
inter planar angles 80, 255
inter planar spacing 259
inter residue distances 113, *114*
INTERCHEM
   analyzing structures 79–83, 109–116
   AUTOBUILD 242–245
   crystal structures 156–162
   docking experiments 236–245
   Dual Display mode screen *52*
   file formats 202–204
   fragment assembly 62–65
   getting started 3–6, 51–52, 267–268
   merging 65–68
   plane groups 149–151
   and QUICKSCAN 236
   set up 267–268
   sketching 53–56
   SMILES 8–9, 57–62
   troubleshooting 268–271
INTERCHEM D files 202–204
internal coordinates 39–41
International Zeolite Association (IZA-SC) 190
Interprobe Chemical Services 3D database 191
inversion axes 153
ionic crystals *see* crystal structures; lattice energy
isomerism *see* stereochemistry
isopropylbenzene *54*
isosteres 219–221

IZA-SC (International Zeolite Association) 190

laptop computers, choosing 14–15
lattice concept 184
lattice energy 186
   background theory 164–166, 168–169
   calculations 169–177
   organic materials 175–177
lattices, crystal
   concept 143—144
   three dimensional 151–155
   two dimensional 144–151
leads (drug design) 211–212, 213
left-handed coordinate systems 204–205
Lennard-Jones potential 33, 166
ligand based design 214
ligands
   docking 215–218, 240–242
   generating 242–245
lighting (for 3D effect) 44
Linux 14, 20–25, 277–278
   history 26
   taxonomy (T) 27
Lipinski's rule (Rule-of-Five) 212
liquid crystals 143
literature, structures from 73–78
L-NMMA.HCl (N^G-monomethyl-L-arginine hydrochloride) 160–161
local alignment 117, *121*, 122–123
low-dimensional structural information 8

Madelung constant 165–166
Maddox, J 141
major groove 100, *103*
malaria drugs 245–248
maleic acid 138
mathematics of molecular modelling 254–260
MDL files 198–199
memory, computer 16, 42
meso compounds 133–134
metals structures 190
methylmalonyl coenzyme-A mutase *112, 114, 115, 116*
Microsoft Windows 14, 270–271, 277
   and Linux 20–25
   comparison of versions of 26

SUA (Subsystem for Unix Applications) 21
Miller planes and indices 158, 177–180, 256–257
MINDO (Modified Intermediate Neglect of Differential Overlap) 37
minor groove 100, *103*
mirror symmetry 145–146
MNDO (Modified Neglect of Differential Overlap) 37
molar refractivity 82(T), 83
molecular formula 81–82
molecular mechanics 32–36, 81
molecular orbital (MO) calculations 36–38
molecular volume 81, 82(T)
molecular weight 81–82
MOLfile *see* MDL files
monoclinic crystals 151, 155(T)
N^G-monomethyl-L-arginine hydrochloride (L-NMMA.HCl) 160–161
monotropes 169
MOPAC program 37–38, 70, 274–275
morphology, crystal 177–182
motherboards 16
multi-structure files 204

natural products, drugs from 248–251
NDB (Nucleic Acid Database) 190
networking 24
non-centrosymmetric crystals 155(T), 162–164
non-enantiomorphous crystals 155(T), 162–164
Nucleic Acid Database (NDB) 190
nucleic acid structures 36, 100, 107–109, 190
Nvidia 28
NWCHEM program 38

'O' notation 42
oblique plane lattice 145
OpenGL 43–45
operating systems 13–14, 20–25
  discs for 17
  file transfer between 278
optical activity 140
optical disc drives 17
optical rotator dispersion ORD 140
optimization *see* energy minimization

organometallic structures 34–35
orthorhombic crystals 151, 155(T)

paclitaxel (Taxol) 249
parallel beta sheets 94, *98*
passive stereo viewing 45, 49
PATA discs 16
PDB (Protein Data Bank) 104–107, 125, 190
PDB files 195–198
peptide bonds 36, 90
Planaria Software 49
plane groups 149–151
plane lattices 144–151, *149*
plastic crystals 143
PM3 (3rd Parameter Model) 38
PM6 (6th Parameter Model) 38
point groups 148, 153, 155(T)
polarity 111
polymorphism 160–161, 168, 169
POSIX 27
power supply, computer 15, 17
PRESTO program 118–124, 272
  set up 271
primitive plane lattices 145–146
priorities (CIP rules) 131–132, 138–139
projectors 20
Protein Data Bank *see* PDB
protein sequences 116–124
protein structures 89–103
  analyzing 109–116
  archive data sources 104–107, 190
  conformation 128
  displaying 107–108
  editing 108–109
  folding 128
  force fields 36
  Greek-key 128
  hairpin 128
  jellyroll 128
  Levinthal paradox 128
  ligand docking (drug design) 215–218, 240–242
  loop 128
  from racemic protein crystals 124–125
  turns in proteins 128
PROTEINS program 106–107, 273, 274(T)
PROTOCHECK program 107
pseudo torsion angles 80

quad buffering 45
Quadro 28
quantum mechanics 36–42, 70–73
QUICKSCAN program 231–235, 273
quinine 63–65, 70–72

racemic proteins 124–125
racemization 134
RAID 28
  disc system 20
Ramachandran plots 113, *114*
random numbers 84
rectangular plane lattice 145–146
reproducibility 84
resolution, projector 20
resolution, screen 18, 19(T), 270(T), 271, 272(T)
results, storing 51, 113, 124
right-handed coordinate systems 204–205
rigid-body rotations 256
ring structures *see* cyclic structures
rotational symmetry 145–146, 151–153
rotations, rigid-body 256
Rule-of-Five (Lipinski's rule; ROF) 212

SATA discs 16
scaling in computational programs 42
scorpion toxin BmBKTx1 124
screens, computer 18–20, 43, 270–271, 272(T)
screw axes 153–154
SDfiles *see* MDL files
security, computer 24
seed random number 84
semi-empirical calculations 37–38
sequence matching 117–124
sheets (protein structure) 94, *98*
SID analysis (Simple Intra-Sequence Distance) 109–110, *112*, 113
single bonds 53, 57
site multiplicity 156
site symmetry 184
sketching 53–56
slice energy 181
SLN (SYBYL Line Notation) 8
small molecules, data sources 189–190
  database DVD 227–232
SMILES strings 8–9, 57–62
  examples 61(T)

rules 60(T)
sodium chloride 164–167
sodium sulfate 169–175, *176*
software 267–276
  additional_items folder 275
  docking software 215, 216–217(T), 218
  graphical display software 42–45
  removing 276
  semi-empirical molecular orbital programs 37–38, 70, 274–275
  updating 276
  see also CONVERT; INTERCHEM; PRESTO; PROTEINS; QUICKSCAN
software suppliers 45
solids 141–143, 164–166
  see also crystal structures
space groups 148–151, 153–154, 162, 262–263(T)
  see also symmetry operations
spirostane 265–266
square plane lattice 145, 146
standard bond lengths 261(T)
STAR files 202
status of drawing process 53
STEER.DAT file 269, 272
stereo viewing 45, 49
stereochemistry 6–8, 83–84, 131–139
  amino acids 90
  chirality 83–84, 131–135, 139, 162–164, 209–210
  conformation 8, 59, 62, 135–137
  E/Z (cis-trans) isomers 84, 138–139
  SMILES 59, 62
  steroid ring systems 266
steroid structures 56, 265–266
storing results 51, 113, 124
strain energy 32–33
structure based design 214
structures
  analyzing 79–83, 109–116
  archive data sources 74–78, 104–107, 189–191, 227–232
  building (*see* building structures)
  correct and incorrect 62
  energies of 32–33, 72–73, 81
  from published literature 73–78
  from racemic protein crystals 124–125
  refining 70–73
  storing 51

Subject Index

supramolecular engineering 159, 161–162
surface energies 182
symmetry
   and internal coordinates 40–41
     three dimensional 151–155
     two dimensional 144–149
symmetry operations 156, 257, 258(T), 260
synthons 159, 161–162

tartaric acid 132–133, 137
Taxol (paclitaxel) 249
Taxotere (docetaxel) 249
Tesla 28
tetracycline *247,* 248
tetragonal crystals 151, 155(T)
thalidomide 209
thenardite (sodium sulfate) 169–175, *176*
therapeutic index (therapeutic ratio) 212
THIS and THAT chains 109–110
three dimensional viewing 44–45
time requirements of calculations 42
torsion angles 32, 39–40, 59, 60(T), 80, *93*, 113
   calculation 255
total cluster polarity 111
transferring files 278
translational symmetry 148–149, 153–154
triclinic crystals 151, 155(T), 259
trigonal crystals 155(T)
triple bonds 58
Tripos International 49
TRIPOS MOL2 files 192, *194*
troubleshooting 275–276
   INTERCHEM 268–271
   PRESTO 271

UFF (Universal Force Field) 36
UniProt 129
uninstalling software 276
united atom approach 36
units 42
UNIX 14, 277–278
   history 26
updating software 276
UPS (Uninterruptible Power Supply) 17

van-der-Waals forces 33
vectors 254
vinblastine 248

vincristine 248
virtual high throughput screening (VHTS) 214–215
VirtualBox 23, 25(T)

Wavefuntion Inc. 49
Windows *see* Microsoft Windows

X-ray crystallography 74–78, 189–190
XR files 74, 192, *193*

Z buffering 44
Z-matrices 39–41
zeolites, archive data sources 190
ZINC database 190
zwitterion structure 90